SP Pvt. Ltd.

Mushroom Cultivation

Development and Perspectives

by

Shweta Kulshreshtha

2019

Studium Press (India) Pvt. Ltd.

Mushroom Cultivation

Development and Perspectives

ISBN: 978-93-85046-62-9

Published by:

Studium Press (India) Pvt. Ltd.
4735/22, 2nd Floor, Prakash Deep Building
(Near Delhi Medical Association)
Ansari Road, Darya Ganj, New Delhi-110 002
Tel.: + 91-11-43240200-15 (15 lines); Fax: 91-11-43240215
E-mail: pubdir@studiumpress.in

Printed at India

About the Editor

Dr. Shweta Kulshreshtha is currently working as Associate professor at Amity University Rajasthan, Jaipur since 2010. She acquired Doctorate degree from University of Rajasthan, Jaipur. Her specialization includes waste characterization, microbial treatment of wastes, mushroom cultivation for the waste remediation, mutagenicity assessment. Currently, she is working on waste remediation approaches and medicinal properties of mushroom and plants. She presented her work in national and international conferences and won many prizes for oral and poster presentations. She has published many papers in SCI listed journals as well as in national journals. She is editorial board member and reviewer of many journals of national and international repute. She delivered invited talks in many conferences. She is life time member of AMI, STOX-India, AFSTI.

Acknowledgements

This is unforgettable moment of contentment on the successful fulfillment of an ambition fostered for long, I prostrate at the feet of Almighty God whose silent blessings gave me the courage, energy and patience to overcome all the hindrances that came in my way to reach the goal.

It is my proud privilege to acknowledge, with deep sense of gratitude, to all authors who contributed to this book. All authors put much effort in writing excellent chapters. The quality of the book could not be possible without their contribution.

I am grateful to Director, Amity Institute of Biotechnology, Amity University Rajasthan, Jaipur for his ceaseless encouragement, constant inspiration and endless co-operation.

I also feel complacent by placing on record my sincere thanks to Studium Press for providing me the opportunity to edit this book.

Shweta Kulshreshtha

Preface

India has agriculture based economy which is related with the production of a variety of crops like spices, pulses, grains and fruits. The export of agricultural produce helps the farmers financially. Nowadays, agriculture had become difficult task in rain-fed areas due to insufficient technological research and innovations. Moreover, there is decrease investment in irrigation and infrastructure facilities required for agriculture. Overuse of pesticides and herbicides are adversely affecting the soil fertility and the quality of crops which further decreasing the export of crops. All these conditions put a mental and financial pressure on the farmer and forcing them to commit suicide.

There is cash crop *i.e.* mushroom which can raise the status of farmers. However, it is not very popular in India. Every farmer doesn't know about it and therefore, cannot cultivate it. This cash crop may be used for their earnings and supporting their family. Besides, it will solve the problem of agricultural waste remediation. Small farmers can also indulge in this business.

Keeping the above point in mind, the book is written in order to provide information on mushroom cultivation. This book discusses the methods of mushroom cultivation. This book will be useful for all the readers who are interested in mushroom cultivation. The description is based on the current knowledge which is definitely a great resource of knowledge on mushroom cultivation. All institutes can establish mushroom cultivation training center and start training programmes for the farmers which will help the farmers to indulge themselves in mushroom cultivation practices in their spare time and earn money. This book will be a boon for them. In one chapter of the book, the role of women in mushroom cultivation is discussed. This shows that the family of farmer can also involve in mushroom cultivations practices and support their family.

The book has chapters on the cultivation of different mushrooms like *Agaricus, Pleurotus,* shiitake, maitake, *Volvariella,* and many medicinal mushrooms. There are chapters on the storage and preservation of mushroom and control of nematodes pests of mushroom crop. We urge to the people to understand the necessity of mushroom cultivation for the farmers and help them by providing necessary support for mushroom cultivation.

With contributions from the renowned experts, this book will be a great knowledge resource on the mushroom cultivation and its perspectives. The book will develop interest in students and researchers, farmers, and general public, policy makers whomsoever involved in mushroom cultivation and improving the conditions of the farmers. The compiled information is expected to stimulate discussion among researchers and scientists who will assist research in formulating new easy and economic methods by using locally available agricultural wastes for mushroom cultivation. Editor will be happy to receive comments and suggestions (if any) at shweta_kulshreshtha@rediffmail.com.

Table of Contents

Table of Abbreviation of Figure

Sl. no.	*Chapter & Fig. no.*	*Figure Title to be read as*	*Figure Legend in brief*	*Page no.*
1.	Chapter 1 Fig. 1	The fruit body of mushroom	Ch. 1, Fig. 1: (MFB)	117
2.	Chapter 1 Fig. 2	Gills of mushroom & Histology of Gills	Ch. 1, Fig. 2: Gills & Histology of Gills	117
3.	Chapter 2.1 Fig. 1	Apparatus used for collection of basidiospores	Ch. 2.1, Fig. 1: Apparatus SC	117
4.	Chapter 2.1 Fig. 2	Basidiospores of *P. florida* collected on black paper under sterile conditions	Ch. 2.1, Fig. 2: Basidiospores *P. florida*	117
5	Chapter 2.1 Fig. 3	Hypha of monokaryon of *P. florida* without clamp connection	Ch. 2.1, Fig. 3: Hypha *P. florida*	117
6	Chapter 2.1 Fig. 4	Mycelial bits from two monokaryons placed together showing a) compatible (b) incompatible mating	Ch. 2.1, Fig. 4: Monokaryons C or IC mating	117
7	Chapter 2.1 Fig. 5	Hypha of dikaryon of *P. florida* with clamp connection	Ch. 2.1, Fig. 5: Dikaryon of *P. florida* (CC)	117
8	Chapter 2.1 Fig. 6	Protoplasts of *Calocybe indica*, Ci-3 strain at 45X	Ch. 2.1, Fig. 6: Protoplasts Ci-3 strain	117
9	Chapter 2.2 Fig. 1	Fungal competitors appeared on mushroom beds. (A) *Rhizopus stolonifer* (B) *Coprinus cinereus* (C) *Aspergillus fumigates* (D) *Alternaria alternate* (E) *Curvularia lunata* (F) *Penicillium* spp.	Ch. 2.1, Fig. 1. Fungal competitors	117
10	Chapter 2.4 Fig. 1	Gills of *Pleurotus florida*	Ch. 2.4, Fig. 1: Gills *P. florida*	118
11	Chapter 2.4 Fig. 2	Appeared pinhead of *Pleurotus* spp. on substrate	Ch. 2.4, Fig. 2: Pinhead *Pleurotus*	118
12	Chapter 2.4 Fig. 3	Initial stage of immature fruiting body formation	Ch. 2.4, Fig. 3: Immature FB (Initial)	118
13	Chapter 2.4 Fig. 4	Immature fruiting body on the substrate	Ch. 2.4, Fig. 4: Immature FB	118
14	Chapter 2.4 Fig. 5	Mature fruiting bodies on the substrate (suitable for harvesting)	Ch. 2.4, Fig. 5: Mature FB	118
15	Chapter 2.4 Fig. 6	Highly mature fruiting bodies on the substrate (not suitable for harvesting)	Ch. 2.4, Fig. 6: Highly mature FB	118
16	Chapter 2.4 Fig. 7	Cultivation of *Pleurotus florida* in plastic bottles	Ch. 2.4, Fig. 7: *P. florida* in plastic bottles	118

Table (*Contd...*)

Table (*Contd...*)

Sl. no.	*Chapter & Fig. no.*	*Figure Title to be read as*	*Figure Legend in brief*	*Page no.*
17	Chapter 2.6 Fig. 1	Fruit body of *Volvariella volvacea.*	Ch. 2.6, Fig. 1: FB *V. volvacea*	118
18	Chapter 2.6 Fig. 2	Paddy straw uninfected (left), and fungal infected (right)	Ch. 2.6, Fig. 2 : Paddy straw UI (left) & I (right)	118
19	Chapter 2.9 Fig. 1	Different varieties of milky mushroom	Ch. 2.9, Fig. 1: Different varieties of milky mushroom	118
20	Chapter 2.9 Fig. 2	Mother spawn filled bottles	Ch. 2.9, Fig. 2: Mother spawn filled bottles.	118
21	Chapter 2.9 Fig. 3	Grains filled polypropylene bag	Ch. 2.9, Fig. 3: Grains filled polypropylene bag.	118
22	Chapter 2.9 Fig. 4	Chemical sterilization of straw	Ch. 2.9, Fig. 4: Chemical sterilization of straw.	119
23	Chapter 2.9 Fig. 5	Spawing bags	Ch. 2.9, Fig. 5: Spawing bags	119
24	Chapter 2.9 Fig. 6	Casing covered with polythene sheet	Ch. 2.9, Fig. 6: Casing covered with polythene sheet.	119
25	Chapter 2.9 Fig. 7	Mushroom initiation after incubation	Ch. 2.9, Fig. 7: Mushroom initiation after incubation	119
26	Chapter 2.9 Fig. 8	Milky mushrooms ready for harvesting	Ch. 2.9, Fig. 8: Milky mushrooms ready for harvesting	119
27	Chapter 2.9 Fig. 9	Harvested milky mushrooms	Ch. 2.9, Fig. 9: Harvested milky mushrooms	119
28	Chapter 2.9 Fig. 10	Contamination by different microbes are shown in this figure.	Ch. 2.9, Fig. 10: Contamination by different microbes are shown in this figure. Contamination in substrate containing bottle (A); *Coprinus* fruit body (B); *Stemonitis* (C), Penicillium (D); *Coprinus* (E); *Trochoderma* (F); *Aspergillus* (G); *Rhizoctonia* (H) Unidentified (I); *Coprinus* (J); *Sclerotium rolfsii* (K).	119
29	Chapter 3.2 Fig. 1	Common nematode pests of button mushroom: Above left- *Aphelenchoides composticola,* Above right- *Aphelenchus avenae,* Below- *Ditylenchus myceliophagus*	Ch. 3.2, Fig. 1. Nematode pests of *Agaricus*	120
30	Chapter 3.2 Fig. 2	Above- *Aphelenchoides swarupi* in (a) cryptobiotic stage in culture plates; (b) Feeding of nematodes on hyphae of *Agaricus bisporus;* (c) *Aphelenchoides swarupi*, (d) *Ditylenchus myceliophagus*	Ch. 3.2, Fig. 2. Different stages of nematodes (a) *A. swarupi* (b) *A. bisporus* (c) *A. swarupi*, (d) *D. myceliophagus*	120

Table (*Contd...*)

Table (*Contd...*)

Sl. no.	***Chapter & Fig. no.***	***Figure Title to be read as***	***Figure Legend in brief***	***Page no.***
31	Chapter 3.2 Fig. 3	Symptoms of nematode damage in button mushroom: Left-infected bed, Right- uninfected bed	Ch. 3.2, Fig. 3. Nematode infected Ch. 3.2, Fig. 3. Nematode uninfected	120
32	Chapter 3.2 Fig. 4	Crop failure of button mushroom due to nematode attack	Ch. 3.2, Fig. 4. Crop failure	120
33	Chapter 3.2 Fig. 5	Predatory nematodes (Left: *Fictor* sp. and Right: *Seinura* sp.) feeding on mushroom nematodes	Ch. 3.2, Fig. 5. Left: *Fictor* sp. and Right: *Seinura* sp. MN	120
34	Chapter 3.2 Fig. 6	Effect of *Paraphelenchus* species on the growth of *Calocybe indica* mycelium grown on Potato Dextrose Agar: Inoculum levels 0, 10, 20, 40 and 100 from left to right	Ch. 3.2, Fig. 6: Effect *Paraphelenchus* on *C. indica*	120
35	Chapter 4.3 Fig. 1	A young *Morchella* ascocarp	Ch. 4.3, Fig. 1: *Morchella* ascocarp	120
36	Chapter 4.3 Fig. 2	Cross section of *Morchella* shows continuous cavity	Ch. 4.3, Fig. 2: *Morchella* continuous cavity	120
37	Chapter 4.3 Fig. 3	Sun dried young morel mushroom	Ch. 4.3, Fig. 3: Morel mushroom	120
38	Chapter 4.3 Fig. 4	*Morchella* sp. Among the ornamental plants	Ch. 4.3, Fig. 4: *Morchella* as ornament	120

Section-1
Introduction

1

Introduction to Mushroom Cultivation

SHWETA KULSHRESHTHA[1*]

1. INTRODUCTION TO MUSHROOM

Mushroom is known as a "macrofungus" which possesses a distinctive fruiting body, large enough to be seen by naked eyes and to be picked up by hand (Chang and Miles, 1992). Mushroom fruit body is also known as carpophores which can be divided into two parts *i.e.*, stalk (stipe) and pileus. Stalk or stipe is the structure that supports the pileus. Pileus is cap like structure which bears gills. Gills are present beneath the cap which possesses basidiospores.

Mushrooms lack chlorophyll and therefore, unable to do photosynthesis. They are heterotrophic in nature and acquire their nutrition from other substrates such as agricultural wastes, industrial wastes, *etc*. In the ancient time, people did not know the art and science of cultivation. They used to collect mushrooms from the forest. After a long time observation, human beings developed the art and science of cultivation. Now, many mushroom species are known which can be cultivated and sold in the market. This is not only the source of income for the farmers but also provide proteinaceous food for the people suffering from protein deficiency. Mushroom cultivation can be successfully implemented to the field because it can be implemented at low cost. Besides this, it can be a potential source of protein to fulfill protein requirement.

2. MORPHOLOGY OF FRUIT BODY

The fruit body of mushroom is known as basidiocarp or carpophore which is mainly divided into six different stages: pinhead, tiny button, button, egg,

[1] Amity Institute of Biotechnology, Amity University Rajasthan, Jaipur.
**Corresponding author:* E-mail: shweta_kulshreshtha@rediffmail.com

elongated fruit body and mature carpophores. At first, a small fruit body appears on the substrate, possesses very tiny stalk with round head which is called as pinhead. At this stage, pileus and stipe are not visible in vertical section.

After this, it forms tiny button stage. At this stage the top of the veil is brown while the rest is white. On vertical section, a thin narrow band on the lower surface of pileus, can be seen which is called as lamellae. In button stage, stipe of fruiting body and pileus is very small and covered by a universal veil. Stipe is visible only in longitudinal section of fruiting body.

Round fruiting bodies become elongated and looks like egg. In egg stage, stipe can be observed only after longitudinal dissection of fruit body. In the egg stage, pileus grows and pushed out of the veil and takes the shape of volva. Gills do not bear basidiospores when observed under microscope. However, the formation of sterigmata, a structure that bear basidiospore, may occur at this stage. This stage is also marketable and mushroom harvested at this stage can also be sold in the market.

The stipe of these fruiting bodies elongated and reached its full length and therefore, this stage is known as elongation stage. Maximum growth in the stipe is found at the attachment site of pileus. When pileus of fruit body open up and take the shape of umbrella, it is called as mature fruiting body. Initially, pileus is covered with thin layer of cells called veil which attaches to the stipe. As the fruit body grows, veil breaks down which allows the pileus to take umbrella like shape and expose the gills. Gills possess basidia that bear oval shaped basidiospores.

At mature stage, the structure is divided into two regions: (i) The pileus or cap, (ii) Stipe or stalk and (Fig. 1).

3. HISTOLOGY OF GILLS

The gills in fruit body and cross-section of gills can be observed under microscope and shown in photomicrograph (Fig. 2). The best stage for observing the histology of gill is an early mature stage in which the pileus is opened up and the gills are exposed. The cross-section of gills reveals that it is composed of three layers- trama, sub-hymenium and hymenium. The innermost layer is made up of loosely packed interwoven hyphae is called trama. On both side of trama layer, a layer of densely interwoven hyphae is present which is called as sub-hymenium. The outermost layer is made up of densely packed hyphae and called as hymenium. In hymenium, terminal cells may be narrow to form paraphyses or swell to form the club-shaped basidia. Basidia are the structures that bear spores. These spores are called as basidiospores. Usually, one basidium possesses four basidiospores. These basidiospores are attached to basidium by the help of sterigmata.

4. CULTIVATION OF MUSHROOM

Mushroom cultivation represents an opportunity for farmers interested in extra-earnings. According to the area available, mushroom cultivation can be done on small scale and large scale. However, there is a need of understanding the conditions and requirements for mushroom cultivation. The most important factor for successful cultivation of edible mushroom is the selection of suitable area. In this regard, there are many factors which need to be considered such as, suitable environmental conditions, availability of raw material and man power for cultivating mushrooms and availability of market for selling fruit bodies. It is advantageous to establish mushroom culture room and facilities in the area where there are some functional mushroom production units are already present or there is any organization which can provide readymade spawn, compost or any technical assistance (Suman and Sharma, 2007). Details of the factors are given below:

4.1. Land for Establishing Mushroom Cultivation Room

One of the most important criteria for cultivating mushroom in economic manner is properly built and equipped farm according to the environmental conditions of the area. It is important to note that mushroom is indoor crop and therefore, it doesn't require agricultural land. Mushroom cultivation can be done on bare land. It will be beneficial to the farmers to establish mushroom cultivation room in the area where spawn providing facilities are available. If these are not available in the area, then spawn preparation rooms along with mushroom cultivation rooms can be established. However, these must be established separated at some distance. Another part of the land must be used for spent compost and substrate disposal. There must be suitable gap in the compost preparation room and spent mushroom compost disposal room.

4.2. Site for Constructing Mushroom Cultivation Room

There are many factors that need to be considered for constructing mushroom cultivation room such as – availability of water, supply of electricity and availability of raw material in the nearby area. This will save the money for import of raw material to the site. It is also suggested to choose the site where labour is easily available because mushroom cultivation is labour intensive job. Mushroom cultivation room must be established in the area where there is a market because mushrooms are perishable food which cannot be preserved for a longer time. Therefore, market for selling mushroom must be available in nearby area and the place must be well connected to the market.

Mushroom cultivation requires moist conditions which can be maintained by spraying water in mushroom cultivation room. Hence, water and electricity must be available in the area. Room must be designed in a manner that water can be recycled. This will prevent the wastage of water.

The whole area can be divided into 6 parts: (i) Raw material storage room (ii) Raw material processing room (iii) Compost making room (optional) (iv) Spawn making room (optional) (v) Mushroom cultivation room (vi) Mushroom preservation room.

First room is used for storing the raw material till it used for cultivation of mushroom. It must not have any type of seepage problem so that raw material can be preserved without contamination.

Second room is required for the processing of raw material. At first, a good quality raw material must be screened and then, sterilized by autoclaving or steaming process in order to remove all the contaminants.

Third room is compost making room which is required for making compost. Compost is the substrate on which mushroom mycelium grows and carpophores (mushroom fruit bodies) are developed. Mushrooms get their nutrition from compost and therefore, quality of compost can be good to yield healthy crop. Compost can be prepared from dead wood or straws or agricultural wastes and industrial wastes (Kulshreshtha *et al.,* 2010) depending upon the choice of mushroom to be cultivated. For *e.g.*, *Lentinula* grows well on dead wood however, *Volvariella* grows well on the compost prepared from straw. Therefore, composting process is very important for successful cultivation of mushroom. Composting involves various microbes and is microbial controlled process and therefore, requires suitable environmental conditions for the succession of different microbial communities. The final population of microbes developed in the compost, supports the growth of mushroom mycelium by providing nutrients and inhibits the growth of other microbes.

The compost preparation depends on the supply of adequate water and oxygen. This can be done by mixing the ingredients evenly and overturn them after definite time. During the composting some structural changes occurs in the straw which makes it soft for cultivating mushroom. It provides proper nutrients to the mushroom crop. Besides, it removes competing microbes which competes for the nutrients. It also helps in maintaining moisture content which acts as water reservoir for growing mushroom.

Composting room is optional to the mushroom growers. If readymade compost is available in the nearby area then this room may not be required. During the composting various types of gases are produced due to the growth of microbes. Microbes utilize the waste and convert it into compost. Compost making room must be away from mushroom growing room in order to avoid the contamination by undesired microbes.

Spawn making room is required for making spawn for cultivating mushroom. Spawn acts as mushroom seed which must be free from contamination in order to get healthy crop of mushroom. Mushroom spores are very small that can be observed under microscope. These microscopic propagules are difficult to handle as seed. Therefore, they are inoculated on some supporting material which supports the growth of mycelium. These mycelium coated substrates are called as spawn which can be preserved for a longer time. Wheat straw, rice straw, agricultural wastes, different types of grains such as can be used as supporting material for preparation of spawn. Spawn affects the quality and quantity of mushroom and therefore, must be prepared in sterile conditions. Any type of contamination with other microbes may destroy the whole crop of mushroom and finally its yield and biological efficiency. Hence, it is of utmost important to prepare and maintain the spawn in sterile conditions.

Substrates are processed by steam sterilization and then cooled to room temperature. After cooling, spore suspension or culture media containing mycelium is added to the substrate in sterile condition and then, incubated at suitable temperature and moisture conditions. Environmental conditions vary according to the mushroom species and mentioned elsewhere in this book. In some cases, calcium carbonate is added to the spawn substrate in order to get good quality spawn.

This room is also optional for mushroom cultivators. If spawn is available as readymade material in the area then, it is not required. Spawn making room must have sterile area or laminar air flow for the inoculation of spawn in sterile conditions.

The most important is mushroom cultivation room. The structure and dimensions of the room can be decided according to the cultivated species and available area. Room must be properly aerated which can be done by installing the ventilation ducts. These ducts can be fitted with filters that are used to filter the air so that mushroom crop can be prevented from infection by other microbes. In these rooms, oxygen must be supplied and gaseous waste products (CO_2) must be removed. This can be done several times in an hour. The blowing air in the room not only removes the waste gases, but also maintains the uniform temperature. This can be achieved by installing the blower at opposite end of the entry, in the lower part of the wall at ground level. This room should also fit with pipes which are used to supply water for irrigating the mushroom crops. It is necessary to install a source of permanent water supply. It is also necessary that this room must be properly sealed to maintain artificial environmental conditions. To maintain heating and cooling, boiler and air conditioners must be installed and walls must be made up of insulated material. Insulated material such as glass wool or thermocol must be incorporated in the wall as suggested by Suman and Sharma (2007). Besides this, doors must be properly fitted without having any space for air exchange. It should possess rubber tubes on the margins which can seal the space between the floor and door. The room must be fitted with artificial light source. The blue print of mushroom cultivation room is given in Fig. 3.

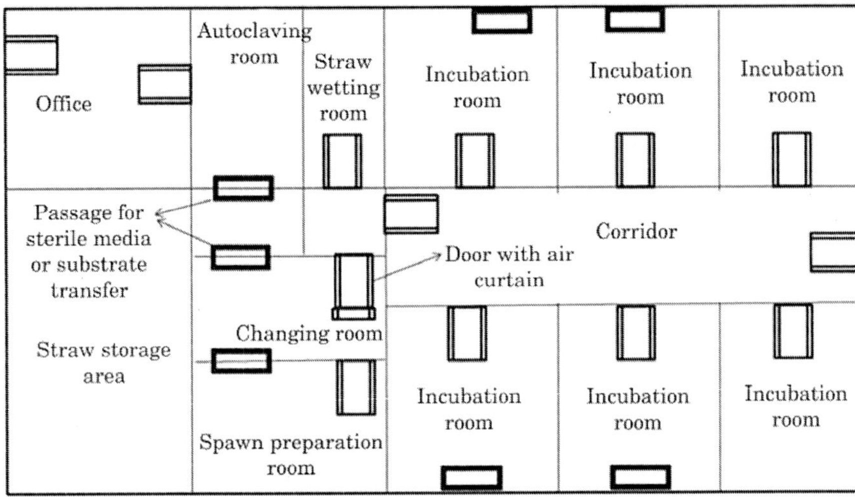

Fig. 3: Layout of mushroom cultivation room.

4.3. Mushroom Preservation Room

This room is used to store the harvested mushroom till their transport to the market. In this room, mushrooms are preserved at low temperature in germ free conditions. This room is equipped with refrigerators which are used to store mushroom. Sometimes, a whole room is used as cooling chamber which is fitted with cooling coils which helps in cooling the room. The walls of this room are also insulated with thermocol or glass wool in order to maintain temperature.

4.4. Requirements for Mushroom Cultivation

4.4.1. *Containers*

Generally, three types of containers – trays, shelves or bags are used for mushroom cultivation. Trays are most commonly used for *Agaricus* cultivation while polythene bags are used for *Pleurotus* cultivation. The plastic bags are widely accepted for mushroom cultivation because it is easy to handle, economic and disposable. However, precautions must be taken in using polythene bags because excessive heat is produced by compost which may further interrupt the spawn run.

The use of tray for mushroom cultivation is preferable than shelves method because these can be easily taken out and cleaned and sterilized. Trays must be made up of wood or rust proof material and withstand rough handling.

Shelves can be prepared by using wood or rust proof metal. These can be made at lower investment than the trays. The use of shelves is not preferable because it is difficult to sterile them.

4.4.2. *Spawn*

A small mushroom can produced millions of spores per hour. However, these spores cannot be used directly for inoculating substrate. Spawn is prepared by inoculating mushroom spores or mycelium on some substrates. Generally, grains act as substrate and are used for preparing spawn. Spawn acts as seed of mushroom which is used to inoculate mushroom compost. The use of mushroom mycelium for inoculating the substrate is preferable method over the use of mushroom spores due to genetic variation. Spores from the same species may be genetically different and therefore, they can produce a new strain which may affect the yield of crop.

4.4.3. *Requirement for spawn production laboratory*

The spawn production is crucial step in mushroom production. Any contamination in spawn can be transferred to mushroom bed and destroy the whole crop. Therefore, the spawn must be prepared in sterile conditions. The requirement for spawn production laboratory is –

(i) Inoculation room

Inoculation room is equipped with laminar air flow which provides sterile area for inoculation of spawn on substrate. Sterile area is also required to maintain germ free conditions to perform the steps of spawn preparation. Several microbes are present in air which can infect the substrate. Therefore, it is necessary to use laminar air flow for the preparation of spawn. Inoculation room must be constructed by using smooth tiles which can be cleaned easily. This room must be fumigated with 40% formaldehyde solution (formalin) which is filled in glass beaker and kept uncovered for overnight in the room. This solution kills the pathogen present in the inoculation room. Laminar air flow chamber can be sterilized with 70% ethyl alcohol and 2% formalin solution.

(ii) Incubation chamber or incubator

Every mushroom species need different temperature to grow. Therefore, equipment is required to control temperature. This can be done by incubator.

(iii) Sterilization room

This is required to sterilize the tools used for mushroom cultivation. This room must be equipped with pressure cooker or autoclave. It can be used for sterilizing different ingredients used for media preparation, substrate (grains, sawdust, tea leaves *etc.*), spawn containers or polythene bags, *etc.* Sterilization can be achieved by autoclave which provides 121°C temperature in closed vessel which is sufficient to kill microbes. For effective sterilization material must be kept at 15lbs/inch^2 pressure for 15 min.

(iv) Tools and glasswares

Spawn preparation needs sterile petri dishes, culture tubes, spirit lamp, inoculating needle, forcep, scalpel etc. Every tool must be sterilized properly in order to avoid contamination. The petri dishes, culture tubes can be put in autoclave for sterilization, however inoculating loop, forcep, scalpel can be sterilized by spirit lamp till these become red hot. Autoclavable plastic bags can be used because these are easy to sterilize by autoclaving procedure.

4.4.4. *Pure culture preparation*

The most important requirement for the mushroom cultivation is mother culture or starter culture. Starter culture can be obtained from a fresh and healthy mushroom fruit body. It can be obtained from mushroom growing centers. If spawn is not available and cultivators are using it first time, then a piece of mycelium can be cut from inner side of mushroom stalk under sterile conditions and then, inoculated on potato dextrose agar or malt extract agar. For this purpose, surface of carpophore is cleaned by spirit or 70% alcohol and cut longitudinally into 2 halves for taking a piece of mycelium. These slants are then incubated at appropriate temperature which varies according to the species of mushroom. After 10-20 days fresh mycelium culture of mushroom is produced. This is called as pure culture which is used for spawn preparation.

4.4.5. *Spawn preparation*

Pure mycelium of mushroom is used for preparing spawn by inoculating it on appropriate substrate. Mostly, grains are used as substrate however, other type of raw materials like agricultural waste (wooden sticks), wheat, rice, rye grains, saw dust, tea leaves *etc.* can also be used. Spawn preparation should be done only on unbroken fresh and healthy grains. These grains must be cleaned for removing undesirable substances like grass, unwanted seeds *etc.* and washed thoroughly in clean water three to four times to remove dust, dirt, soil particles. After washing, grains are soaked overnight and then boiled in such a way that they must not burst. After boiling, excessive water is removed by sieving process and left them in sieve for few hours for maintaining the moisture content. Now, grains are mixed with calcium sulphate and chalk powder in such a way that their pH must be in neutral range *i.e.,* 7-7.8.

4.5. Inoculation for Spawn Preparation

There are three methods of inoculating substrate for spawn preparation which are as follows:

4.5.1. *Broadcast spawning*

This is a traditional method not in use nowadays. In this method, spawn is broadcast over the surface of compost beds and allowed to grow there. At first, spores germinate and cover the surface whole bed with their mycelium and later reach the depth of compost bed. The advantage of this method is time saving approach. Besides this, the limitation is related to the loss of nutritional value of grains and drying of mushroom bed. Therefore, care should be taken to avoid drying. This method is not used nowadays.

4.5.2. *Mixed spawning*

In this method, spawn is mixed in such a way that it is evenly distributed throughout the compost. This is done by the help of specially designed device. The advantage of this method is reduction in spawn running period as mycelium need not to spread for a longer distance. Short run of mycelium leads to the shorter time for covering the substrate. Therefore, it is more advantageous than the other methods. The only precaution need to be taken is the cooling down of compost to 25°C before inoculation (Royse, 2010).

4.6. Characteristics of Compost Used for Spawn Growth

It is important to note that – there are many factors that need to be considered for good yield of mushroom. Although growth rate of the mushroom is their own genetic characteristics depending on strain of the mushroom used, yet in some cases growth rate is affected by the environmental conditions and quality of compost. Therefore, it is important to know the factors affecting the spawn quality and mushroom growth which are as follows:

4.6.1. *pH*

The pH of the compost 6.5 to 8.2 is acceptable. It is observed that pH of the compost changes from the time of inoculation till the production of fruit bodies. Therefore, it is also observed that pH is not an important factor that affects the quality of compost.

4.6.2. *Nitrogen content*

Nitrogen is provided by ammonia which is an important factor to be considered for obtaining good quality compost. Ammonia concentration must be lower than 0.05 percent. If ammonia is higher than 0.1% then the growth of the mycelium is retarded and further, higher concentration of ammonia (higher than 0.2%) inhibits the growth of mycelium.

Nitrogen sources must be chosen carefully because they affect the yield of mushroom. Material with low nitrogen contents such as chipboard, cocoa shell, ammonium sulfate produce lower yield of mushroom compared to the materials with higher nitrogen content such as poultry manure, urea, brewer's grains, hop, molasses, cocoa meal *etc.* (Noble *et al.,* 2002).

The use of poultry manure must be avoided in order to reduce odorous sulfur containing compounds. Urea and cocoa meal not only reduced the odour but also acts as prewetting agent. Rape straw is better option to form compost because it results in the reduction of odorous emissions and maintain porosity of substrate for gaseous exchange. It is better option than wheat straw (Noble *et al.*, 2002).

4.6.3. *Lipids requirement*

Mushroom compost, if supplemented with vegetable oils, linoleic and oleic acid esters, provides good yield of mushroom. These are stimulating agents that stimulate the growth of mushroom. This was observed for the growth of *Agaricus bisporus*. This shows that mushrooms are able to utilize lipids (Moore, 1998).

4.7. Physical Conditions for Mushroom Growth

4.7.1. *Temperature*

Every mushroom species can grow at the favorable temperature that varies for different species of mushroom. The optimum temperature of the compost bed for the growth of the mycelium of the mushroom is 23°C to 25°C. Any deviation in the range of optimum temperature may affect the yield of mushroom. When appropriate temperature is provided, mushroom mycelium grows rapidly and covers the substrate and hence, reduces the time required for spawn run. Mushrooms cannot grow at higher temperature. The growth of mycelium, formation of fruit bodies are completely retarded when substrate is kept at or above 30-40°C. This range also varies for different species of mushroom. The compost temperature rises during the mycelial growth.

4.7.2. *Humidity*

Humidity is an important factor required for the growth of mycelium. Water is sprayed in the room and frequency of spraying water depends upon environmental conditions. Room must be closed tightly to maintain temperature and humidity conditions. In the mushroom cultivation room, 95% relative humidity is required.

4.7.3. *Gas concentration*

Mushrooms require CO_2 rich atmosphere to grow. This can be achieved by keeping the doors closed tightly. Initially, little oxygen is required for spawn run which can be provided by ventilators but later mushroom mycelium starts spreading on the substrate and produce CO_2 which supports its growth. CO_2 concentration cannot be maintained in open room and this may affect the growth of mushroom.

4.7.4. *Moisture content*

Substrate is soaked overnight in water containing bavistin and formalin solution. Then, take it out from the solution by pressing with hands and spread on the ground to remove excessive moisture. Pressing must be done in such a way that little amount of water is removed. If force is applied then the substrate becomes too dry to use. At the time of spawning moisture content should be 65-75%.

4.7.5. *Hygienic condition*

Hygienic conditions are required for cultivating mushroom due to the delicacy of mushroom crop. Mushroom crop is susceptible to various insects, fungal and bacterial infections. This room must be fumigated after each crop in order to avoid the growth of insects. Walls and floor of room must be smooth to avoid the binding of pathogens. If hygienic conditions are not maintained, there is a chance of infection in the whole crop because low temperature and humid conditions supports the growth of pathogens. Therefore, floor, walls must be cleaned properly by using any disinfectant solution. During the installation of ventilator, it must be fitted with antibacterial filters. Keeping unusable extra material in the culture room must be avoided.

4.8. Casing

Casing is layer of moist material that is made on the top of compost layer prior to allowing substrate to fruiting bodies formation. The usage of casing layer is recommended to prevent the loss of moisture from the substrate compost. The recommended thickness of casing layer is 3-4 cm. After complete spawn run, compost should be covered with casing material. The material that has high porosity, water holding capacity and neutral pH, can be used for casing purpose. Peat moss is considered as the best casing material. However, its availability at some specific place restricted its wide application. There are some alternatives that were used successfully such as loam soil and sand (ratio 4:1); decomposed cow dung and loam soil (ratio 1:1); 2-3 years old spent compost; sand and lime. These compositions were applied for *Agaricus* species.

Researcher also tried the combination of cow dung and loamy soil (ratio 3:1); farm yard manure, loamy soil and sand (ratio 3:1); and old spent oyster mushroom substrate containing saw dust as casing material (Amin *et al.,* 2010). This method was found to be suitable for the cultivation of *C. indica.* All casing substrates were sterilized by pasteurization at 65°C for 4 hrs. It was inoculated in the packets in the form of 5 different level of thickness *i.e.*, 1-5 cm layers. The bags or trays were then maintained at 30-35°C temperature and relative humidity at 70-80%. However, immediately after casing of the compost, the temperature of the room should be maintained at 23-28°C; relative humidity (85-90%) and low CO_2 concentration for 8-10 days. These conditions were found to be favourable for reproductive growth of mycelium. The best growth and biological efficiency were observed for 3 cm thick layer of casing material composed of cow dung and loamy soil (Amin *et al.,* 2010).

In an another experiment, 4 different casing layers were made using peat in different combinations such as peat (100%), peat perlite (75%: 25%), and peat perlite clinoptilolite (60%:20%:20%) and found that peat perlite clinoptilolite combination provide good yield of mushroom (Adanacioglu *et al.*, 2015).

Pardo and colleagues evaluated the potential of composted pine bark, coconut fiber pith and wood fiber as an alternative to peat and reported that there are no significant differences in the mushroom production when these materials were used as casing material (Pardo *et al.*, 2004). Casing material may possess harmful microbes that are able to destroy the whole crop. Therefore, it is recommended that casing soil must be sterilized before use. Chemical or physical sterilization methods can be used for the purpose. Chemical treatment must be done at least 15 days before application while steam treatment and pasteurization can be done on the same day.

Physical treatment can be given in the form of steam exposure or autoclaving. Autoclaving is done at 121°C for 2 hrs. This time may vary from 30 min to 2 hrs depending upon the material used for casing. Autoclaving can be done by using pressure cooker and autoclave. Steam sterilization without pressure can be done in a drum which is filled with water (1/ 4 amount of water according to the total capacity of drum) and the wire net or porous plate is fixed in the drum for keeping casing or composting material for sterilization. Wire net or porous plate is required, so that steam can reach and penetrate the materials for killing pathogens. This drum is placed on boiler to heat the water for steam production. Again, time requirement varies and depending on the substrate. The minimum boiling time is 1 hr and maximum is 6 hrs.

Chemical method of sterilization of substrate is based on the use of chemical insecticides and pesticides. These can be incorporated into casing or in compost. This type of chemical treatment is associated with the reduction in mushroom yield. To evaluate the effect of incorporation of insecticides in the compost or casing, experiment was performed by Shamshad *et al.* (2009). In this experiment, casing material was treated with organophosphate (diazinon), pyrazole (fipronil), and an insect growth regulator (triflumuron). Another insect growth regulator cyromazine was incorporated into compost. This incorporation led to the reduced growth and further, yield of mushroom. Therefore, the effect of these insecticidal agents was evaluated. It was reported that cyromazine at dose 10 mg/kg into compost and triflumuron at dose 20 mg/kg into casing provided complete protection against pest without any significant loss in yield (Shamshad *et al.*, 2009). It is important to use chemical pesticides or insecticides in appropriate amount; otherwise it will affect the mushroom yield and level of toxicants in it.

4.8.1. *Effect of microbes present in casing layer on mushroom growth*

Generally, any types of microbial contamination in compost or casing lead to the reduction in mushroom productivity. In contrast to this, Cho *et al.* (2008) reported the presence of some beneficial bacteria present in casing layer. The microbial communities present in sterile compost were different than that of present in non-sterile compost. Detailed analysis of the bacterial community revealed that *Pseudomonas* was predominant in unsterilized soil while *Flavobacterium* is predominant in sterile casing layer soil. *Pseudomonas* was found to associate with the formation of fruit body and yield enhancement. This study showed that specific population of bacteria in the casing layer plays an important role in the formation fruit body of mushroom. This proves that every microbial population present in casing layer is not harmful for mushrooms (Cho *et al.*, 2008).

4.8.2. *Casing management*

This involves the practices and techniques associated with yield enhancement, better management of flushes of carpophores, size of carpophores. This can be done by scratching, ruffling and compost added at casing (CACing) method.

In the scratched technique, casing layer is mechanically removed and redistributed on the compost layer without disturbing the mycelium growth. This is done to reduce the surface compaction and to maintain CO_2 level. This will maintain the proper aeration of the substrate for the growth of fruit body.

When mycelium has grown into the casing layer and covered its three-fourth part, it is collected and redistributed on compost layer by the help of specially designed rototiller- like machine. Ruffling method disturbs the entire casing layer and mycelium that spread on it. In this method care should be taken that compost must not be mixed with casing layer or mycelium growth on the compost must not be disturbed.

Compost added to casing (CACing) technique involves the incorporation of the small amount of fully colonized compost (after spawn run when mushroom mycelium colonizes the substrate) into the casing layer at the time of casing. It can be done by using readymade compost covered with mushroom mycelium. This is mixed with casing material and then applied on the layer of compost. The application of readymade CACing is considered as sterile and safe, however, increases the cost of production. This compost should be mixed uniformly with the casing layer because success of technique depends on it. If the compost is contaminated with microbes, then it should be discarded to prevent spread of the infection. When this technique is applied properly, it reduces the growing cycle and time and increases the yield and weight of cultivated fruit bodies. CACing is better than the other method of casing (Zied *et al.*, 2010)

4.8.3. *Maintenance of casing by ruffling method*

Like scratched method, mycelium of mushroom grows down towards the compost in this method and finally, reaches the bottom layer. However, it removes all drawback associated with scratched method. Casing layer is poured on the surface of compost. When mycelium covers three fourth of the casing layer then, it is mechanically mix with the layer. Rototiller machine can also be used for the purpose. This method disturbs the entire casing layer and mycelium that grows into it. This is required to avoid the suffocation of the substrate. In this method, care should be taken so that compost layer should not be disturbed and mixed with casing; otherwise, mycelium will grow in casing layer and form the clumps which finally reduced the production. This method is most widely used method than the broadcasting method if mechanized spawning equipment is unavailable. Like broadcasting method, mushroom mycelium can grow down and penetrate the compost layer and spread in it.

4.9. Environmental Crop Management

Environmental conditions vary for different species of mushroom and hence, appropriate conditions must be maintained during the cropping season. Generally, there is need to maintain compost temperature at 16-20°C for which air temperature must be maintained at 14-18°C. In winter, the room must be properly ventilated with filtered air. However, fast currents of air must be avoided because it may cause the drying of compost beds. In summer season, cooling system must be used to maintain desired temperature. In the room, humidity must be maintained between 85-90%.

4.10. Harvesting

When mushroom fruit bodies become mature, these are harvested. The fruit bodies take usually 8-10 days to mature after the formation of pinheads. Harvest the mushroom when the margins of carpophores are downward towards the stipe. Cap of mushroom should be unopened cup shaped structure at the time of harvesting. In over-mature mushrooms, cap is fully expanded and loses it cup shape and gills are fully exposed.

Harvesting of the fruit bodies can be done by picking- up them with stipe. This can be done by twisting the mushroom fruit body for detaching them from the substrate. After this, mushroom fruiting bodies are cleaned by trimming the roots part embedded in the substrate. Two or three flushes can be harvested from one bundle of substrate.

4.11. Packaging and Marketing

After harvesting fresh and healthy mushrooms, they are packed in the plastic packets and immediately, transferred to the market in cold conditions.

Mushrooms can also be marketed in dried form. Therefore, mushroom fruit bodies can be dried in an oven at 45°C. This is followed by packing in plastic packets. Drying process removes the moisture content and then increases the shelf-life of the product. These can be cultured and marketed throughout the year which can be source of income for the mushroom growers.

5. REFERENCES

Adanacioglu, N., Boztok, K. and Akdeniz, R.C. 2015. The effects of light intensity, casing layers, and layering styles on royal sun medicinal mushroom, *Agaricus brasiliensis* (higher basidiomycetes) cultivation in Turkey. *Int. J. Med. Mushrooms*, 17(4): 397–405.

Amin, R., Khair, A., Alam, N. and Lee, T.S. 2010. Effect of different substrates and casing materials on the growth and yield of *Calocybe indica*. *Mycobiology,* 38(2): 97–101. doi: 10.4489/MYCO.2010.38.2.097.

Cho, Y.S., Weon, H.Y., Joh, J.H., Lim, J.H., Kim, K.Y., Son, E.S., Lee, C.S. and Cho, B.G. 2008. Effect of casing layer on growth promotion of the edible mushroom *Pleurotus ostreatus*. *Mycobiology,* 36(1): 40–44. doi: 10.4489/MYCO.2008.36.1.040.

Moore, D. 1998. Front matter. *In: Fungal Morphogenesis* (*Developmental and Cell Biology Series*). Cambridge: Cambridge University Press, pp. I–VI. Doi: 10.1017/CBO9780511529887.

Noble, R., Hobbs, P.J., Mead, A. and Dobrovin-Pennington, A. 2002. Influence of straw types and nitrogen sources on mushroom composting emissions and compost productivity. *J. Ind. Microbiol. Biotechnol.*, 29(3): 99–110.

Pardo, A., de Juan, A.J., Pardo, J. and Pardo, J.E. 2004. Assessment of different casing materials for use as peat alternatives in mushroom cultivation. Evaluation of quantitative and qualitative production parameters. *Span. J. Agric. Res.*, 2(2): 267–272.

Royse, D.J. 2010. Effects of fragmentation, supplementation, and addition of phase II compost to 2nd break compost on mushroom (*Agaricus bisporus*) yield. *Bioresour. Technol.,* 101: 188–192.

Shamshad, A., Clift, A.D. and Mansfield, S. 2009. Effect of compost and casing treatments of insecticides against the sciarid *Bradysia ocellaris* (Diptera: Sciaridae) and on the total yield of cultivated mushrooms, *Agaricus bisporus*. *Pest Manag. Sci.,* 65(4): 375–80. doi: 10.1002/ps.1700.

Suman, B.C. and Sharma V.P. 2007. *Mushroom Cultivation in India*. Daya Books, Cultivated mushroom, pp. 179.

Zied, D.C., Minhoni, M.T.A., Pardo-González, J.E. and Pardo-Giménez, A. 2010. A study of compost added to a casing technique in *Agaricus bisporus* cultivation from phase III bulk compost. *Hort. Science*, 45: 1649–1653.

Section-2

Principles of Mushroom Production and Cultivation

2.1

Methods and Strategies for Strain Improvement of Commercial Mushroom Varieties

JATINDER KAUR[1]* AND H.S. SODHI[2]

ABSTRACT

Productivity and quality of widely cultivated mushrooms are mainly dependent on strain make up, therefore, many strains have constantly been produced, aiming at higher yield and improved quality attributes such as accumulation of nutrients of interest, pathogen resistance, sporelessness and adaptability to wide temperature range. For strain improvement, a complete knowledge of the life cycle of species is desirable with the ability to manipulate it genetically. Genetic manipulations can be achieved in a number of ways. Cross breeding (anastomosis) is an important tool to combine characteristics of two parent strains into one entity. The investigation of the molecular basis of mating provides access to the use of mating type genes in order to facilitate breeding. Protoplast fusion has been used as an alternate method to create mushroom hybrids, especially when conventional method cannot be used. Unlike conventional hybridization, protoplast fusion can be performed intraspecifically, intergenerically and even interheterogenerically. However, for the same reason as implied to the conventional method, the greater the distance in genetic relationship between the two mating isolates, the less successful will be protoplast fusion. In recent years, mutagenesis has proved as an appropriate method for strain improvement of mushrooms. Chemical mutagens like N-methyl-N'-nitro-N-nitrosoguanidine, sodium nitrite, 5'-bromouracil result in single nucleotide substitution while ultraviolet rays can cause deletion and translocation. These alterations can produce beneficial effects on mushroom yield and characteristics.

[1 2] Department of Microbiology, Punjab Agricultural University, Ludhiana, Punjab.
**Corresponding Author*: E-mail: jatinder_sandhu@rediffmail.com

Key words: Strain improvement, Anastomosis, Protoplast fusion, Mutations.

INTRODUCTION

Mushroom industry is growing tremendously as the culinary and pharmaceutical values of different mushrooms are being discovered and reported frequently. With increase in demand of mushroom there has been an increasing trend of research on strain improvement of mushrooms. New strains with increased yield, resistance to pests, improved fruiting quality, wider substrate utilization, sporelessness, environmental bioremediation, and stress tolerance to biotic and abiotic factors are being researched upon (Kapoor *et al.,* 1996). *Agaricus bisporus*, the most widely produced mushroom has been subjected to number of strain improvement techniques followed by other speciality mushrooms like *Pleurotus*, *Lentinus*, *Volvariella*, *Calocybe etc.* Numbers of different techniques are being employed to bring together desirable characters or bring about some useful mutations. Simple techniques such as selection (Monospore and Multispore) and cross breeding (hybridization) can be easily adopted to improve the characters of the cultivated mushrooms. Other techniques involve protoplast fusion and mutations. For strain improvement there is a need to study exact life cycle and mating system of mushrooms.

Life Cycle of Basidiomycetes

Basidiomycetes are characterized by septate mycelium. Like most fungi, basidiomycetes are haploid. The life cycle begins with the germination of haploid basidiospores under appropriate environmental conditions. A single basidiospore germinates to form a monokaryon mycelium which contains usually one individual haploid nucleus or sometimes two genetically identical haploid nuclei (Waters *et al.,* 1975; Kües, 2000; Kües *et al.,* 2002). Two compatible monokaryons with opposite mating type anastomose and nuclear exchange leads to formation of dikaryons. A fertile dikaryotic mycelium is characterized by vigorous growth of mycelium, the presence of two distinct haploid nuclei per septa and clamp connections (Kües and Casselton, 1992; Badalyan *et al.,* 2004; Skrede *et al.*, 2013). The dikaryotic mycelium gives rise to fruiting bodies under defined conditions. In the specialized basidium, present on pileus of fruiting body, there occurs karyogamy (fusion of distinct nuclei) followed by meiosis results in the formation of four basidiospores per basidium (Moore, 1998; Kües, 2000). Each basidiospores receives one haploid nucleus resulting from meiosis and the cycle goes on (Srivilai *et al.,* 2009).

Mating System

Basidiomycetes mainly possess two sexes, but have multiple mating types. Individuals of basidiomycetes fungi can only mate if their mating types are different and the mating type is determined by the loci (James *et al.,* 2006). Mating system of basidiomycete fungi is termed as tetrapolar or bifactorial incompatibility. In this system mating type is determined by two unlinked loci A and B. Factor A is known to control nuclear pairing, clamp cell formation, co-ordinate cell division and clamp cell septation while factor B leads to nuclear migration, septa dissolution and clamp cell fusion (Kothe, 2001). Compatibility at both the loci is essential for mating and fruiting. Tetrapolar mating system *i.e.,* involvement of two mating type factors (A and B) which heterozygosity at both loci required for formation of dikaryotic mycelium, in basidiomycetes has been observed in at least two of the three major lineages, the Ustilaginomycetes, or smut fungi, and the Hymenomycetes, the mushroom fungi (Burnett, 1975). Multiple alleles are present at the mating-type loci which allow individuals within a population to be mating compatible with each other. The homobasidiomycetes possess a large number of alleles at both loci (Whitehouse, 1949; Raper, 1966).

Approximately 65% of the homobasidiomycetes possess tetrapolar mating system while 25% species have a bipolar system which is controlled by a single locus with multiple alleles (Raper, 1966). This results in higher inbreeding rate in bipolar system as compared to tetrapolar system.

The bipolar mating system is supposed to have evolved from a tetrapolar mating system. Raper (1966) has put forward three plausible hypotheses regarding the origin of the bipolar mating system. According to one of the hypotheses, there occurs a primary mutation at one of the mating-type loci which display a self-compatible phenotype and result in bipolar mating behaviour of tetrapolar strains possessing the mutated alleles. When such mutated self-compatible mating-type alleles get fixed in a population, the population will be subsequently considered as bipolar. The second hypothesis suggests translocation of chromosomal segment having one of the mating-type loci into close vicinity of the other mating type loci, ultimately leading to fusion of the two into a single non recombining region. The third hypothesis states that as A and B mating type loci control different, but interlinked functions during dikaryotic growth and fruiting. Therefore, during evolution the function of either of the mating-type loci could be gradually taken over by the other.

TECHNIQUES USED FOR STRAIN IMPROVEMENT

Conventional Breeding

Mushroom breeding is comparatively a new approach (Sonnenberg *et al.,* 2005). In this technique, there are two critical steps involved: germination of spores and cross breeding/hybridization/ anastomosis.

Collection and germination of spores

Collection and germination of spores are of utmost importance. Bukowski (1954) devised the technique for collecting spores of the cultivated mushroom by placing sterile dish under growing basidiocarp. A modified collecting chamber for basidiospores was devised by Phutela (1986) (Figs. 1 and 2). The spore collecting technique remains the same for different mushrooms, but the germination frequency and germination techniques for spores of various mushroom species are quite different. Ferguson (1902) described the detail process for germinating mushroom spores and found that mass spore transfer maintained in a slightly acidic medium showed good germination initially only for a few spores, but later there was an enhancement in germination frequency due to production of some volatile compounds by germinated spores. He also pioneered the spore germination studies of *A. campestris* on a defined medium which was adopted by Falck (1924) in *Psalliota* sp. using distilled water, dung decoction, synthetic and semi-synthetic media, but he recorded the best germination on 10% malt extract supplemented with 0.2% succinic acid. Other media like potato dextrose agar (Bretzloff *et al.,* 1962; Kneebone *et al.,* 1974) wheat agar and Lambert agar (Fritsche, 1978) have been reported for spore germination and raising monospore and multispore cultures.

Sometimes temperature shock can also prove beneficial for germination of mushroom spores. According to Mills and Eilers (1973) heating spores to 45°C for 4 h resulted in 23% germination in case of *Coprinus radiate* on media. It can also prove beneficial in germination of other fungal spores.

The association of growing mycelium with the basidiospores is known to enhance their germination (Ferguson, 1902; Falck, 1924), which is due to diffusion of certain metabolic products from the mycelium (Stanek, 1959). Some such compounds have been identified as 2,3-dimethyl-1-pentene (Mc Teague *et al.,* 1959) 7C olefin, isovaleric acid and isoamyl alcohol (Losel, 1964).

Cross breeding

Cross breeding (anastomosis/hybridization) is an important tool to combine characteristics of two parent strains into one entity, the scientific basis of which was provided by Raper *et al.* (1972) in *A. bisporus*. Breeding high yielding strains has traditionally been accomplished by trial and error method. Large numbers of hybrids can be obtained and cultivated to evaluate the production characteristics (Eger, 1978). In heterothallic basidiomycetes, basidiospores, a meiotic product, can be collected and germinated individually to give rise to single spore monokaryotic cultures (Fig. 3). These single spore isolates are then inter-crossed (Fig. 4) to identify their mating type compatibility to form hybrid dikaryons (Fig. 5) followed by characterization aided by RAPD and RFLP analysis. If two mycelia involved in a cross are similar in both *A* and *B*, then the *A*- and the *B*-controlled pathways of dikaryon development remain turned off. There is a lack of interaction between the two mycelia due to incompatibility and both will continue to grow as haploid mycelial entities (Fig. 2). If either *A* genes or *B* genes are different between the two mycelia, then only the processes induced by the *A* pathway or *B* pathway of development, respectively, are turned on after the mycelia fusion. These are semi-compatible interactions, which are incapable of producing fertile dikaryons. A fully compatible mating occurs when both *A* genes and the *B* genes are different, leading to the formation of stable and fertile dikaryon with vigorous aerial mycelium growth. Thereby, nuclear migration is induced, nuclei are exchanged between both mycelia and clamp connections are formed in every new cell (Casselton and Olesnicky, 1998; Kothe, 1999; Kües, 2000). The case is different for homobasidiomycetes (*A. bisporus). A. bisporus* spores carry two haploid nuclei with different mating types and, therefore, the germinating spore directly forms a dikaryon (Raper *et al.,* 1972).

Moessner (1962) was first to report the possibility of the interstrain breeding in *A. bisporus* by anastomosis between hyphae of two fertile cultures. Two very successful strains U-1 and U-3 of *A. bisporus* were developed by the cross breeding of infertile monospore culture at Mushroom Experimental Station, Horst, Netherlands (Fritsche, 1983). Since then a number of interstrain and interspecies hybridizations have been attempted (Yu and Chang, 1987; Eichlerova and Homolka, 1999; Kapoor *et al.,* 1996; Kaur *et al.,* 2008; Chauhan, 2009; Kumar, 2010).

Kapoor *et al.* (1996) performed cross hybridization among the monokaryons of *P. florida*. Out of eight dikaryons developed, there was one dikaryon (Biological efficiency 61.0%) which out yielded the parent culture (biological efficiency 49.7%). Eichlerova and Homolka (1999) were able to develop a dikaryon with high laccase and ligninolytic enzyme production by mating between compatible monkaryons of *P. ostreatus*. Bao *et al.* (2004) tested the mating compatibility of 25 *Pleurotus* spp. by mon-mon mating and di-mon mating. The results revealed 5 intersterilty groups (*P. ostreatus, P. cystidiosus, P. cornucopiae, P. pulmonarius* and *P. salmoneo stramineus* complexes) and 8 independent intercompatible spp. (*P. calyptratus, P. corticatus, P. dryinus, P. eryngii, P. nebrodensis, P. ostreatus* var columbines, *P. djamor and P. flabellatus*). In another case, Sawashe and Sawant (2005) developed hybrid cultures S-51, S-55 and S-56, which required significantly shorter period for spawn run as compared to the parents, *P. florida* and *P. eous*. Moreover, the average weight of the fruit body of the hybrid cultures S-55 and S-56 was found to be higher than that of both the parents. Kaur *et al.* (2008) attempted cross breeding of monokaryons of *P. florida* and obtained a fertile grey fruiting dikaryon, PFJ4. According to Jaswal *et al.* (2014) laccase and protease activity was higher in hybrids 37 and 46, respectively which were produced by interspecific breeding between *P. florida* and *P.*

sajor-caju. Cross breeding may not also produce fruiting dikaryons with desirable characters. For *e.g.,* Kumara and *Edirimanna* (2009) attempted hybridization between monospore cultures of Lanka Oyster and American Oyster (*Pleurotus ostreatus*). Five hybrid cultures were selected on the basis of their growth in laboratory conditions but none of them showed higher yield when compared with the parent strains.

Protoplast Fusion

Protoplast fusion has been used as a method to create mushroom hybrids, especially when conventional method could not be achieved. The use of protoplast fusion for development of new strains could be traced back to 1980s. As conventional hybridization, protoplast fusion can be performed intraspecifically (Kiguchi and Yanagi, 1985; Toyomasu and Mori, 1987) interspecifically (Takehara *et al.,* 1993; Matsumoto *et al.,* 1997) intergenerically (Eguchi *et al.,* 1993; Zhao and Chang, 1996) and even interheterogenerically (Toyomasu and Mori, 1987). However, for the same reason as that implied to the conventional method, the greater the distance in genetic relationship between the two mating isolates, the less successful will be protoplast fusion (Anne and Peberdy, 1976).

Genetic manipulation using protoplast fusion involves 3 major steps: Protoplast liberation, protoplast fusion and protoplast regeneration.

Protoplast liberation

The cell wall of fungi is made up of chitin, which needs to be dissolved for liberation of complete protoplast. A number of factors are to be kept in mind while liberating protoplast from fungal cells. They are being discussed below:-

Factors affecting protoplast liberation

Mycelial age

Mycelia age can affect protoplast isolation (Fig. 6) and its regeneration, although both juvenile and older mycelia of basidiomycete can be used for protoplast isolation and regeneration, the protoplasts from older mycelia can be regenerated into mycelia more rapidly and easily. The probable reason could be that older mycelia have a more intact nucleus and organelles (Santiago *et al.,* 1982a; Santiago *et al.,* 1982b; Peng *et al.,* 1993; Wang *et al.,* 2005).

Lytic enzymes

Lytic enzyme concentration is one of the most important factors affecting protoplast release. If too low, mycelia cannot be digested sufficiently; too high, protoplast membranes will be injured in spite of high yields, which will lead to a low regeneration frequency (Kitamoto *et al.,* 1988; Zhang and Li, 1993; Rao and Prakash, 1995; Naseema *et al.,* 2008).

Lywallzyme, a lytic enzyme produced by Guang Dong Institute of Microbiology, P.R. China, has been used to liberate the protoplasts in a number of studies (Sun *et al.,* 2001; Wang *et al.,* 2004; Li *et al.*, 2006). A Novozym 234 solution (5 mg/ml prepared in osmotic stabilizer solution) was used by Mukherjee and Sengupta (1986) for the liberation of protoplasts of *Volvariella*.

Osmotic stabilizers

The osmotic stabilizers, used for rinsing mycelia, diluting lytic enzyme and preparing regeneration media, are also important to influence protoplast liberation and regeneration. As for protoplast liberation, an osmotic stabilizer is only suitable for a few species owing to the differences of cell wall compositions in different species (Cheng and Bélanger, 2000). For example, mannitol was best for *Agrocybe aegerita* (Zhang *et al.,* 2004) while $MgSO_4$ proved best one for *Sclerotium rolfsii* (Morinaga *et al.,* 1985; Faria *et al.*, 2004) while in case of *C. comatus* 0.6 M KCl provided the highest yield. An osmotic stabilizer solution was developed by Mukherjee and Sengupta (1986) for the liberation of protoplasts of *Volvariella* which contained 0.6 M NaCl in 0.01 M phosphate buffer (pH 6.0). 0.6M $MgSO_4$ was used as an osmotic stabilizer by Djajanegara and Masduki (2010) in case of protoplast isolation of white and brown oyster.

Protoplast fusion

Spontaneous protoplast fusion is very rare due to the negative charge on the protoplast membrane (Masekesa, 2007). This leads to strong repulsive forces which inhibit fusion. Therefore, there is need of fusogens to increase the frequency of induced fusion (Navratilova *et al.,* 2007). Calcium chloride at high concentration and high pH neutralize the overall negative charge of the protoplast resulting in their close proximity and ultimate fusion. Polyethylene glycol (PEG) is another fusogen which is being used since 1974. It increases the frequency of fusion and it has very low toxicity. It helps in fusion as it induces protoplast agglutination. The most widely adopted method to produce fusion events between protoplasts has been the use of PEG polyethylene glycol in presence of $CaCl_2$. Development of fusants between *P. ostreatus* and *P. djamor* using 40% PEG has been reported (Dhitaphichit and Pornusuriya, 1995). Electrofusion is also being frequently used. Protoplasts are placed in a small culture cell along with electrodes. A potential difference is applied across the electrodes which lead to an alignment of protoplasts and ultimately their fusion. Numerous reports are available pertaining to the use of electrofusion in *Agaricus bisporus* (Sonnenberg and Wessels, 1987) and *Pleurotus ostreatus* (Magae *et al.,* 1986; Kaur and Kapoor, 2014). Fusion events brought about by calcium chloride and PEG are very random in contrast to electrofusion where more control can be achieved.

Protoplast regeneration

Once the fusion process is over, the protoplast need to be regenerated whereby their cell wall is formed and they continue to divide and reproduce. A number of factors need to be taken into account for successful regeneration. The two most important being: the type of osmotic stabilizer and regeneration medium.

Factors affecting protoplast regeneration

Osmotic stabilizer

In absence of cell wall, the cell becomes susceptible to surrounding osmotic pressure and is prone to shrinkage or bursting depending on whether the surrounding environment is hypertonic or hypotonic, respectively. Osmotic stabilizers like sorbitol, mannitol, inositol *etc.* ensure that water does not flow either into or out of the protoplast. Regarding protoplast regeneration, the optimum osmotic stabilizers supplemented into regeneration media also vary for different species. For instance, sucrose was reported best for *A. bisporus* and *A. bitorquis* (Yanagi *et al.,* 1985; Sonnenberg *et al.,* 1988) while in case of *P. brevicompactum*, 0.8M KCl gave better results (Varavallo *et al.,* 2004).

In the experiments carried out by Li *et al.* (2010) CYM medium containing 0.6 M manitol proved to be the best, with 1.4% of regeneration percentage, followed by the CYM appended with 0.6 M $MgSO_4.7H_2O$. Mannitol, as osmotic stabilizer, has been used for protoplast regeneration in plant and other fungi. According to speculation, it is the direct precursor of cell wall synthesis or the indirect one by metabolism and transformation, and thus, could speed up the cell wall synthesis and make protoplast regenerate more easier and quicker (Lau *et al.,* 1985; Peng *et al.,* 1993; Kim *et al.,* 2000; Chitnis and Deshpande, 2002; Balasubramanian *et al.*, 2003).

Regeneration media

Besides osmotic stabilizers, media can also influence significantly protoplast regeneration. A regeneration medium contains required nutritional components which induce the formation of cell wall of the protoplast and help in their further proliferation. Numbers of regeneration media have been tested for regeneration of protoplast fusants by numerous scientists. In the studies by Faria *et al.* (2004) on protoplast regeneration of *Sclerotium rolfsii* ATCC 201126, the MYG medium supplemented 0.6M sucrose (26%) or 0.7M NaCl (17%) displayed the highest regeneration rate. Dhitaphichit and Pornsuriya (2005) demonstrated *Pleurotus* fusants protoplasts regeneration by culturing them on a plate of regeneration medium which contained 30 g agar, 20 g malt extract, 20 g glucose, 1 g peptone and 1000 ml of 0.6 M sucrose followed by overlaying with the same kind of medium but with a concentration of agar of only 5 g. The media used by Mukherjee and Sengupta (1986) for regeneration of mutated protoplasts of *Volvariella* contained (wt/vol) 1% glucose and 0.5% yeast extract in osmotic stabilizer solution.

The CYM and MYG media were tested for usefulness in protoplast regeneration by Sun *et al.* (2001). The average regeneration percentage was 1.5% in CYM medium and 0.1% in MYG medium. Colonies appeared after incubation for 3 d on CYM plate and 6 d on MYG plate. These findings suggested that the CYM medium was much more efficient than the MYG medium for protoplast regeneration.

Minimum Regeneration Media (MRM) containing 10 g dextrose, 2.8 g $(NH_4)_2SO_4$, 50 ml A solution (10 g KCl, 10 g $MgSO_4.7H_2O$ and 0.2 g $FeSO_4.7H_2O$), 50 ml B solution (20 g K_2HPO_4) and 15 g bacteriological agar followed by overlaying with the same kind of medium but with a concentration of agar of only 5 g was used for regeneration of fusants by Djajanegara and Masduki (2010). The strain of *A. bisporus* (ACCC50658) was used for protoplast isolation and regeneration was maintained on PDYA medium added with 0.6 M sucrose (Begin and Spear, 1991; Nazrul *et al.*, 2010).

Mutagenesis of Protoplasts

Mutations are permanent changes in genetic makeup of an organism which are passed on to the offsprings. Mutations may be silent, useful or even lethal mutations. Numbers of physical and chemical induced mutagens are available which change the gene sequence and may bring about desired changes. Physical agents like ultraviolet light leads to formation of T-T dimer and this further brings about change in replicated strand of DNA. Alkylating agents are usually used during mutagenic studies. They are electrophilic compounds with affinity for nucleophilic centers in organic macromolecules. These include a wide variety of chemicals like nitrous acid, hydroxylamine, N-methyl- N'-nitro-N-nitrosoguanidine (NTG) and ethyl methanesulfonate (EMS). These lead to transition mutations. Base analogues are yet another group of induced mutagen. Base analog is a substance other than a standard nucleic acid base that can be incorporated into a DNA molecule by the normal process of polymerization. Such a substance must be able to pair with the base on the complementary strand. For example, 5-bromouracil is an analogue of thymine and might cause an A-T to G-C transition mutation. Substances whose dimensions are roughly the same as those of a purine-pyrimidine pair may act as intercalating agents. In aqueous solutions, these substances form stacked arrays, and are also able to stack with a base-pair by insertion between two base-pairs. This may result in frameshift mutation. The above mutations lead to formation of putative mutations which are confirmed as mutants using molecular techniques which are discussed further.

Fungi are multicellular organisms; therefore it is difficult to go for mutation of each and every cell. Therefore, protoplast acts as ideal material for mutagenesis, which can be passed throughout the organisms on regeneration of mutated protoplast.

Numbers of experiments have also conducted on basidiomycetes for strain improvement using mutagenesis. The protoplasts of *Volvariella* were irradiated at 254 nm at a dose rate of 1 J/m^2 per s. After irradiation, the protoplasts were kept for 30 min in the dark to allow full expression of the damage induced by ultraviolet (UV) light. The irradiated protoplasts and untreated control, in parallel, were suitably diluted with osmotic stabilizer solution and plated for regeneration on solid medium. For N-methyl-N'-nitro-N-nitrosoguanidine (NTG) treatment, protoplasts were exposed to the action of NTG for different periods from 15 to 120 s. NTG was used at two concentrations, 15 and 30, µg/ml of osmotic stabilizer solution. The protoplasts were immediately washed three times with osmotic stabilizer solution to remove NTG, diluted, and plated on regenerating solid medium (Mukherjee and Sengupta, 1986).

To induce the lignocellulolytic mutants of *Pleurotus ostreatus*, the mycelia were irradiated by gamma-ray to doses of 1–2 kGy. Five strains were isolated by the criteria of clamp connection, fruiting body formation, growth rate and activities of extracellular enzymes. All isolated strains were able to produce fruiting bodies and grew similarly to the control. The extracellular enzymes activities in liquid media of isolated strains were up to 10 times higher than the control. Genetic similarities of the isolated strains ranged from 64.4% to 93.3% of the control. From these results, it seems that the genetic diversity of *P. ostreatus* could be changed and useful strains can be induced by gamma-ray radiation to recycle or reuse biowastes (Lee *et al.*, 2000).

In another experiment Kaur *et al.* (2011) performed an experiment on mutagenesis of protoplast of *Calocybe indica* using four mutagenic treatments (UV, NTG, Ethidium bromide and bromouracil). Four high yielding mutants were obtained (CMN-3, CMN-9, CMN-11 and CMB-4) which gave significantly higher yield that the parent strain (Ci-3). Kang *et al.* (2013) performed mutagenesis on *Flammulina velutipes* and were able to develop new strain which were high temperature tolerant.

MOLECULAR CHARACTERIZATION

Molecular characterization is an efficient, effective and important tool for analysis of new hybrids or mutants. It is essential to determine the similarity coefficient and variation of newly developed strains from the parent strains. Numbers of molecular techniques are available. Out of these, for fungus characterization, use of Randomly Amplified Polymorphic DNA (RAPD) and Restriction Fragment Length Polymorphism (RFLP) is more prominent. These use the primers for amplification of specific DNA segments and ultimately run on a gel. Using various authorized softwares the DNA sequences are related with the parent strain.

CONCLUSIONS

Numbers of different approaches are being used for strain improvement of mushrooms. All the techniques may not be successful for a particular genus but a complete study of mating system of a fungus may prove beneficial for selecting the appropriate technique for its improvement. Mushroom breeding and manipulation needs lots of patience and hard work. Every experiment may not lead to desired results. There are still number of mushroom genus whose life cycle, mating system and strain improvement are in their infant stages and in some cases they have not even be explored to their slightest.

REFERENCES

Anne, J. and Peberdy, J.F. (1976). Induce fusion of fungal protoplast following treatment with polyethylene glycol. *Journal of General Microbiology,* 92: 413–417.

Badalyan, S.M., Polak, E., Aebi, M. and Kues, U. (2004). Role of peg formation in clamp cell fusion of homobasidiomycete fungi. *J. Basic Microbiol.*, 44: 167–177.

Balasubramanian, N., Julietg, A., Srikalaivani, P. and Lalithakumari, D. (2003). Release and regeneration of protoplasts from the fungus *Trichothecium roseum. Canadian Journal of Microbiology*, 49: 263–268.

Bao, D., Kinugasa, S. and Kitamoto, Y. (2004). The biological species of oyster mushrooms (*Pleurotus* spp.) from Asia based on mating compatibility. *Journal of Wood Science*, 50: 162–168.

Begin, M. and Spear, M. (1991). A novel method for inducing the expression of sectors in *Agaricus bisporus. In: Science and cultivation of edible fungi,* Maher M.J. (*ed.*), Balkema, Rotterdam, pp. 105–109.

Bretzloff, C.W., Robbins, W.A. and Curme, J.H. (1962). Observation on multisporus isolates from the cultivated mushroom *Agaricus bisporus* (Lange) Sing. *Mushroom Science*, 5: 188–196.

Bukowksi, T. (1954). Technique for collecting spores of cap fungi for germination purposes. *Acta Microbiologica Polonica*, 3: 399–401.

Burnett, J.H. (1975). *Mycogenetics*. John Wiley & Sons, London.

Casselton, L.A. and Olesnicky, N.S. (1998). Molecular genetics of mating recognition in basidiomycete fungi. *Microbiol. Mol. Biol. Rev.*, 62: 55–70.

Chauhan, S. (2009). *RAPD marker assisted development of improved strains of Agaricus bisporus (Lange) Sing.* Ph.D. Dissertation, Punjab Agricultural University, Ludhiana.

Cheng, Y. and Bélanger, R.R. (2000). Protoplast preparation and regeneration from spores of the biocontrol fungus *Pseudozyma flocculosa*. *FEMS Microbiological Letters*, 190: 287–291.

Chitnis, M.V. and Deshpande, M.V. (2002). Isolation and regeneration of protoplasts from the yeast and mycelial form of the dimorphic zygomycete Benjaminiella poitrasii: Role of chitin metabolism for morphogenesis during regeneration. *Microbiological Research*, 157: 29–37.

Dhitaphichit, P. and Pornsuriya, C. (2005). Protoplast fusion between *Pleurotus ostreatus* and *P. djamor*. *Songklanakarin Journal of Science and Technology*, 27: 975–982.

Djajanegara, I. and Masdukia, A. (2010). Protoplast fusion between white and brown oyster mushrooms. *Journal of Agricultural Science,* 11: 16–23.

Eger, G. (1978). Biology and breeding of *Pleurotus*, pp. 497–519. *In*: *The biology and cultivation of edible mushrooms* (Chang, S.T. and Hayes, W.A. (*eds.*)) AOP. New York, p. 819.

Eguchi, F., Fukuzumi, T. and Higaki, M. (1993). Production of new species of edible mushrooms by protoplast fusion and analysis of fusant between *Pleurotus sajor-caju* and *Mycoleptodonoides aitchisonii*. *Mokuzai Gakkaishi,* 41: 342–348.

Eichlerova, I. and Homolka, L. (1999). Preparation and crossing of basidiospores-derived monokaryons- A useful tool for obtaining laccase and other ligninolytic enzyme higher producing dikaryotic strains of *Pleurotus ostreatus.* Antonie van Leeuwenhoek, 75: 321–327.

Falck, R. (1924). Uber die sporen Kei mung des champignons. *Mykol Untersuchungen and berichte, Beihft.I.*

Faria, J.I., Molina, O.E. and Figueroa, L.I.C. (2004). Formation and regeneration of protoplasts in *Sclerotium rolfsii* ATCC 201126. *Journal of Applied Microbiology*, 96: 254–262.

Ferguson, M.C. (1902). A preliminary study of the germination of spores of *Agaricus campestris. U.S. Dept. Agric. Bur. Pl. Ind. Bull.*, 16: 1–40.

Fritsche, G. (1978). Breeding work, *In:* Chang, S.T. and Hayes, W.A. (*eds.*), *The Biology and Cultivation of Edible Mushrooms,* Academic Press, London, pp. 239–50.

Fritsche, G. (1983). Breeding *Agaricus bisporus* at the Mushroom Experimental Station, Horst. *Mushroom Journal,* 122: 49–53.

James, T.Y., Srivilai, P., Kues, U. and Vilgalys, R. (2006). Evolution of the bipolar mating system of the mushroom *Coprinellus disseminatus* from its tetrapolar ancestors involves loss of mating-type-specific pheromone receptor function. *Genetics*, 172: 1877–1891.

Jaswal, R.K., Sodhi, H.S. and Sharma, S. (2014). Characterization of interspecific hybrid dikaryons of the oyster mushrooms, *Pleurotus florida* PAU-5 and *P. sajor-caju* PAU-3 (higher Basidiomycetes) from India. *Int. J. Med. Mushrooms,* 16: 355–364.

Kang, L.Z., Han, F., Lin, J.F., Guo, L.Q. and Bai, W.F. (2013). Breeding of new high-temperature-tolerant strains of *Flammulina velutipes. Scientia Horticulturae*, 151: 97–102.

Kapoor, M., Sodhi, H.S. and Dhanda, S. (1996). Strategies for strain improvement in *Pleurotus* species. *Mushroom Research*, 5: 57–66.

Kaur, J., Sodhi, H.S. and Kapoor, S. (2008). Breeding of *Pleurotus florida* (oyster mushroom) for phenotypic pigmentation and high yield potential. *Journal of the Science of Food and Agriculture*, 88: 2676–2681.

Kaur, J., Sodhi, H.S., Kapoor, S., Khanna, P.K. and Jaswal, R.K. (2011). Strain improvement of speciality mushroom, *Calocybe indica*, through mutagenesis. *Applied Biological Research*, 13: 62–69.

Kaur, L. and Kapoor, S. (2014). Protoplast electrofusion for development of somatic hybrids between *Pleurotus florida* and *Pleurotus sajor-caju*. *International Journal of Pharma and Bio Sciences*, 5: 507 –519.

Kiguchi, T. and Yanagi, S.O. (1985). Intraspecific heterokaryon and fruit body formation in *Coprinus macrorhizus* by protoplast fusion of auxotrophic mutants. *Applied Microbiology and Biotechnology*, 22: 121–127.

Kim, B.K., Kang, J.H., Jin, M., Kim, H.W., Shim, M.J. and Choi, E.C. (2000). Mycelial protoplast isolation and regeneration of *Lentinus lepideus*. *Life Sciences*, 66: 1359–67.

Kitamoto, Y., Mori, N., Yamamoto, M., Ohiwa, T. and Ichikawa, Y. (1988). A simple method for protoplast formation and improvement of protoplast regeneration from various fungi using an enzyme from *Trichoderma harzianum*. *Applied Microbiology and Biotechnology*, 28: 445–450.

Kneebone, L.R., Hwang, S.W., Schultz, P.G. and Patton, Jr. T.G. (1974). Comparative production and performance of stock cultures of eight strains of *Agaricus bisporus* preserved by liquid nitrogen freezing and by repeated vegetative transfer. *Mushroom Science*, 9: 229–35.

Kothe, E. (1999). Mating types and pheromone recognition in the homobasidiomycete *Schizophyllum commune*. *Fungal Genet. Biol.,* 27: 146–152.

Kothe, E. (2001). Mating-type genes for basidiomycete strain improvement in mushroom farming. *Appl. Microbiol. Biotechnol.*, 56: 602–612.

Kües, U. (2000). Life history and developmental processes in the basidiomycete *Coprinus cinereus*. *Microbiol. Mol. Biol. Rev.*, 64: 316–353.

Kües, U. and Casselton, L.A. (1992). Homeodomains and regulation of sexual development in basidiomycetes. *Trends Genet.,* 8: 154–155.

Kües, U., Walser, P.J., Klaus, M.J. and Aebi, M. (2002). Influence of activated *A* and *B* mating-type pathways on developmental processes in the basidiomycete *Coprinus cinereus*. *Mol. Genet. Genomics,* 268: 262–271.

Kumar, R. (2010). Interspecific hybridization between *Pleurotus florida* and *Pleurotus sajor- caju* to develop hybrid dikaryon(s) and their molecular characterization. Ph.D. Dissertation, Punjab Agricultural University, Ludhiana, India.

Kumara, K.L.W. and Edirimann, I.C.S. (2009). Improvement of strains of two oyster mushroom cultivars using duel culture technique. *World Applied Sciences Journal*, 7: 654–60.

Lau, W.C., Dhillon, E.K.S. and Chang, S.T. (1985). Isolation and reversion of protoplasts of *Pleurotus sajor-caju*. *Mircen J.*, 1: 345–53.

Lee, Y.K., Chang, H.H., Kim, J.S., Kim, J.K. and Lee, K.S. (2000). Lignocellulolytic mutants of *Pleurotus ostreatus* induced by gamma-ray radiation and their genetic similarities. *Radiation Physics and Chemistry*, 57: 145–50.

Li, L., Qingyu, Yin Q., Liu, X. and Yang, H. (2010). An efficient protoplast isolation and regeneration system in *Coprinus comatus*. *African Journal of Microbiology Research,* 4: 459–65.

Li, G., Li, R.X.., Liu, Q.Y., Wang, Q., Chen, M. and Li, B.J. (2006). A highly efficient polyethylene glycol-mediated transformation method for mushrooms. *FEMS Microbiological Letters*, 256: 203–08.

Losel, D.M. (1964). The stimulation of spore germination of *Agaricus bisporus* by living mycelium. *Annals of Botany,* 28: 541–54.

Magae, Y., Kashiwagi, Y., Senda, M. and Sasaki, T. (1986). Electrofusion of giant protoplasts of *Pleurotus cornucopiae. Applied Microbiology and Biotechnology,* 24: 509–11.

Masekesa, R.T. (2007). Developing a method for the enhancement of the Beta carotene content of sweet potato (cv Brondal) using protoplast technology. Master thesis, University of Zimbabwe.

Matsumoto, M., Eguchi, F. and Higaki, M. (1997). Polyethylene glycol centrifugal fusion between *Pleurotus ostreatus* and *Pleurotus sajor-caju. Mokuzai Gakkaishi*, 43: 215–19.

Mc Teague, D.M., Hutchinson, S.A. and Reed, R.I. (1959). Spore germination in *Agaricus campestris. Comum. Bot. Bucuresti*, 60: 240–49.

Mills, G.L. and Eilers, F.I. (1973). Factors influencing the germination of basidiospores of *Coprinus radiates. The Journal of General Microbiology*, 77: 393–401.

Moessner, E.J. (1962). Preliminary studies of the possibility of obtaining improved cultures through mycelial fusion (anastomosis). *Mushroom Science*, 5: 197–203.

Moore, D. (1998). Fungal morphogenesis. Cambridge University Press, Cambridge, UK, ISBN: 0-521-55295-8, p. 485.

Morinaga, T., Kikuchi, M. and Nomi, R. (1985). Formation and regeneration of protoplasts in *Coprinus pellucidus* and *Coprinus cinereus. Agriculture and Biological Chemistry*, 49: 523–24.

Mukherjee, M. and Sengupta, S. (1986). Mutagenesis of protoplasts and regeneration of mycelium in the mushroom *volvariella volvacea. Applied and Environmental Microbiology*, 52: 1412–14.

Naseema, A., Dhanya, B., Anjanadevi, I.P., Sheena, K.G. and Alex, S. (2008). Isolation and regeneration of protoplasts from the mycelium of *Fusarium pallidoroseum*-A potential biocontrol agent of water hyacinth [*Eichhornia crassipes* (Mart.) Solms]. *Journal of Tropical Agriculture*, 46: 67–69.

Navratilova, B., Greplova, M., Vyvadilova, M. and Klima, J. (2007). Protoplast fusion in selected vegetables of *Brassica, Communis* and *Solanum* genera. *ISHS Acta Horticulturae,* 725

Nazrul, M.I., Lin, F.X. and Yin-Bing, B. (2010). Screening of homokaryotic protoclones of *Agaricus bisporus* (J. Lge) imbach by colony characters and ISSR markers. *Bangladesh Journal of Botany*, 39: 119–22.

Peng, M., Lemke, P.A. and Shaw, J.J. (1993). Improved conditions for protoplast formation and transformation of *Pleurotus ostreatus. Applied Microbiology and Biotechnology*, 40: 101–06.

Phutela, R.P. (1986). *Strain improvement and development in the cultured mushroom Agaricus bisporus.* Ph.D. thesis, Punjab Agricultural University, Ludhiana.

Rao, K.S. and Prakash, A.H. (1995). A simple method for the isolation of plant protoplasts. *Journal of Biosciences*, 20: 645–55.

Raper, C.A., Miller, R.E. and Raper, J.R. (1972). Genetic analysis of the life cycle of *Agaricus bisporus. Mycologia*, 64: 1088–117.

Raper, J.R. (1966). *Genetics of Sexuality in Higher Fungi.* Ronald Press, New York.

Santiago, C.M. (1982a). *Studies on the Physiology and genetics of Volvariella volvacea* (Bull. ex. Fr.) Sing. Ph.D. Thesis, University of Nottingham, England.

Santiago, C.M. (1982b). Production of *Volvariella* protoplasts by use of *Trichoderma* enzyme. *Mushroom Newsletter for the Tropics*, 3: 3–6.

Sawashe, S.G. and Sawant, D.M. (2005). A study on monoculture isolation and interspecies hybridization of *Pleurotus*. *Journal of Maharashtra Agricultural University*, 30: 49–52

Skrede, I., Maurice, S. and Kauserud, H. (2013). Molecular characterization of sexual diversity in a population of *Serpula lacrymans*, a tetrapolar basidiomycete. *G3: Genes Genomes Genetics,* 3: 145–152.

Sonnenberg, A.S.M. and Wessels, J.G.H. (1987). Heterokaryon formation in the basidiomycete *Schizophyllum commune* by electrofusion of protoplasts. *Theoretical and Applied Genetics*, 74: 654–58.

Sonnenberg, A.S.M., Baars, J.J.P., Hendrickx, P.M. and Kerrigen, R.W. (2005). Breeding mushrooms: State of art. *Mush. Biol. Mush. Prod. Eds.*, 5: 163–173.

Sonnenberg, A.S., Wessels, J.G. and Vangriensven, L.J. (1988). An efficient protoplasting/regeneration system for *Agaricus bisporus* and *Agaricus bitorquis*. *Current Microbiology*, 17: 285–91.

Srivilai, P., Loutchanwoot, P. and Sukha, J. (2009). Blue light signaling inactivates the mating type genes-mediated repression of asexual spore production in the higher basidiomycete *Coprinopsis cinerea*. *Pak. J. Biol. Sci.*, 12: 110–118.

Stanek, M. (1959). The germination of the basidiospores of cultivated mushroom, *Agaricus bisporus. Ceska Mykologia*, 13: 241–51.

Sun, L., Cai, H.Q., Xu, W.H., Hu, Y.L., Gao, Y. and Lin, Z.P. (2001). Efficient transformation of the medicinal mushroom *Ganoderma lucidum*. *Plant Molecular Biology Reporter*, 19: 383–87.

Takehara, T., Kumata, A. and Aono, S. (1993). Interspecific protoplast fusion between some *Pleurotus* species using auxotrophic mutants. *Mokuzai Gakkaishi*, 39: 855–59.

Toyomasu, T. and Mori, K.I. (1987). Intra- and interspecific protoplast fusion between some *Pleurotus* species. *Agricultural Biology and Chemistry*, 51: 935–37.

Varavallo, M.A., de Queiroz, M.V., Pereira, J.F. and de Araújo, E.F. (2004). Isolation and regeneration of *Penicillium brevicompactum* protoplasts. *Acta Scientiarum*, 26: 475–79.

Wang, B.T., Xiong, Y. and Jiang, L. (2004). Studies on protoplast preparation and regeneration of *Agaricus bisporus. Southwest China Journal of Agricultural Science*, 17: 215–16.

Wang, W.M., Wu, R.Y. and Ko, W.H. (2005). Variation and segregation following nuclear transplantation in *Antrodia cinnamomea*. *Botanical Bulletin of Academia Sinica*, 46: 217–22.

Waters, H., Moore, D. and Butler, R.D. (1975). Structure of aerial and submerged sclerotia of *Coprinus lagopus*. *New. Phytol.*, 74: 199–205.

Whitehouse, H.L.K. (1949). Multiple allelomorph heterothallism in the fungi. *New Phytol.*, 48**:** 212–244.

Yanagi, S.O., Monma, M., Kawasumi, T., Hino, A., Kito, M. and Takebe, I. (1985). Conditions for isolation of and colony formation by mycelial protoplasts of *Coprinus macrorhizus*. *Agricultural and Biological Chemistry*, 49: 171–79.

Yu, M.Y. and Chang, S.T. (1987). Effect of osmotic stabilizers on the activity of mycolytic enzymes used in fungal protoplasts liberation. *Mircen Journal*, 3: 161–67

Zhang, Y., Wang, Q., Zhang, X.M. and Zhang, H.Y. (2004). A Study on the construction and UV-inducing of *Agrocybe aegerita* protoplast. *Edible Fungi of China,* 23: 15–17.

Zhang, Z.G. and Li, D.P. (1993). Studies on protoplasts technologies of filamentous fungi-effects of dewallzyme on the formation from several filamentous fungi (I). *Acta Scientiarum Naturalium Universitatis Normalis Hunanensis,* 16: 172–74.

Zhao, J. and Chang, S.T. (1996). Intergeneric hybridization between *Pleurotus ostreatus* and *Schizophyllum commune* by PEG-induced protoplast fusion and electro-fusion. *World Journal of Microbiology and Biotechnology,* 12: 573–78.

2.2

Is Pasteurization Really Needed for Small Scale Mushroom Farming?

SIDDHANT[1*], RUCHIRA SINGH[2], C.S. SINGH[3]
AND R.S. KANAUJIA[4]

ABSTRACT

Various treatments, viz., hot water application (T1), autoclaving (T2), radiation (T3) and chemical treatment (T4) were tried for the pasteurization of fresh and 12 months stored wheat straw to evaluate different pasteurization techniques and to find out their necessity in cultivation of mushrooms at small scale. In present investigation, the mushroom beds were examined for yield potential of Pleurotus eous. The crop of mushroom was harvested in three flushes where yield and biological efficiency ranged 255-300 gm, 51-60% and 90-285 gm, 18-57% respectively, under different pasteurization practices in both fresh and stored straw substrate. Among the techniques employed, chemically treated substrate proved most effective in respect of various parameters of mushroom production. The occurrence of competitor fungi was also noticed during the cultivation. Six fungal species belonging to Phycomycetes (Rhizopus stolonifer), Basidiomycetes (Coprinus cinereus) and Deuteromycetes (Aspergillus fumigatus, Alternaria alternata, Curvularia lunata and Penicillium sp.) were detected from mushroom beds. The overall results revealed the importance of pasteurization against yield loss due to presence of competitors. However, it did not seem compulsory in fresh substrate and optimum yield could be achieved without pasteurization.

Key words: Biological efficiency, Fungal competitors, Fresh and stored wheat straw, Pasteurization, *Pleurotus eous*, Yield loss.

[1] Department of Botany, Durgesh Nandini Degree College, Faizabad, Uttar Pradesh.
[2,3,4] Department of Botany, K.S.S. P.G. College, Ayodhya, Faizabad, Uttar Pradesh.
**Corresponding author:* E-mail: siddhant.ani@gmail.com

INTRODUCTION

The *Pleurotus* mushroom is generally referred to as 'Oyster mushroom' throughout the globe and 'Dhingri' in India. The term itself originated from Greek word "Pleuro" which means formed laterally or in a side way position, reflecting to the lateral position of the strip in relation to pileus. It occupied third rank among cultivated ones in respect of production at global level (Gyorfi and Hajdu, 2007). They are highly nutritious food rich in crude fibre and protein. They also contain low fat, low calories and good vitamins, therefore, considered as a pool of nutrients and popularly called "Vegetable meat" in respect of current energy food crisis. In addition, they possess multi-functional medicinal properties. For growing this mushroom, wheat straw is found to be most common and efficient substrate (Bhatti *et al.*, 1987) which is attacked by various contaminants of fungal, bacterial and pest nature, as a result decreased yield and biological efficiency are observed. To diminish the damages produced by competitors on mushroom development and yield, it is recommended to disinfect the substrate before spawning which is achieved by various methods (Arya and Arya, 2003; Bano *et al.,* 1979; Kumar *et al.*, 1990; Champawat and Chitale, 2003; Ram and Thakur, 2005; Tewari and Pandey, 1988). Removal of competitors will permit a faster and better, and more uniform spawn run and will assure a better resistance to further contaminations. As the consequence the mushroom yield will be higher and qualitatively superior.

In present communication, fresh and 12 months stored wheat straw was used as a cultivation medium for the farming of pink oyster mushroom *Pleurotus eous* to evaluate the different pasteurization techniques and find out their necessity in preparation of mushroom beds from fresh and stored wheat straw.

MATERIALS AND METHODS

Mushroom Culture

The pure culture of *Pleurotus eous* was obtained from the Mushroom Section of Plant Pathology Department, Chandra Shekhar Azad University of Agriculture and Technology, Kanpur (U.P.), India. It was maintained on potato dextrose agar medium (peeled, sliced and boiled potato, 200g; dextrose, 20g; agar, 20g l^{-1}) by using serial subculture method (Naraian *et al.,* 2009).

Spawn Strategy

a) Spawn substrate

Wheat grain (*Triticum aestivum*) was used as a spawn substrate which was purchased from the seed market of Faizabad. The spawn was prepared by the conventional method (Naraian *et al.*, 2009).

b) Spawn dose

Inoculation of substrate was made with spawn of *P. eous* at the rate of 15% w/w on dry weight basis under aseptic conditions.

Substrate for Mushroom Cultivation

a) *Substrate*

The substrate used for cultivation for *Pleurotus eous* was fresh and 12 months stored wheat straw. The fresh wheat straw was collected at the time of threshing of harvested crop from village Madarahia.

b) *Substrate preparation (Pasteurization)*

The wheat straw substrate (500 gm) was washed separately in fresh water. When excess water was drained off, it was pasteurized by following treatments:

1. ***Hot water application (T1):*** In this method, substrate was boiled in the water for one hour at 100°C.
2. ***Autoclaving (T2):*** This method included sterilization of substrate at 15 lbs/inch^2 pressure for 60 minutes.
3. ***Radiation treatment (T3):*** In this treatment, the substrate was exposed to UV light for 45 minutes.
4. ***Chemical treatment (T4):*** Wheat straw substrate was pasteurized in the solution of formaldehyde (500 ppm) and bavistin (75 ppm) for 18 hours.
5. ***Control (T5):*** Substrate was not pasteurized at all and it was soaked in water for 24 hours.

Method of Cultivation

Plastic bag technology was employed in this experiment. The beds were prepared from pasteurized substrate by multilayered spawning method according to the procedure described by Bano (1971). In this method, the wheat straw was placed at the bottom of polythene bag. Later, a layer of mushroom spawn was sprinkled. In the same way, four layers of straw and three layer of spawn were kept in the polythene bag. The mouth of bags was then tightened with fibre thread. The treatments were replicated thrice. These were incubated in cultivation room at 22-30°C temperature for spawn run. When mycelia had completely covered the beds, the polythene covering were removed and relative humidity was maintained 85-95 percent with the help of humidifier.

Concerning Data

a) *Data regarding mushroom production and yield parameters*

It included time lapsed in spawn run, pin head initiation and maturity of fruit bodies, number of flushes, total yield, biological efficiency, number and weight per sporocarp on the fresh and stored substrate subjected to different pasteurization techniques. The biological efficiency of mushroom was worked out as percentage yield of fresh mushroom in relation to dry weight of the substrate as suggested by Chang and Miles (1989).

b) *Encountered microorganism*

The competitors encountered on the mushroom beds were examined under light microscope. These were identified on the basis of morphological characters.

STATISTICAL ANALYSIS

Completely Randomized Design (CRD) was employed for yield data of the experiments. These were statistically analyzed. The Critical Difference (CD) was worked out at five percent probability level.

RESULTS

Different techniques were employed for the pasteurization of substrate. Their effect on cultivation statistics of *Pleurotus eous* with fresh and stored straw is presented in Tables 1-3 and plate.

Mushroom Production

All the practices employed for the substrate preparation, *viz.,* T1, T2, T3 and T4 showed almost similar response among themselves for substrate colonization, primordial initiation and fruit body maturation (15, 18 and 22 days, respectively) with both fresh and stored wheat straw. Results also indicated that they produced early sporophore production in comparison to unpasteurized set (T5) [Table 1].

Mushroom Yield

The crop of mushroom was harvested in three flushes where yield and biological efficiency ranged 255-300 gm, 51-60% in case of fresh straw substrate under different pasteurization process while in case of stored straw substrate, yield and biological efficiency ranged 90-285 gm, 18-57 %, respectively. In both the cases, chemical treatment gave highest yield and biological efficiency over others, although, in case of fresh straw it was at par to control. In contrast to fresh substrate, yield data of stored straw represent a highly significant treatment response over control where chemical pasteurization (285 gm and 57%) was found most effective followed by hot water treatment (240 gm and 48%), autoclaving (225 gm and 45%) and radiation treatment (165 gm and 33%). The unpasteurized sets gave negligible results in aforesaid manifestations (90 gm and 18%) [Table 2].

Table 1: Effect of pasteurization technique on substrate colonization, primordial development and fruit body maturation in fresh and stored wheat straw substrate.

Pasteurization technique	*Fresh wheat straw as substrate*			*Stored wheat straw as substrate*		
	Substrate colonization (Days)	*Primordial development (Days)*	*First harvest (Days)*	*Substrate colonization (Days)*	*Primordial develop ment (Days)*	*First harvest (Days)*
Hot water (T1)	15	18	22	15	18	22
Autoclave (T2)	15	18	22	15	18	22
UV- radiation (T3)	15	18	22	16	19	23
Chemical treatment (T4)	15	18	22	15	18	22
Control (T5)	17	20	24	23	27	32

Figures are average of three replications.

Percentage Increase in Yield Over Control

Data on yield increase (%) revealed that actual requirement of pasteurization practices. Promising results were not obtained with fresh wheat straw whereas the stored wheat straw exhibited the respective increase in yield (%) which showed the efficiency of pasteurization against stored straw substrate (Table 2).

Appearance of Competitors

During the course of study, a total of 06 fungal species were encountered on the mushroom beds (Plate) that hampered the exact potential of the crop. Among them five, *viz., Rhizopus stolonifer* (Phycomycetes), *Aspergillus fumigatus, Alternaria alternata, Curvularia lunata* and *Penicillium sp.* (Deuteromycetes) appeared mainly on pasteurized beds (Figs. 1A to 1F) but their occurrence was much lesser in fresh straw beds in early phase of spawn run. Later, their growth was suppressed by growing mushroom mycelium, therefore, they did not show much harmful effect on cropping of mushroom. While *Coprinus cinereus* (Basidiomycetes) were appeared on unpasteurized beds made from stored straw which resulted significant reduction in mushroom yield (Table 2) and sporophore quantity (Table 3). The higher decomposition of substrate was also detected in these beds.

Table 2: Mushroom production on fresh and stored wheat straw substrate under different pasteurization practices.

Pasteurization technique	*Fresh wheat straw as substrate*			*Stored wheat straw as substrate*		
	Total yield from three flushes [gm/500 gm dry substrate]	*Biological efficiency (%)*	*% increase over control*	*Total yield from three flushes [gm/500 gm dry substrate]*	*Biological efficiency (%)*	*(%) increase over control*
Hot water (T1)	260	52.0	–7.14	240	48.0	+166.67
Autoclave (T2)	255	51.0	–8.92	225	45.0	+150.00
UV- radiation (T3)	272	54.4	–2.85	165	33.0	+83.34
Chemical treatment (T4)	300	60.0	+7.14	285	57.0	+216.67
Control (T5)	280	56.0	0.0	90	18.0	0.0
SE	14.03	2.80	–	10.64	2.12	–
CD (P=0.05)	31.29	6.25	–	23.74	4.74	–

Figures are average of three replications.

Table 3: Quality and quantity parameters of sporophores.

Pasteurization technique	*Fresh wheat straw as substrate*		*Stored wheat straw as substrate*	
	Number of sporocarp	*Weight per sporocarp (gm)*	*Number of sporocarp*	*Weight per sporocarp (gm)*
Hot water (T1)	50	5.20	42	5.71
Autoclave (T2)	48	5.31	44	5.11
UV- radiation (T3)	51	5.33	32	5.15
Chemical treatment (T4)	53	5.66	57	5.00
Control (T5)	49	5.71	19	4.73
SE	5.86	0.54	3.68	0.64
CD (P=0.05)	13.07	1.21	8.22	1.43

Figures are average of three replications.

DISCUSSION

The pasteurization techniques reduce the natural microflora of substrate and improve substrate colonization and yield of mushroom. Due to this reason, pasteurized substrates showed quick mycelial run and higher sporophore formation. In terms of yield performance, chemical treatment gave better response with both fresh and stored wheat straw. This was because of the fact that chemical treatment permitted the beneficial microorganisms in the substrate which consume little or no celluloses or lignins from the material on which the mushroom use to grow, but they do use 'hemicelluloses' the natural material in most substrates that favour the growth of harmful organisms. They also favour the production of more cellulases by *Pleurotus* spp. and thereby increased the yield (Krishnamoorthy *et al.,* 1991). Production of this enzyme is directly proportional to the yield of *Pleurotus in vitro* (Kochuthresiamma *et al.,* 1991; Nallathambi and Marimuthu, 1994). Besides this, bacterial population actively corrodes the surface of the substrate providing more suitable site for the colonization of fungus. *Pleurotus* spp. are also reported to utilize the nitrogen fixed by N-fixing bacteria present in the substrate. In addition, chemical treatments slightly modified the nature of substrate and favoured the growth of fungal mycelium. As compared to chemical treatment, autoclaving and hot water treatments gave insignificant response. This was due to higher temperature which eliminated all the beneficial microorganisms present in the substrate (Bano *et al.,* 1979). The pasteurization is not found compulsory in washed fresh straw substrate due to its low contamination and the reduction in substrate disinfection costs (Moda *et al.*, 2005) but in case of stored substrate, they showed significant biological efficiency of mushroom. It is well known that stored straw is an easy source of contaminants which competes with mushroom mycelium for the available nutrients. When it was pasteurized, growth of competitors, present in the substrate was suppressed resulting into healthy mushroom production. In contrast to this, unpasteurized stored straw gave highly significant reduction in yield and biological efficiency. This loss was either due to depletion of food material in substrate required for growing mushroom mycelia or through the production of toxic substances by the microorganisms (Atkins, 1949). Although, Doshi and Singh (1991) stated that toxic metabolites produced by microorganisms were very mild and had little inhibitory effect on growth of *Pleurotus sajor-caju* and whatever effect is seen was because of depletion of nutrients from the medium due to contaminants. The decomposition of substrate was also visible in unpasteurized beds. This might be due to the combined action of contaminating microflora and cultivated fungus.

During the cultivation of *Pleurotus eous* different types of competitor fungi were encountered on mushroom beds. This result was supported by the conclusions given by different workers which revealed the presence of weeds associated with oyster mushroom cultivation (Vijay and Sohi, 1987; Siddique *et al.,* 2004; Siddhant *et al.,* 2014).

CONCLUSIONS

The results revealed that pasteurization is needed to protect the mushroom beds from competitors and produced higher yield of mushroom. However, in case of fresh substrate, pasteurization was not found compulsory and optimum yield was achieved without it at small scale mushroom cultivation.

ACKNOWLEDGEMENTS

The authors are thankful to Dr. B.P. Kanaujia, C.S.A. University of Agriculture and Technology, Kanpur for providing the pure culture of *Pleurotus eous.*

REFERENCES

Arya, C. and Arya, A. 2003. Effect of acid hydrolysis of substrate on the yield of oyster mushroom (*Pleurotus sajor-caju* (Fr.) Singer). *Mush. Res.,* 12(1): 35–38.

Atkins, F.C. 1949. Major diseases and competitors of the cultivated white mushroom. Yaxley, Peterborough: Mushroom Growers' Association, p. 23.

Bano, Z. 1971. Cultivation of *Pleurotus flabellatus*. *Second Int. Symp. Pl. Pathol.* IARI, New Delhi, p. 135.

Bano, Z., Rajarathnam, S. and Nagaraja, N. 1979. Some aspect on the cultivation of *Pleurotus flabellatus* in India. *Mush. Sci.,* 10: 597–608.

Bhatti, M., Mir, F.A. and Siddiq, M. 1987. Effect of different bedding material on relative yield of oyster mushroom in successive flushes. *Pak. J. Agric. Res.,* 8: 256–259.

Champawat, R.S. and Chitale, K. 2003. Effect of different sterilants and farm waste on yield of *Pleurotus citrinopileatus. J. Mycol. Pl. Pathol.,* 33(2): 292–293.

Chang, S.T. and Miles, P.G. 1989. *Edible mushrooms and their cultivation*. CRC Press, Boca Raton, Florida, pp. 256–274.

Doshi, A. and Singh, R.D. 1991. Toxic effect of culture filtrates of different contaminant fungi on *Pleurotus sajor-caju*. *Indian Phytopath.,* 44(2): 231–232.

Gyorfi, J. and Hajdu, C.S. 2007. Casing –material experiments with *P. eryngii. Int. J. Horticult. Sci.,* 13: 33–36.

Kochuthresiamma, J., Kothandaraman, R. and Mathew, J. 1991. Comparative studies on cellulolytic enzymes production and yield of three species of oyster mushroom. *In*: Nair (*ed.*), *Proceedings of National Symposium on Mushrooms, Kerala Agriculture University, Kerala,* India, pp. 187–189.

Krishnamoorthy, A.S., Habbebullah, B. and Marimuthu, T. 1991. Studies on chemical sterilization of paddy straw for oyster mushroom cultivation. *In:* Marimuthu Krashnamoorthy and Jayarajan (*eds.*), Glimpses of oyster mushroom research, Tamil Nadu Agriculture University, Coimbatore, India, pp. 27–29.

Kumar, S.M., Shukla, C.S. and Dixit, A.K. 1990. Modified hot-water substrate treatment for better yield of oyster mushroom (*Pleurotus sajor-caju* (Fr.) Singer). *Indian J. Mycol. Pl. Pathol.,* 20(2): 154–155.

Moda, E.M., Horii, J. and Spoto, M.H.F. 2005. Edible mushroom *Pleurotus sajor-caju* production on washed and supplemented sugarcane bagasse. *Sci. Agric. (Piracicaba, Braz.),* 62(2): 127–132.

Nallathambi, P. and Marimuthu, T. 1994. Effect of various substrate treatments on enzyme activities of *Pleurotus* spp. in correlation with yield. *Indian J. Mycol. Pl. Pathol.,* 24(3): 167–171.

Naraian, R., Sahu, R.K., Kumar, S., Garg, S.K., Singh, C.S. and Kanaujia, R.S. 2009. Influence of different nitrogen rich supplements during cultivation of *Pleurotus florida* on maize cobs substrate. *Environmentalist,* 29: 1–7.

Ram, R.C. and Thakur, D. 2005. Evaluation of new substrate preparation techniques for oyster mushroom cultivation. *Mush. Res.,* 14(1): 37–39.

Siddique, A.B., Gogoi, R. and Puzari, K.C. 2004. Evaluation of phyto-extracts against contaminants of oyster mushroom. *Indian J. Mycol. Pl. Pathol.,* 34(2): 329–332.

Siddhant, Yadav, S., Mishra, R. and Singh, R. 2014. Effect of substrate disinfection on biological efficiency of *Pleurotus sajor-caju* (Fr.) Singer. *Plant Archives,* 14(1): 205–209.

Tewari, R.P. and Pandey, M. 1988. An important method of oyster mushroom (*Pleurotus sajor-caju*) cultivation. *Indian J. Mycol. Pl. Pathol.,* 18(1): 104.

Vijay, B. and Sohi, H.S. 1987. Cultivation of oyster mushroom *Pleurotus sajor-caju* (Fr.) Singer on chemically sterilized wheat straw. *Mush. J. Tropics,* 7: 67–75.

2.3

Production Technology of White Button Mushroom

RAM CHANDRA[1*]

ABSTRACT

The white button mushroom (Agaricus bisporus (Lange) Sing.) is very popular throughout the world and most important mushroom on commercial scale. This mushroom is extensively cultivated throughout the world and contributed about 60 percent of the total world production of the mushroom. It can be successfully cultivated in places where the environmental conditions are favorable. The optimum temperature for mycelial growth is 20–25°C and that for fruit body formation 18–24°C and also needs a high percentage of relative humidity (85–95%). The compost is prepared by long method of composting in which wheat straw is spread on cemented floor and wetted thoroughly by sprinkling water. For spawning, completely colonized and fresh spawn should be used. The upper surface of compost is covered with newspaper sheet within the polythene bag. These spawned bags are placed in growing chamber where temperature ranges between 20–25°C. After completion of spawn run, the newspaper sheet is removed and surface of compost is covered (3–4 cm thick layer) by casing soil. Mushroom beds are sprayed regularly with water to keep the casing layer adequately moist. The yield of white button mushroom depends on the compost ingredients, supplements used in the compost, casing materials, temperature and relative humidity. Approximately 15–20 kg fresh button mushroom can be raised from one quintal compost.

Key words: *Agaricus bisporus*, Compost, Casing materials and yield.

[1] Department of Mycology & Plant Pathology, Institute of Agricultural Sciences Banaras Hindu University, Varanasi – 0221005 (U.P.)
**Corresponding author:* E-mail: rcrbhumpp@yahoo.com

INTRODUCTION

The white button mushroom (*Agaricus bisporus* (Lange) Sing.) is very popular throughout the world and most important mushroom on commercial scale. This mushroom is extensively cultivated throughout the world and contributed about 60 percent of the total world production of the mushroom. It can be successfully cultivated in places where the environmental conditions are favorable.

The optimum temperature for mycelial growth is 20–25°C and that for fruit body formation 18–24°C and also needs a high percentage of relative humidity (85–95%). Cultivation of white button mushroom requires technical skill.

NUTRITIONAL VALUE OF MUSHROOM

Mushrooms are considered to be healthy food because of their relatively high and qualitatively good protein content and because of their good vitamins, minerals and low fat content. Mushrooms have been recommended by FAO as food that contributes to the protein nutrition of developing countries which depends largely cereals. Mushrooms are important constituents of minor forest produce, that grow on the most abundant biomolecule of this biosphere, that is, cellulose. Presently, mushrooms are regarded as a macro-fungus with a distinctive fruiting body which can be either epigeous or hypogeous and large enough to be seen with the naked eyes and to be picked by hand (Chang and Miles, 1992). Only fruit body of the mushroom can be seen whereas the rest of the mushroom remains underground as mycelium. Geologically, mushrooms existed on the earth even before man appeared on it, as evidenced from the fossil records of the lower cretaceous period. Thus anthropologically speaking, there is every possibility that man used the mushrooms as food when he was still a food gatherer and hunter on the chronology of cultural evolution. Mushrooms offer tremendous applications as they can be used as food and medicines besides their key ecological roles. The proximate composition of cultivated mushroom is shown in Table 1. They represent as one of the world's greatest untapped resources of nutrition and palatable food of the future.

Mushrooms have been found effective against cancer, cholesterol reduction, stress, insomnia, asthma, allergies and diabetes (Bahl, 1983). Due to high amount of proteins, they can be used to bridge the protein malnutrition gap. Mushrooms as functional foods are used as nutrient supplements to enhance immunity in the form of tablets. Due to low starch content and low cholesterol, they suit diabetic and heart patients. One third of the iron in the mushrooms is in available form. Their polysaccharide content is used as anticancer drug. Even, they have been used to combat HIV effectively (Namba, 1993; King, 1993).

Table 1: Proximate composition (Percent fresh weight) of the cultivated mushrooms.

Mushrooms/ Vegetable	*Moisture*	*Protein*	*Fat*	*Carboh-ydrate*	*Fibre*	*Ash*	*Calorie*
Agaricus bisporus	90.1	2.9	0.3	5.0	0.9	0.8	36
Pleurotus sajor-caju	90.2	2.5	0.2	5.2	1.3	0.6	35
Volvariella volvacea	90.1	2.1	1.0	4.7	1.1	1.0	36
Cabbage	91.9	1.8	0.1	4.6	1.0	0.6	27
Cauliflower	90.8	2.6	0.4	4.0	1.2	1.0	30
Potato	74.7	1.6	0.1	22.6	0.4	0.6	97

Source: Sohi, 1988

Mushrooms as a Source of Food

The mushrooms are consumed basically for their texture and flavor, and Mushrooms have from nutrition point of view, a distinct place in human diet which otherwise consist of items either of plant or animal" origin. Mushrooms are perhaps the only fungi deliberately and knowing consumed by human being and they complement and supplement the human diet with various ingredients not encountered in or in deficient in food items of plant and animal origin. Chang and Buswell (1996) have coined the term mushroom nutraceuticals and pharmaceuticals. Nutraceuticals are the functional food enriched / modified and consumed as normal diet to provide health giving benefits.

Man has been hunting for the wild mushrooms since antiquity (Cooke, 1977). Thousands of years ago, fructifications of higher fungi have been used as a source of food (Mattila *et al.*, 2001) due to their chemical composition which is attractive from the nutrition point of view. During the early days of civilization, mushrooms were consumed mainly for their palatability and unique flavors (Rai, 1994; 1997). Present use of mushrooms is totally different from traditional because, lot of research has been done on the chemical composition of mushrooms, which revealed that mushrooms can be used as a diet to combat diseases.

Proteins

The value of protein is determined by the kinds of amino acids that form protein. Mushrooms contain all the essential amino acids as well as the most commonly occurring non-essential amino acids and amides. Mushrooms are rich in lysine and tryptophan, the two essential amino acids that are deficient in cereals. The most nutritious mushrooms are almost equal in nutritional value to meat and milk. Protein is the main body building constituent of our food. Protein content of mushrooms depends on the composition of the substratum, size of pileus, harvest time and species of mushrooms (Bano and Rajarathnam, 1982).

Verma *et al.* (1987) reported that mushrooms are very useful for vegetarian because they contain some essential amino acids which are found in animal proteins (Table 2). The digestibility of *Pleurotus* mushrooms proteins is as that of plants (90%) whereas that of meat is 99% (Bano and Rajarathnam, 1988). Rai and Saxena (1989a) observed decrease in the protein content of mushroom on storage. The protein conversion efficiency of edible mushrooms per unit of land and per unit time is far more superior compared to animal sources of protein (Bano and Rajarathnam, 1988).

Table 2: Essential amino acids (% crude protein) in edible mushrooms.

Amino acid	*Agaricus bisporus*	*Pleurotus sajor-caju*	*Volvariella volvacea*
Leucine	7.5	7.0	4.5
Isoleucine	4.5	4.4	3.4
Valine	2.5	5.3	5.4
Tryptophan	2.0	1.2	1.5
Lysine	9.1	5.7	7.1
Threonine	5.5	5.0	3.5
Phenylalanine	4.	5.0	2.6
Methionine	0.9	1.8	1.1
Histidine	2.7	2.2	3.8

Source: Bano and Rajarathnam (1982); Li and Chang (1982).

Carbohydrates and Fats

The carbohydrate content of mushrooms represents the bulk of fruiting bodies accounting for 50 to 65% on dry weight basis. The mannitol, also called as mushroom sugar constitutes about 80% of the total free sugars, hence it is dominant (Tseng and Mau, 1999; Wannet *et al.*, 2000). Mc-Connel and Esselen (1947) reported that a fresh mushroom contains 0.9% mannitol, 0.28% reducing sugar, 0.59% glycogen and 0.91% hemicelluloses.

In mushrooms, the fat content is very low as compared to carbohydrates and proteins and it is important for patients of obesity. The fats present in mushroom fruiting bodies are dominated by unsaturated fatty acids. Total fat content in *A. bisporus* was reported to be 1.66 to 2.2/100 g on dry weight basis (Maggioni *et al.*, 1968). Ogundana and Fagade (1981) indicated that mushrooms have 4.481% fats on dry weight basis.

Vitamins

Mushrooms are good source of vitamins (Table 3) such as vitamin 'B_1' (Thiamine), vitamin 'B_2' (Riboflavin), niacin, biotin and vitamin 'C' (Ascorbic acid). Mushrooms are one of the best sources of vitamins especially Vitamin

B (Breene, 1990; Mattila *et al.*, 1994; Zrodlowski, 1995; Chang and Buswell, 1996; Mattila *et al.*, 2000). Vitamin content of edible mushrooms has been reported by Esselen and Fellers (1946) and Litchfield (1964).

Table 3: Vitamins content in some edible mushrooms.

Mushrooms	*Content mg/100g D. wt.*			
	Thiamine	*Riboflavin*	*Niacin*	*Ascorbic acid**
i. *Agaricus bisporus*	1.1	5.0	55.7	81.9
ii. *Lentinus edodes*	7.8	4.9	54.9	0.0
iii. *Volvariella volvacea*	0.32 – 0.35	1.63 – 2.97	64.8	20.2
iv. *Pleurotus* spp.	1.16-4.8	4.7*	46.1	0.0

*Adapted from Crisan and Sands (1978).
Source: Chadha and Sharma (1995).

Minerals

Mushrooms are also good source of minerals such as potassium, phosphorus and sodium and contain low but available form of Iron, Potassium and Sodium ratio is very high which is desirable for patients of hypertension. Minerals and heavy metals content of *Pleurotus* spp. are given in Table 4.

Table 4: Minerals content of *Pleurotus* mushrooms.

Species	*K*	*P* (mg/100g. D wt.)	*Mg*	*Na*	*Ca*	*Fe*	*Cd*	*Zn* (ppm)	*Cu*	*Pb*	*Hg*
P. sajor-caju	3260	760	221	60	20	12.4	0.3	29	12.2	3.2	0
P. sajor-caju	2240	565	156	256	40	2.8	–	2.6	0.5	–	–
P. eous	4570	1410	242	78	23	9.0	0.4	82.7	17.8	1.5	0
P. flabellatus	3760	1550	292	75	24	12.4	0.5	56.6	21.9	1.5	0
P. florida	4660	1850	192	62	24	18.4	0.5	11.5	15.8	1.5	0

Source: Rai (1995).

The fruiting bodies of mushrooms are characterized by a high level of well assimilated mineral elements. Major mineral constituents in mushrooms are K, P, Na, Ca, Mg and elements like Cu, Zn, Fe, Mo, Cd form minor constituents (Bano and Rajarathanum, 1982; Bano *et al.*, 1981; Chang, 1982). K, P, Na and Mg constitute about 56 to 70% of the total ash content of the mushrooms (Li and Chang, 1982) while potassium alone forms 45% of the total ash. Abou-Heilah *et al.* (1987) found that content of potassium and sodium in *A. bisporus* was 300 and 28.2 ppm, respectively. *A. bisporus* ash analysis showed high amount of K, P, Cu and Fe (Anderson and Fellers, 1942). Varo *et al.* (1980) reported that *A. bisporus* contains Ca (0.04 g), Mg (0.16), P (0.75 g), Fe (7.8 g), Cu (9.4 mg), Mn (0.833 mg) and Zn (8.6 mg) per kilogram fresh weight.

Medicinal Importance

Numerous reports have shown that mushroom and mushroom products have significant medicinal properties such as immunomodulation, anticancer, antioxidant, blood pressure lowering, cholesterol lowering, liner protective antifibrotic, anti-inflammatory, anti-diabetic and antimicrobial activities. *Ganoderma lucidum* is the most popular medicinal mushroom in China and has been used for a wide range of health benefits, from preventive measures and maintenance of health to regulation and treatment of chronic as well as acute ailments (Chang, 1995). One of the most interesting mushroom derived products is the polysaccharides which exhibit promising immunomodulatory and anticancer effects. Numerous anti-tumour polysaccharides have been discovered from mushroom. These anti-tumour agents of polysaccharides nature in *Lentinus, Schizophyllan* and polysaccharides proteins complexes (PSK and PSP) were isolated from *Lentinus edodes, Schizophyllum commune* and *Coriolus versicolor*, respectively. Many mushrooms polysaccharides have shown antiviral activities including human viruses.

Cholesterol is negligible, fat and calories are found less amount in the mushroom. So, it is comfortable for patients of hypertension and blood pressure and for diabetes patient, due to lack of sugar in mushroom. Some genera of mushrooms have medicinal quality. Medicinal mushroom *Ganoderma lucidum* have been traditionally used in Japan and China. Cosmetic product and tonic beverages have been produced in China from *Ganoderma lucidum.* Antitumour effect have been reported by the extract of various edible fungi including *Lentinus edodes, Flammulina valutipes, Pleurotus ostreatus, Agaricus bisporus, Pholiota nemak, Tricholoma matsutake and Auricularia auricula.*

The compounds extracted from *Agaricus bisporus, Lentinus edodes, Coprinus comatus and Oudemansiella mucida* have been reported to have antifungal and antibacterial properties. Various pharmaceutical compounds have been isolated from several mushrooms (Buswell and Chang, 1993). Some mushroom genera are given in Table 5 which have medicinal properties.

The mushrooms are consumed basically for their texture and flavour, and there is a little awareness about their nutritional and medicinal attributes. Mushrooms have also been investigated by the nutritionists after the Second World War when food shortage stimulated the research on nutritional value of the food items.

ANTIOXIDANT ACTIVITY

As in India, mushrooms have traditionally been used in China and Japan also for the medicinal and tonic properties and in recent years there has been an upsurge interest in this aspect. Anti-tumor effects have been reported by the extracts of various edible fungi including *Lentinus edodes*, *Flammulina velutipes*, *Pleurotus ostreatus*, *Agaricus bisporus*, *Pholiota nameko*, *Tricholoma matsutake* and *Auricularia auricula*. Recently, maitake (*Grifola frondosa*) and shiitake (*Lentinus edodes*) mushrooms have been reported to be inhibitory to AIDS virus in USA and Japan (Namba, 1993). *Coriolus versicolor* is a medicinal mushroom widely prescribed for the prophylaxis and treatment of cancer and infection in China (Chu *et al.*, 2002).

Table 5: Various pharmaceutical components isolated from several mushroom are listed in table.

Pharmacodynamic	***Component***	***Species***
Antibacterial effect	Hirsuitic acid	Many species
Antibiotic	E-B-methoxyacrylate	*O. radicata*
Antiviral effect	Polysaccharide, Protein	*L. edodes* and *Polyporaceae*
Cardiac tonic	Volvatoxin, Flammutoxin	*Volvariella, Flammulina*
Decrease cholesterol	Eritadenine	*Collybia velutipes*
Decrease blood pressure	Triterpene	*G. lucidum*
Decrease level of blood	Peptide glycogen, *Ganoderma glucan*	*G. lucidum*
Anti thrombus	5' AMP, 5'GMP	*Psalliota hortensis*
Inhibition of PHA	r-GHP	*P. hortensis, L. edodes*
Anti-tumour	B-glucan RNA complex	Many spp. *Hpysizygus marmoreus*
Increase secretion of bile	Armillarisia A	*Armillaria tabescens*
Analgesic and sedative effect	Marasmic acid	*Marasmius androsaceus*

Source: Chadha and Sharma (1995).

Antioxidants are chemical compounds that protect cells from the damage caused by unstable molecules known as free radicals. Free radicals are powerful oxidants and those chemical entities that contain unpaired electrons. They are capable of randomly damaging all components of the body, *viz.* lipids, proteins, DNA, sugars and are involved in mutations and cancers (Przybytniak *et al.*, 1999).

PRODUCTION TECHNOLOGY

The compost is prepared by long method of composting in which wheat straw is spread on cemented floor and wetted thoroughly by sprinkling water. This moistened straw is mixed with other ingredients and again with water. A heap (stack) is made of this mixture and covered with polythene sheet. The compost is de-composted by total seven turnings and each turning is done at 4 days interval. Gypsum is mixed during 3rd turning and insecticide dust powder is mixed at the last turning for prevention of insect pests.

For spawning, completely colonized and fresh spawn should be used. The amount of spawn is 2–2.5 kg/quintal compost. The 5 kg spawned compost is filled in one polythene bag. The upper surface of compost is covered with news paper sheet within the polythene bag. These spawned bags are placed in growing chamber where temperature ranges between 20–25°C. After completion of spawn run, the newspaper sheet is removed and surface of compost is covered (3–4 cm thick layer) by casing soil. The casing soil is prepared from 2 years old farm yard manure and loam soil (1:1 ratio). Mushroom beds are sprayed regularly with water to keep the casing layer adequately moist. The yield of white button mushroom depends on the compost ingredients, supplements used in the compost, casing materials, temperature and relative humidity. Aproximately15–20 kg fresh button mushroom can be raised from one quintal compost.

Mushroom cultivation presents an economically important bio-technological industry that has expanded all over the world in the last few years. Mushroom cultivation mainly depends on the agricultural crop residues. These crop residues are abundantly available in our country. It is primarily used as cattle feed and remaining part is burnt or spreaded in the fields. Agro-wastes contain about 80% cellulose, hemi-cellulose and lignin which are not easily degradable, but most of the edible fungi possess enzyme system which can degrade these components. Thus, their cultivation can avoid environmental pollution by recycling agricultural wastes and can also convert straw into easily digestible animal feed. The white button mushroom *Agaricus bisporus*, which is an important commercial mushroom being grown in many parts of our country.

Mushroom Spawn Production

The mushroom spawn of button mushroom (*Agaricus bisporus*) is prepared on wheat grains (*Triticum aestivum* Linn). Well cleaned wheat grains are boiled for 30 minutes or until the grain become soft but firm. The excess water is drained off and the grains cooled on a cheese cloth and turned for several times with a spoon for quick cooling. These cooled grains are mixed with 2% Chalk (calcium carbonate) and 2% Gypsum (calcium sulphate), Gypsum and chalk powder are added to prevent the grains from sticking together and correct the pH, respectively. The prepared grains (500 g/ bag) are filled into polypropylene bags and plugged with non-absorbent cotton plug. These grains filled bags are sterilized in an autoclave for 1 hour at 121°C temperature (15 lbs pressure per square inch). After sterilization, bottles are allowed to cool at room temperature and inoculated with bits of mushroom mycelium (7 days old cultures) of *Agaricus bisporus*. Then inoculated bags are incubated at 25 ± 2°C temperature in a B.O.D. incubator for 16 days. These bags are shaken at 4 days interval to spread the mycelium completely impregnated the grains.

Compost Preparation

The basic materials for synthetic compost, wheat straw like are wheat bran, urea, potassium (Muriate of potash), phosphorus (Single super phosphate), gypsum molasses and lindane are required for compost of button mushroom (Table 6).

Table 6: Formulation of the compost for button mushroom.

Ingredients		*Quantity in kg*	
	Lowest	*Medium*	*Highest*
Wheat straw	300	600	1200
Wheat bran	30	60	120
Urea	3.5	7.5	15
Muriate of potash	3.0	6.0	12
Single super phosphate	3.0	6.0	12
Molasses	4.0	8.0	16
Gypsum	30	60	120
Insecticide	0.250	0.500	1.00

Procedure

The compost is prepared by Long Method of Composting (LMC), which was first advocated in India by Mantel *et al.* (1972). If pasteurization facilities are available, then the short method of composting is applied, however, where pasteurization are lacking, the long method of composting is applied. There are three quantities (lowest, medium and highest) are given in Table 6 and farmer can select any quantity for preparation of compost for cultivation of white button mushroom.

Wetted wheat straw had spread thinly over entire floor of the composting yard and then gradually wetted by sprinkling water. The straw is then turned for even wetting at the stage and water content is maintained at 75 percent.

The moist straw is mixed with wheat bran and fertilizers uniformly scattered over the straw. A heap was made after each turning but not compressed tightly so as to maintain the aerobic condition in the compost heap. Gypsum is mixed at the third turning and water is sprayed at each turning to make up the loss of water due to evaporation. Insecticide is mixed at 7th turning for prevention of insect pests. Total eight turning is done and each turning at four days interval. The compost is then ready for spawning when it appeared as dark brown in colour and without any smell of ammonia and had sufficient moisture content (68–70%) when pressed between palms.

Spawning

The compost is mixed with mushroom spawn @ 2.5 kg/quintal compost and spawned compost is filled (4 kg/bag) in a cylindrical polythene bag (size 40 x 40 cm) and compost filled bags are covered with newspaper sheets to prevent loss of moisture content in mushroom beds are placed in growing chamber, where temperature range between 22–28°C.

Casing

The basic material for preparation of casing soil such as coir pith, saw dust, vermicompost and sand.

The white button mushroom needs casing it is an overlay of neutralized above casing mixtures applied to colonized compost to stimulate pinning and fruiting of mushrooms after completion of the spawn run. The newspaper sheets are removed and the surface of compost is uniformly layered about 3.5 cm by casing formulations. New casing technology has been developed for taking higher yield of button mushroom and detail are given in Table 7.

Table 7: Effect of casing mixtures on yield of button mushroom (*Agaricus bisporus*).

Casing mixtures	*Yield (g) of diff. flushes* I^{st}	IInd	IIIrd	IVth	*Total yield (g)*
CCP + FYM + SD (1:1:1)	266.67	302.00	298.3	216.6	1033.67
CCP + FYM (1:1)	183.33	200.00	256.6	196.6	736.67
VC + SD + FYM + sand (1:1:1:1)	213.33	246.67	226.6	200	886.00
CCP + VC + sand (1:1:1)	196.67	233.3	216.6	186.6	833.34
VC + FYM (1:1)	216.67	250.0	233.3	183.3	883.33
CCP + VC + FYM + SD + sand (1:1:1:1:1)	270.33	320.00	305.33	216.6	1112.26
CCP + VC + FYM (1:1:1)	190.33	205.00	250.0	190.67	836.00
$SEm_{\pm}$	56.42	19.03	17.30	16.29	
C.D. (P= 0.05)	24.23	41.47	27.33	35.50	

Each observation is the average of three replications. CP = Coconut coir pith, FYM = Farm yard manure, SD = Saw dust, VC = Vermicompost.

The highest yield of first flush obtained from the casing mixture of CCP + VC + FYM + SD + sand (270.33 g) followed by CCP + FYM + SD, VC + FYM, VC + SD + FYM + Sand and CCP + VC + Sand (266.67, 216.67, 213.33 and 196.67) respectively. Casing mixture CCP + VC + FYM + SD + Sand showed the highest (320 g) yield whereas CCP + FYM (250 g) showed lowest yield in the harvesting of second flush. In the 3rd flush, maximum yield was obtained from casing CCP + VC + FYM + SD + Sand (305.33 g) where as lowest yield was obtained from CCP + FYM (196.6 g). The result was significantly difference between casing mixture VC + FYM and CCP + FYM and between CCP + FYM + SD and VC +SD + FYM + Sand.

The total yield (1112.26 g) was obtained from casing mixture, CCP + VC + FYM + SD + Sand and lowest yield (736.67 g) was obtained from CCP + FYM. Casing mixture of second highest yield (1033.67 g) obtained from CCP + FYM + SD.

This finding confirm with the result of Dhar *et al*. (2003) were used eight commonly available casing materials in India *viz*, FYM, SMC, CCP, MG, VC, Terracare-A, Terracare-B and FYM + SMC to identify the suitable casing materials for use in button mushroom cultivation and coir pith result as early pinning and significantly higher number of fruit bodies and total yield.

This result also confirms to the findings of Pardo *et al*. (2004) evaluated different casing materials for the cultivation of button mushroom.

Cropping

Pin heads are appeared between two-three weeks after casing and they became ready for harvesting within next one week. The mature unopened mushroom buttons are harvested at most once a day. They are picked by gently twisting of buttons without causing disturbance. If small pits are formed in the casing layer during harvesting of fruit body which are immediately re-cased with respective casing mixtures. Mushroom beds are sprayed regularly with water to keep the casing soil adequately moist. The temperature 18–24°C should be required during cropping period.

Yield

The yield of white button mushroom depends on the compost ingredients, supplements used in the compost, casing materials, temperature and relative humidity. Approximately 15–20 kg fresh button mushroom can be raised from one quintal compost.

CONCLUSIONS

In developing countries malnutrition is one of the major problems because most of the population remains under the economic line. Mushrooms are considered to be healthy food because of their relatively high and qualitatively good protein content and because of their good vitamins, minerals and low fat content. India being vegetarian country requires protein for pulse, vegetable and mushrooms. Mushroom are in polyvalent use which have many useful facets *viz*. food use, bioconversion of agricultural wastes in to precious biomolecules, aesthetic value, symbol of religion, environmental cleaner in nature, pharmaceutical, nutraceutical, medicine, nutrition, agriculture, spent substrate for fodder use, biodiversity and white revolution through mushroom cultivation. So, this technology for growing on different agricultural residues not only provide protein rich foods to vegetarian population, also this will minimized the uses of excessive chemical, pesticides, fertilizer in mushroom production technology. India will become mushroom producer country in the world if proper scientific focus on agricultural wastes and local easily available casing mixtures used for mushroom production.

REFERENCES

Abou-Heilah, A.N., Kasionalsim, M.Y. and Khaliel, A.S. 1987. Chemical composition of the fruiting bodies of *Agaricus bisporus*. *Int. J. Expt. Bot.,* 47: 64–68.

Anderson, E.E. and Fellers, C.R. 1942. The food value of mushrooms *(A. campestris)*. *Proc. Am. Soc. Hort. Sci.,* 41: 301.

Bahl, N. 1983. Medicinal value of edible fungi. *In: Proceeding of the International Conference on Science and Cultivation Technology of Edible Fungi. Indian Mushroom Science II*, pp. 203–209.

Bano, Z., Bhagya, S. and Srinivasan, K.S. 1981. Essential amino acid composition and proximate analysis of mushroom *Pleurotus florida. Mushrooms News Lett. Trop.,* 1: 6–10.

Bano, Z. and Rajarathanam, S. 1982. *Pleurotus* mushrooms as a nutritious food. *In*: *Tropical Mushrooms – Biological Nature and Cultivation Methods* (Chang, S.T. and Quimio, T.H. (*eds.*)), The Chinese University press, Hong Kong, pp. 363–382.

Bano, Z. and Rajarathanum, S. 1988. *Pleurotus* mushroom part II. Chemical composition nutritional value, post harvest physiology, preservation and role as human food. *Crit. Rev. Food Sci. Nutr.,* 27: 87–158.

Breene, W.M. 1990. Nutritional and medicinal value of specialty mushrooms. *J. Food Protect.,* 53: 883–894.

Buswell, J.A. and Chang, S.T. 1993. Edible mushrooms attributes and applications. *In: Genetics and Breeding of Edible Mushrooms* (Chang, S.T.J., Buswell, J.A. and Miles, P.G. (*eds.*)), Gordon and Breach, Philadelphia, pp. 297–394.

Chadha, K.L. and Sharma, S.R. 1995. Mushroom Research in India. History, Infrastructure and Achievements. *In: Advances in Horticulture*. Chadha, K.L. and Sharma, S.R. (*eds.*), Malhotra Publish House, New Delhi, 13: 1–33.

Chang, S.T. 1982. Prospects for mushroom protein in developing countries. *In: Tropical Mushroom – Biological Nature and Cultivation Methods* (Chang, S.T. and Quimio, T.H. (*eds.*)), Chinese University Press, Hong Kong, pp. 463–473.

Chang, S.T. and Buswell, J.A. 1996. Mushroom nutriceuticals. *World J. Microbiol. Biotechnol.,* 12: 473–476.

Chang, S.T. and Miles, P.G. 1992. *Mushroom Biology – A New Discipline. Mycologist*, 6: 64–65.

Chang, S.T. 1995. *Ganoderma* the leader in production and technology of mushroom nutraceuticals. Recent Advances in *Ganoderma lucidum* Research. Seoul, Korea; The Pharmaceutical Society of Korea, pp. 43–52.

Chu, K.K., Ho, S.S. and Chow, A.H. 2002. *Coriolus versicolor*: A medicinal mushroom with promising immunotherapeutic values. *J. Clin. Pharmacol.*, 42: 976–984.

Cooke, R.C. 1977. Fungi, Man, and his Environment, Largman, London, New York, p. 144.

Crisan, E.W. and Sands, A. 1978. A nutritional value. *In:* Chang, S.T. and Hayes, W.A. (*eds.*), *The Biology and Cultivation of Edible Mushrooms.* Academic Press, New York, pp. 172–189.

Dhar, B.L., Ahlawat, O.P. and Gupta Yash 2003. Evaluation of agro-industrial wastes as casing materials in *Agaricus bisporus* cultivation in India. *Mushrooms International*, 92: 5–9.

Esselen, W.B. and Fellers, C.R. 1946. Mushrooms for food and flavor. *Bull. Mass. Agric. Exp. Sta.*, 434p.

King, T.A. 1993. Mushrooms, the ultimate health food but little research in U.S. to prove it. *Mushroom News,* 41: 29–46.

Li, G.S.F. and Chang, S.T. 1982. Nutritive value of *Volvariella volvacea*. *In: Tropical Mushrooms – Biological Nature and Cultivation Methods* (Chang, S.T. and Quimio, T.H. (*eds.*)), Chinese University Press, Hong Kong, pp. 199–219.

Litchfield, J.H. 1964. Nutrient content of morel mushroom mycelium: B vitamin composition. *J. Food Sci.,* 29: 690–691.

Maggioni, A., Passera, C., Renosto, F. and Benetti, E. 1968. Composition of cultivated mushrooms (*Agaricus bisporous*) during the growing cycle as affected by the nitrogen source in compositing. *J. Agr. Chem.,* 16: 517–519.

Mantel, E.F.K., Agarwala, R.K. and Seth, P.K. 1972. A guide to mushroom cultivation. Ministry of Agriculture, Farm Information Unit, Directorate of Extension, New Delhi, *Farm Bull.,* 2.

Mattila, P., Konko, K., Eurola, M., Pihlawa, J.M., Astola, J., Vahteristo, Lietaniemi V., Kumpulainen, J., Valtonen, M. and Piironen, V. 2001. Contents of vitamins, mineral elements, and some phenolic compounds in cultivated mushrooms. *J. Agric. Food Chem.*, 49: 2343–2348.

Mattila, P.H., Piironen, V., Uusi, R. and Koivistoinen, P.E. 1994. Vit. D content in edible mushroom. *J. Agric. Food Chem.,* 42: 2449–2453.

Mattila, P.K., Konko, M., Eurola, J., Pihlava, J., Astola, L., Vahteristo, V., Hietaniemi, J., Kumpulainen, N., Valtonen, V. and Piironen, V. 2000. Contents of vitamins, mineral elements and some phenolic compounds in the cultivated mushrooms. *J. Agric. Food Chem.,* 49: 2343–2348.

Mc-Connel, J.E. and Esselen, W.B. 1947. Carbohydrate in cultivated mushrooms. *Food Res.,* 12: 118–121.

Namba, H. 1993. Maitake mushroom the king mushroom. *Mushroom News,* 41: 22–25.

Ogundana, S.K. and Fagade, O. 1981. The nutritive value of some Nigerian edible mushrooms. *In: Mushroom Science XI, Proceedings of the Eleventh International Scientific Congress on the Cultivation of Edible Fungi, Australia,* pp. 123–131.

Pardo, A., Juan, A.J., de Pardo, J. and Pardo, J.E. 2004. Assessment of different casing materials for use as peat alternatives in mushroom cultivation. Evaluation of quantitative and qualitative production parameters. *Spanish J. Agril. Res.*, 2(2): 267–272.

Przybytniak, G. and Ambroz, H. 1999. Free radicals, their identification and determination. *In:* Marciniec, B. and Zak, S. (*eds.*), Analytical methods in studies of pollutants and hazardous. Bydgoszcz: BTN, p. 17.

Rai, R.D. 1994. Nutritional and medicinal values of mushrooms. *In: Advances in Horticulture.* (Chadha, K.L. and Sharma, S.R. (*eds.*)), Malhotra Publishing House, New Delhi, pp. 537–551.

Rai, R.D. 1995. Nutritional and medicinal values of mushrooms. *In*: *Advances in Horticulture* (Sharma, S.R. and Chadha, K.L. (*eds.*)), Malhotra Publishing House, New Delhi, 13: 537–557.

Rai, R.D. 1997. Medicinal mushrooms. *In:* Advances in Mushroom Biology and Production (Rai, R.D., Dhar, B.L. and Verma, R.N. (*eds.*)) Mushroom society of India. NRCM, Solan, H.P. pp. 355–368.

Rai, R.D. and Saxena, S. 1989. Biochemical changes during the post harvest storage of button mushroom (*Agaricus bisporus*). *Curr. Sci.,* 58: 508–10.

Sohi, H.S. 1988. Mushroom culture in India, Recent research findings. *Indian Phytopath.*, 41: 313–326.

Tseng, Y.H. and Mau, J.L. 1999. Contents of sugars free amino acids and free 5-nucleotides in mushroom, *Agaricus bisporus* during the post harvest storage. *J. Sci. Food Agric.,* 79: 1519–1523.

Varo, P., Lahelman, O., Nuurtamo, M., Saari, E. and Koivistoinen, P. 1980. Mineral element composition of finish food. VII Postal, vegetables, fruits, berries, nuts and mushrooms. *Acta Agric. Scandinavica Supplement*, 22: 107–113.

Verma, R.N., Singh, G.B. and Bilgrami, K.S. 1987. Fleshy fungal flora of N.E.H. India- I. Manipur and Meghalaya. *Indian Mush. Sci.,* 2: 414–421.

Wannet, W.J.B., Hermans, J.H.M., Vander Drift, C. and Op den Camp, H.J.M. 2000. HPCL detection of soluble carbohydrates involved in mannitol and trehalose metabolism in the edible mushroom, *Agaricus bisporus*. *J. Agaric. Food Chem.*, 48: 287–291.

Zrodlowski, Z. 1995. The influence of washing and peeling of mushrooms *Agaricus bisporus* on the level of heavy metal contaminations. *Pol. J. Food Nutr. Sci.*,

2.4

Cultivation of *Pleurotus* (*Dhingri Mushroom*)

Shweta Kulshreshtha[1]*

ABSTRACT

Pleurotus is commonly known as Dhingri mushroom which is one of the well-known cultivable mushroom species throughout the world. Pleurotus can be cultivated on variety of waste residues like wheat straw, rice straw, maize straw and different types of industrial wastes like cardboard industrial and paper industrial wastes. The nutraceutical properties of Pleurotus make it an attractive option to remove waste and to fight against the health issues related to protein deficiency. Here, in this chapter, the cultivation methods of Pleurotus have been described.

Key words: *Pleurotus*, Dhingri cultivation, Lignocellulosic wastes.

1. INTRODUCTION

Mushrooms are a type of higher fungi, belongs to basidiomycetes or ascomycetes group, which produce characteristic spore-forming fruiting bodies (carpophores) as a means of sexual reproduction. Several edible mushrooms can be cultivated at commercial level and are readily available at grocery stores and food markets. They have a unique flavor and aromatic properties; and nutritionally, these are considered to be rich in protein, fiber, carbohydrates, vitamins and minerals (Cohen *et al.*, 2002). Mushrooms are eaten as meat substitutes and are known as an excellent food source to alleviate malnutrition in developing countries (Eswaran and Ramabadran, 2000). Malnutrition is a problem in developing third world countries and

[1] Amity Institute of Biotechnology, Amity University Rajasthan, 14-Gopal Bari, Ajmer Road, Jaipur-302006, Rajasthan.
**Corresponding author:* E-mail: shweta_kulshreshtha@rediffmail.com

due to rising prices nutritious food is becoming a limiting factor for the peoples. Mushrooms can be cultivated on waste economically and without extensive efforts and hence, can be a very reliable and profitable option to alleviate hunger and malnutrition (Pathmashini *et al.*, 2008; Kulshreshtha *et al.*, 2014).

Mushrooms are also known as an important resource of useful, easily accessible, natural compounds with various biological activities that may promote human well-being (Elisashvili, 2012). The various cellular components and secondary metabolites (polysaccharides, proteins and their complexes, phenolic compounds, polyketides, triterpenoids, steroids, alkaloids, nucleotides, etc.) have been identified as a source of therapeutic compounds. These compounds possess cholesterol-lowering, anti-diabetic, antioxidant, antitumor, immunomodulating, antimicrobial, and antiviral activities. Some of these compounds are ready for clinical trials, industrial trials and further commercialization, while others are in various stages of development (Elisashvili, 2012).

Pleurotus spp. are gilled macrofungi (Fig. 1) which belongs to basidiomycetes group. Many species of *Pleurotus* are very popular and easily cultivable species and therefore, species of interest for the whole world. In 90's, *Pleurotus* spp. were at third place in worldwide production of edible mushrooms, after *Agaricus bisporus* and *Lentinula edodes* (Chang, 1999). But now, these are the second most popular mushroom after button mushroom all over the world (Royse, 2014). The cultivation of *Pleurotus* spp. is an economic and important approach worldwide, which has expanded in the past few years. Some species of *Pleurotus* are expected to become an important exportable food source due to its low moisture content. Many species of *Pleurotus* are edible and can be used for medicinal purposes. The medicinal properties are found in various cellular components of its fruiting body (carpophore). It possesses many secondary metabolites which have been isolated and identified from the fruiting bodies, culture mycelium, and culture broth of mushrooms (Elisashvili, 2012).

Pleurotus spp. not only possess medicinal properties but also possess important biotechnological and environmental applications due to their ligninolytic abilities. *Pleurotus* spp. can remediate agricultural waste, forest waste and waste of food processing industries because of their ability of colonization of a huge variety of lignocellulosic substrates. Other benefits of using *Pleurotus* spp. are – requirement of shorter growth time in comparison to other edible mushrooms. The substrate used for their cultivation may be sterilized or pasteurized, which will not affect their yield. *Pleurotus* spp. require control of few environmental parameters and their fruiting bodies are not often attacked by diseases and pests. Besides this, they can be cultivated in a simple and cheap way. All this makes *Pleurotus* spp. cultivation an excellent alternative for the production of mushrooms when compared to other mushrooms (Sánchez, 2010). In this way, *Pleurotus* cultivation reduces lignocellulosic wastes from the environment and provides profit to the cultivator by producing edible fruiting bodies. It may become an important useful approach to treat solid agricultural and industrial wastes in both developed and developing countries (Kulshreshtha *et al.*, 2013; Kulshreshtha *et al.*, 2010).

Pleurotus mushroom is being cultivated from ancient times and nowadays becoming popular throughout the world due to its medicinal, nutritional and many other beneficial properties. Mushroom cultivation is a combination of art and science techniques. In this chapter, some parameters are discussed which need to be considered by the farmers for getting good yield of mushroom.

2. CULTIVATION OF *PLEUROTUS*

Mushroom cultivation requires the four basic requirements: (i) Culture room, (ii) Spawn, (iii) Strains and (iv) Substrate.

2.1. Cultivation Rooms and Culture Conditions

The mushroom cultivation room can be built on arable land or on cemented floor depending on the level of investment. It should be designed in such a way that every space should be used efficiently. It should provide suitable environmental conditions for *Pleurotus* cultivation and should prevent contamination on the farm. If a mushroom farm has many rooms, then rooms should not be connected to one another. This is necessary to avoid the spread of contamination.

An important thing to consider is floors of the room. Mushroom cultivation process requires water spray to maintain the humidity in the cultivation room. Keeping this in mind, floors should be slightly inclined to drain the excessive water during irrigation of room. Water can be collected in a bucket, after filtering the coarse debris, for recycling purpose.

Hygiene is important for every culture room. Doors and windows of the room should have a fine wire mesh, cloth or any other type of simple filter to keep the insects away from the cultivation room.

The spawn preparation should be done in sterilized conditions in order to avoid the contamination. However, readymade spawn is available which can be used directly for inoculation purpose. Therefore, spawn preparation is an optional requirement for this purpose. It depends on the availability of readymade spawn. This facility can be avoided if cultivator gets already prepared spawn from any nearby centre. Spawn preparation room must possess a laminar air flow or wooden hood for spawn preparation. This wooden hood or laminar air flow is used to provide sterile condition.

2.2. Spawn

Mushroom cultivation requires spawn which is generally known as mushroom seed. Spawn is used as propagation material for mushroom cultivation. The yield of mushroom depends on the quality of spawn.

Spawn may be available in the market or may be prepared in the spawn preparation laboratory. Spawn can also be prepared from the mature fruiting bodies of cultivated mushroom (Karam *et al.*, 2010) by collecting the spores or mycelial tissues from sporocarp, in sterile conditions and then inoculate them on appropriate medium for mycelial run. This mycelial culture will be used for the production of spawn. Spawn preparation requires a lot of efforts to maintain its quality. If mushroom strain is degenerated, then prepared spawn affects the yield of mushroom.

Considering these points, cultivator should produce the spawn carefully. Selection of best strain for mushroom cultivation is must and all protocol must be followed under aseptic condition to avoid contamination. During the spawn preparation and inoculation all aseptic transfers are performed in laminar air flow or wooden hood.

The most important requirement for spawn preparation is sterilized grains or any substrate which can be inoculated with fresh mycelium of mushroom. A variety of grains can be used for the preparation of spawn such as wheat, rye, sorghum grains *etc.* The grains are cleaned with water and then boiled in water for 30 min. The boiled grains are filled in sterile polythene bags or in with flasks and then mixed with 2% calcium carbonate. These bags are sterilized in autoclave at 15 psi for one hour. After sterilization, bags are cooled to room temperature and inoculated with 1cm diameter discs. These discs are cut from the periphery of the 7-10 days old mycelium, grown in petri dishes. Sometimes a piece of mycelium is inoculated by the help of needle. These bags can be inoculated by one of the three ways: (i) Top layer inoculation, (ii) Inoculation by mixing and (iii) Alternate layer inoculation. The inoculated bags are kept undisturbed at 20°C for 20-25 days for mycelial run. As the mycelium covered the whole grains, these flasks are stored at 4°C for further use. These mycelium covered grains are called spawn which can be used to inoculate the substrate for mushroom cultivation.

For mushroom cultivation, at first substrates are filled in bottles or in autoclavable plastic bags and then sterilized by autoclaving at 15 psi for one hour. This sterilized substrate is cooled and mixed with calcium carbornate (2% solution). This substrate is now mixed with spawn. About 2% (w/w) spawn is used to inoculate the substrate. The spawn inoculation is also done in laminar air flow under sterilized condition by the help of sterilized inoculating needle. Media disks of uniform size can be cut from the actively growing mycelium and can be used for substrate inoculation.

Pathmashini *et al.* (2008) conducted a study to examine the effect of different types of spawns on oyster mushroom (*Pleurotus ostreatus*) production. Locally available grains of kurakkan (*Eleusine coracana*), maize (broken) (*Zea mays*), sorghum (*Sorghum bicolor*), and paddy (*Oryza sativa*) were used for spawn production. This spawn was further used for inoculation of sawdust filled bags for the cultivation of *P. ostreatus* and inoculated bags were examined for spawn running (mycelia development), pinhead formation and fruit body formation, mean yield, and biological efficiency. Kurakkan spawn was found to be the most efficient spawn type for growth in the sawdust medium compared to other spawns tested. Besides this, Kurakkan spawn produced the highest mean yield of fresh mushroom (Pathmashini *et al.*, 2008).

2.3. Oyster Mushroom Strains and Its Verification

There are many species of *Pleurotus* known for their edibility. Some of them are domesticated and some of them are cultivated commercially at small scale or large scale. Strain with high yield potential is required for increasing the yield of cultivable mushroom. In order to identify, strains with high yield potential and other traits of interest, assessment of genetic and phenotypic diversity is necessary (Alam *et al.*, 2009). The genetic distance can be analyzed for the germplasm materials to take the distantance advantages by using assays based on biochemical or molecular biology and best strain can be selected by comparing the results with cultivated strain.

Pleurotus fruiting bodies are simple in morphology. These fruiting bodies may be changed under the influence of various environmental conditions which results in confusion for the taxonomy identification based on morphology. It became more difficult for mushroom genus that has many related species such as *Pleurotus, Auricularia* and *Armillaria* (Zhang *et al.*, 2010). Mostly, isolated mushroom species is found to be contaminated with fungi having same morphological characteristics. In these cases, taxonomy and classification based on morphology of fruiting body is also a difficult task. Therefore, strain selection based on the morphology may not be successful in selecting the high yield providing species of mushroom. Besides strain selection, germplasm evaluation and strain verification of the selected strain should also be done. In both of these cases molecular tools and techniques will be useful (Zhang *et al.*, 2010). Routine Molecular tools such as Random Amplified Polymorphic DNA (RAPD), Inter-Simple Sequence Repeats (ISSR) and Internal Transcribed Spacers (ITS) sequencing can be used to identify high yielding isolates, strain verification and strain characterization (Zervakis *et al.*, 2001; Tuckwell *et al.*, 2005; Ro *et al.*, 2007; Zhang *et al.*, 2010). Ro *et al.* (2007) used these tools for the characterization of *Pleurotus* isolates. According to Alam *et al.* (2009) the physical and chemical conditions for optimal mycelial growth should also be done along with strain verification.

Molecular tools to observe strain degeneration

Sometimes, yield of mushroom is not found to be increased even after the selection of high yield strain which is due to strain degeneration. Recently, Lee *et al.* (2014) reported that a strain of *Pleurotus eryngii* was isolated from commercial farm. RAPD analysis of isolated DNA revealed the differences in the DNA banding pattern of normal and isolated strain. This showed that isolated strain was a degenerated strain which occasionally emerged due to chromosome-level rearrangements that possibly occur through somatic recombination (Lee *et al.*, 2014). This degenerated strain affects the yield of edible mushroom and need to control before it enters the commercial process. The most effective method to control the loss caused by degenerated strain is – its removal at spawn development stage to prevent further loss.

Improvement by using and designing new strains (hybrids)

Mushroom production can be optimized by improving the culture process and by using high yield varieties of *Pleurotus*. A lot of work has been done on mushroom culture process improvement. However, there is scarcity of reports on genetic breeding of *Pleurotus* strains. Increase in mushroom yield depends on the spawn strain quality – whether it is high yielding strain or not. In order to obtain high yield variety, new isolates or low yield strains can be bred with high yield providing strains of mushroom. Hybridization or genetic breeding process is useful to combine desirable characteristics from different strains and to produce a variety of germplasm from the existing germplasm.

Two compatible monokaryotic hyphae are able to fuse and give rise to a dikaryotic mycelium in which two parental nuclei remain independent during the vegetative growth of mycelium and fruiting body development. In the basidia, karyogamy takes place immediately before the meiotic division and therefore, it is the only site where true diploidy occurs. Later, these diploid basidia produce four uninucleate basidiospores. In the growing hyphae tip, conjugate division of the nuclei in the growing hyphae tip forms clamp connections which are the indicative of the sexual compatibility of strains and dikaryotic condition (Goltapeh *et al.*, 2007).

The 3 weeks old culture of monosporous isolates of *Pleurotus ostreatus* can be used for heterokaryon production. Two monosporous isolates can be placed 15 mm apart in the center of petri dish containing PDA. The inoculated plates are incubated in dark at 24°C till the formation of large contact zone in both mycelia. It takes approximately 14 days. A strip of 2 mm size was cut from the contact zone, and transferred on PDA containing new culture plate. This culture is again incubated at 25°C for 7 days or till the formation of large colonies. This culture is observed under microscope for confirming the presence of clamp connection (Goltapeh *et al.*, 2007; Ha *et al.*, 2015).

Gharehaghaji *et al.* (2007) produced hybrids of *Pleurotus ostreatus* by using breeding approach. At first, basidiospores were collected and suspended in sterile distilled water. The basidiospores, present in the suspension, were counted with the help of haemocytometer. These basidiospores were inoculated on Potato Dextrose Agar (PDA) medium for colony formation. Growth rate of the mycelium on PDA and type of colony of each isolate was determined and then 17 monokaryons were selected. This selection was based on some characteristics such as morphological interaction in the contact zone of mycelium, increasing in growth rate of hybrid, change of colony morphology and the presence of clamp connections between dikaryotic cells. These characteristics were used to differentiate monokaryons from dikaryons. Further, screened monokaryons were allowed to breed. A total of 27 hybrids were produced by this method (Gharehaghaji *et al.*, 2007). It is not necessary that every hybrid possess better characteristics than the parental strain. Therefore, in the next step, hybrid with desirable characteristics is screened among the developed one. The paired isolates are incubated at 25-27°C for 7-10 days for mating. Small piece is cut from the meeting point of the paired monokaryons. The spawn was made by standard method and then used for mushroom cultivation. The different characteristics of sporophore like colour, size, yield and temperature requirement for cultivation can be compared with their parent strains. Hybrids can be grown on suitable substrate and then high yielding and early fruiting strains or any desirable properties possessing strains, can be selected (Bahukhandi and Sharma, 2002).

Strain maintenance

Strains can be maintained at low temperature on various agars or cereal grains. This can be revived by periodic sub-culturing process which can be done by inoculating mycelium on a fresh medium. This is labour intensive process and not very successful due to strain degeneration. Strain can be maintained by cryogenic preservation, after adding cryopreservant in vials. Many vials can be preserved easily and producer can check their sterility and for presence of superior lines for spawn. The spawn producer can reproduce a particular strain easily at any time.

This cryopreservation has the advantage that likelihood of changes in the genetic material is lower but has some limitations as all fungi cannot respond similarly to freezing conditions for nitrogen preservation. Therefore, it is important to conduct more extensive research to determine optimal conditions for nitrogen preservation and viability time for specific culture.

The wheat or any other grains can be used for the preparation of spawn. At first, these grains are soaked in water for overnight. These grains are placed in petri dishes and then sterilized at 121°C for 45 minutes. After cooling, these petri dishes are inoculated with desired strains containing 1

cm^2 agar disc in the laminar air flow under flame area. These petri dishes are incubated at appropriate temperature and humidity conditions required for spawn run (Table 1). After spawn run, some grains are collected from these petri dishes and put in a sterile glycerol containing tubes. These polypropylene tubes are prepared by placing 1.2 ml of glycerol at 10% v/v and then sterilized at 121°C for 15 minutes. After inoculation with mycelium containing grains, these tubes are sealed tightly, placed in boxes and then introduced into the liquid nitrogen container. To assess the viability, one tube can be taken out after 6 months and mycelium containing grains are inoculated on petri dishes and incubate at appropriate temperature and conditions (refer to Table 1) (Lara-Herrera *et al.*, 1998).

Table 1: Different species of *Pleurotus* and their requirement for growth.

Points to be considered	***Pleurotus florida***	***Pleurotus cornucopiae***	***Pleurotus eryngii***	***Pleurotus eryngii var. ferulae***	***Pleurotus pulmonarius***	***P. salmoneostramineus***
Common name	Florida oyster mushroom	Yellow oyster mushroom	King oyster mushroom	White Elf mushroom	Summer oyster mushroom	Pink oyster mushroom
Qualities	Summer strain which gives good cultivation results at temperature above 20°C	Easy to grow	Thick-stemmed fruit bodies, strong texture, can be preserved	White and fleshy mushroom	Shaded light-beige to light brown fruiting bodies, very tasty fruit bodies, gives good cultivation results at high temperature	Over 19°C, this mushroom clusters of pink fruit bodies
Substrate (Recommended)	Hardwood/ wheat straw	Hardwood/ wheat straw	75% hardwood, mixed fine + course 10% cereals 5% oil seeds 10% bran	75% hardwood, mixed fine + course 10% cereals 5% oil seeds 10% bran	Hardwood/ wheat straw	Hardwood/ wheat straw
Incubation						
Room temperature	20–22°C	22–24°C	23–25°C	10–25°C	20–22°C	20–22°C

Table 1: *Contd...*

Points to be considered	*Pleurotus florida*	*Pleurotus cornucopiae*	*Pleurotus eryngii*	*Pleurotus eryngii var. ferulae*	*Pleurotus pulmonarius*	*P. salmoneostramineus*
Substrate temperature	25–30°C	27–32°C	25–28°C	12–28°C	25–30°C	25–30°C
Duration	12–14 days	12–13 days	10–15 days	10–15 days	14–16 days	13–16 days
Fruiting induction						
Temperature	15–20°C	15–20°C	18°C	0–4°C	15–20°C	18–24°C
Relative humidity	90–95%	90–95%	65–70%	65–70%	90–95%	90–95%
Fruiting conditions						
Room temperature	Varies between 11–28°C, better fruiting at 13–20°C	Varies between 13–28°C, better fruiting at 17–22°C	12–15°C	12–15°C	Varies between 11–28°C, better fruiting at 15–18°C	18-24°C
Relative humidity	85%	85%	95–97%	95–97%	85%	90- 95 %
CO_2 concentration	<1000 ppm	<1000 ppm	<1200 ppm	< 1200 ppm	<1000 ppm	<1000 ppm
Light	800–1500 lux	800–1500 lux	800–1500 2000 lux	1200–2000 lux	800–1500 lux	800–1500 lux
Flushes						
Number	2–3	2–3	2	1–2	2–3	2–3
Interval	8–10 days	8 days	1–2 weeks	1–2 weeks	8–10 days	10–12 days
Total production cycle	2.5 months	2 months	2 months	2 months	2 months	2 months
Similar conditions required for	*Pleurotus citrinopileatus*			*Pleurotus nebrodensis*		

In the cryopreservation techniques, many other grains such as oat, rice grains, grains without hulls, millet grains, brown rice grains etc. Procedure for grains preparation and mycelium inoculation is same as mentioned in aforesaid paragraph. The cryopreservant solutions (mentioned in Table 2) and the cryotubes were autoclaved separately at 121°C for 20 min and then cryopreservant is filled in cryotubes. These cryotubes are inoculated by grains covered with mushroom mycelium (Mantovani *et al.*, 2012).

Table 2: Different cryopreservant and their sterilization.

Sl. no.	*Cryopreservant solution*	*Mode of action of cryopreservant*	*Sterilization of solution*
1	1 % Dimethyl sulfoxide (DMSO)	Penetrates the cellular wall and the plasma membrane	DMSO solution is filtered through 0.22 μm filter
2	5 % glycerol	Penetrates the cellular wall and the plasma membrane	Sterilized by autoclaving at 121°C for 20 min.
3	10 % saccharose	Penetrates only the cellular wall	Sterilized by autoclaving at 121°C for 20 min.
4	4 % glucose	Penetrates only the cellular wall	Sterilized by autoclaving at 121°C for 20 min.
	6 % polyethylene glycol–6000 (PEG)	Does not penetrate the wall or the plasma membrane	Sterilized by autoclaving at 121°C for 20 min.
	5 % malt extract	Does not penetrate the wall or the plasma membrane	Sterilized by autoclaving at 121°C for 20 min.

2.4. Substrate

The production of mushroom substrate is the most critical stage of cultivation, having dramatic consequences on yield also crop quality. The substrate not only provides physical support but also acts as a source of nutrients for the mushrooms needed to complete their life cycle.

Pleurotus spp. possess a variety of enzymes which are required to utilize a large variety of agricultural or agro-industrial wastes. Therefore, *Pleurotus* can be cultivated on various substrates such as paddy straw, wheat straw, ragi straw, stalk and leaves of maize, millets and cotton, used citronella leaf, sugarcane bagasse, saw dust, jute and cotton waste, dehulled corncobs, peanut shells, dried grasses, dried populus leaves, sunflower stalks, used tea leaf waste, discarded waste paper, coffee byproducts, tobacco waste, apple pomace etc. (Upadhyay and Sohi, 1988; Upadhyay *et al.*, 1996; Upadhyay and Verma, 2000). Recently, it was cultivated on date palm leaves by Alananbeh *et al.* (2014). It can also be cultivated by using industrial wastes like paper mill sludges, handmade paper and cardboard industrial waste (Kulshreshtha *et al.*, 2010; Kulshreshtha *et al.*, 2013) etc. However, their use relies on the type, culture, quality, availability of substrate locally which is important to consider in order to reduce the cost of production. Standards of mushroom cultivation substrates are- (1) It should possess some antimicrobial compound to inhibit the growth of microbes, (2) It should be selective for the growth of mushroom, (3) The pH of the substrate should be in the range of 6 to 7.5, (4) It should be able to retain moisture, (5) It should be sufficiently porous to allow the gas exchange which is produced during the cultivation and (6) It should possess appropriate amount of carbon and nitrogen sources because these help in obtaining highest yield. Agostini *et al.* (2011) reported that *Pleurotus* can be grown on soybean hulls to which urea (U), ammonium sulfate (AS) or mixture of AS:U (1:1) were added to achieve carbon/nitrogen (C/N) ratios of 5, 15, 20 or 30. The average longitudinal mycelial growth was measured and laccase activity was determined and found that higher C/N ratios increased mycelial growth and decreased laccase production. The highest activities were obtained with a C/N ratio of 5. C/N ratios lower than 30 induced laccase syntheses, inhibited mycelial growth and were a better condition for prehydrolysis of plant residues. The carbon/nitrogen ratio varies with the species of mushroom. However, its physicochemical spawn running time, pinhead formation, mature fruiting body formation time and yield varies with the variation in substrate or their combination and supplements provided.

Substrate influences the physicochemical parameters and productivity of mushroom. Therefore, it is necessary to prepare substrate carefully in such a way that should provide all optimum conditions for the growth of mushrooms. The substrate should be prepared for the mycelial growth of selective mushroom species. This can be done by the application of various heat treatments to eliminate competitive fungi or microbes (Badham, 1991; Jandaik and Guleria, 1999). The popular methods of heat treatment are steam sterilization, pasteurization and hot water treatment. Besides this, composting and chemical sterilization methods can be used for substrate preparation.

3. MUSHROOM CULTIVATION TECHNOLOGY

3.1. Production in Bags

The most important substrates for *Pleurotus* spp. cultivation are wheat straw (*Triticum aestivum*) and cottonseed hulls (*Gossypium hirsutum*). Wheat straw needs chopping to reduce its length to 2-6 cm, however, cottonseed hulls can be used without chopping. After this, these substrates are wetted for overnight and then spread on clean and decontaminated floor to maintain 70% moisture. 1.25 ml/l of 40% formalin solution (v/v) and 0.75 g/l (w/v) of anti-fungal substance bavistin are used for decontaminating of substrate. This is followed by pasteurization of substrate which can be done by passing the steam at 65°C for 1 hour or by putting it into steam sterilizer with loosen pressure releasing knob. After cooling the mixture is supplemented with calcium carbonate. This chemical method of decontamination for cultivation of mushroom is not acceptable worldwide due to toxicity consideration. Therefore, the most preferable method of decontamination of substrate is heat treatment by dipping substrate in hot water, steam sterilization, pasteurization etc.

In large scale production unit, wheat straws are fed in large capacity scale-mounted feed mixer and then it is filled with water up to the desired level and then the slurry is supplemented with calcium carbonate. After few hours, moistened substrate is filled into galvanized metal boxes with perforated floors. These perforated floors are needed to drain off the excessive moisture and later for the supply of air steam mixer through it from top to bottom direction for the period of one hour.

Now, this pasteurized substrate is filled in porous polythene bags after cooling and inoculated with spawn. Generally, 2-5% amount, according to the weight of the substrate, is used for inoculation. Increase in spawn inoculation from 2% of the wet weight of the substrate to 5% may result in increased yield. It is observed that – when spawn running rate is high, it covers the whole substrate rapidly which help in reducing the time of exposure of non-colonized substrate to competitors such as bacteria and fungi. This is the reason of increasing productivity of mushroom on increasing the amount of spawn. After incubation at appropriate temperature and humid conditions, pinhead fruiting bodies (Fig. 2) begin to form around the edges of bag perforations and become mature gradually (Figs. 3 to 5). The fruit bodies are harvested from the substrate after maturation (Fig. 5) when their margins are folded downward. Maturation time is approximately 3 to 4 weeks after the spawning and it is depending on the strain, temperature, humidity, substrate, any type of supplement if used. As they grow, their margins turns upward due to increasing age and decreasing content of water, they acquire uplifted shape of basidiocarp. These are called as over-matured fruit body. The over-matured fruiting bodies (Fig. 6) lose their taste, sometimes pleasant aroma, spores formation and are not considered as suitable for harvesting.

3.2. Cultivation in Bottles

The technique of *Pleurotus* cultivation in bottles was developed in Japan and nowadays becoming popular in the other countries. In this technique, at first, bottles are arranged in trays and then sterilized. These bottles are filled with substrate and put in autoclave at 121°C for 1 hr. After cooling, these bottles are inoculated with spawn of *Pleurotus* mushroom. Further, these are incubated at appropriate temperature and humidity for spawn running. Upon completion of spawn run, whole substrate is covered with white mycelium of *Pleurotus* spp. At this stage lid of bottles is removed and surface of substrate should be scratched (removal of little bit of mycelium from the substrate) in order to induce primordial formation. These primordia develop into pinhead, immature fruiting body and then mature fruiting body. Mature fruiting bodies are harvested and packed for sale.

Lee *et al.* (2009) cultivated *P. eryngii* commercially in wide-mouthed polypropylene bottles in semi-automated confined facilities. A complex substrate which consists of sawdust, rice bran, powdered corncob, beet pulp, wheat bran, cottonseed hull, and shell powder was used for cultivation. Such nutritionally rich complex media provides rapid mycelial propagation and high-quality mushroom production.

In India, a new technique of cultivation of *Pleurotus* in mineral water bottles is developed at Howly (Barpeta) Assam Agricultural University, Assam, India. In this technique, 1 litre mineral water bottle is used which is sterilized by any disinfectant solution. The bottles are made porous by the help of sterile needles. Now, these porous bottles are filled with pasteurized straws to make a layer. Spawn is inoculated on this layer and again a layer of wheat straw is made. This alternate method of spawning is done till the bottle is filled up. The mouth of the bottle is sealed with cap. These inoculated bottles are kept in dark room and maintain the temperature (18-21°C) and humidity (80-90%). These bottles are sprayed with water mist to maintain humidity. Fruiting bodies appeared in the bottle are shown in the Fig. 7.

3.3. Cultivation in Plastic Containers

Generally plastic bags are used for the cultivation of *Pleurotus* species but it is reported that these bags are not suitable for gaseous exchange and may suffocate the culture because in the bags substrate is filled compactly and aeration is received by only peripheral culture. In the center of bag, proper aeration may be interfered by over compaction of bundles. Another problem related to plastic bag is – farmers cannot recycle these bags due to which these are discarded after one time use. This results in increase in cost to the farmers because they have to purchase plastic bags every time and also cause pollution due to accumulation of discarded bags. To solve this problem, recently, Mamiro *et al.* (2014) developed the technology of cultivation of mushroom in re-usable containers. These buckets possessed a lid and were modified by making holes at appropriate distance by piercing the side surface of container by red hot nail. Each side of bucket had 6 holes and the distances between holes were 3.5-4.0 cm. These holes are required for emerging the pinheads. A container should be of 5 L capacity in order to avoid the problem of substrate heating at the center related to large containers. However, optimum size of container and material of the container is point of interest of the future research. The yield obtained in these containers is comparable to that obtained from plastic bags. This proves that recyclable containers can be used successfully for the cultivation of *Pleurotus* spp. (Mamiro *et al.*, 2014). One important consideration in this technique is – cleaning of containers before reuse them for *Pleurotus* cultivation. After first use, these containers are cleaned with the detergent solution and then filled with fresh substrates and then sterilize in autoclave at 121°C for 1 hr. Containers are sterilized with the sterilization of substrates.

4. SUBMERGED CULTURE

Recently, the submerged cultivation of mushrooms has received a great deal of attention because it is an efficient method for producing the bulk amount of mycelium. This mycelium can be further use and process for the production of various metabolites. Submerged cultivation of mushrooms has significant industrial potential due to increased production of mycelium. However, it can be successfully used when a novel production system is established which should possess solution to the problem associated with mushroom cultivation on solid substrates (Elisashvili *et al.*, 2012) production of different metabolites is easy in submerged cultivation technique, however, the physiological and engineering aspects of this process is yet to be determined.

In contrast to above, it is reported that *Pleurotus ostreatus* produces higher amount of ligninolytic enzymes in solid substrate than in liquid culture. This experiment was performed on isolates of *Pleurotus ostreatus* and the ligninolytic abilities were compared with the parental strain. No significant difference among the isolates was found and all strain was found to produce high amount of enzymes on solid substrates compared to liquid substrate (Eichlerová *et al.*, 2000).

5. HARVESTING, PACKAGING AND MARKETING

Harvesting period varies and depends on the strains used for cultivation. Moreover, it is also affected by the variation in substrate, environmental parameters. Therefore, exact time of mature fruiting body formation cannot be stated.

The best time to harvest the mushrooms is when the mushroom caps are slightly curled downward and before their up-fold. Flattening of cap is an indication that it is in sporulating stage. The spores are harmless however; these reveal that the mushroom has passed the maturation stage. To harvest mushroom, the stem of the mushrooms is cut with a clean knife or may twist to take out them from the substrate. The mushrooms may mature at different times, so cut them individually when they are ready.

Harvested mushrooms are sealed in clean polythene bags and refrigerated immediately. Sometimes these fruiting bodies are dried in sunlight or in oven at 45°C. These packets are ready to sell in markets.

5.1. Factors Affecting the Process

A number of factors, such as pH, contamination, aeration of substrate etc., are found to affect the mushroom cultivation process. Mushroom yield depends on aeration of the substrate which relies on the type of raw material used for substrate preparation, texture and porosity of substrate (Bech and Rasmussen, 1968; Hayes, 1972). Kulshreshtha *et al.* (2010, 2011, 2013) used handmade paper industrial waste for the cultivation of *Pleurotus* spp. and found very low yield due to lack of aeration. However, on mixing the handmade paper industrial waste with wheat straw, porosity of the substrate is increased which further results in increase of yield.

Recently, it is reported that – use of calcium chloride containing water for irrigation purpose during the cultivation helps in improving the postharvest storage of mushroom by reducing the surface bacterial population (Chikthimmah *et al.*, 2005). These examples revealed the influence of various factors on cultivation process. Therefore, it is necessary to consider all the factors that affect the yield of mushroom.

5.1.1. *Effect of substrate*

The quality of substrate and the material used for substrate preparation affects the yield of mushroom. Long sized straws cannot provide good yield and therefore, need to be cut or milled before use to improve the yield of mushroom. Additional physical processing can be done depending on the requirement and type of substrate used (Royse *et al.*, 2004).

The yield of mushroom was found to be affected by cellulose, lignin and fibre contents of the substrates. According to Obodai *et al.* (2003) rice straw was found to be the best alternate substrate for growing oyster mushrooms after wheat straw. Royse *et al.* (2004) reported that yields were three times higher on switch grass that was harvested after the grass had senesced (winter; beige color) compared to grass that was harvested when it was green (summer; time of flowering). This shows the effect of cellulose and lignin content in the dried grass which ultimately affects the mushroom productivity.

5.1.2. *Influence of temperature*

The yield of mushroom is also affected by temperature because every step of mushroom cultivation requires appropriate temperature. Delay in spawn run, pinhead formation, and mature fruiting bodies formation is observed if temperature is not controlled. In order to find out the effect of temperature on mycelial growth of *Pleurotus ostreatus, P. eryngii, P. pulmonarius,* an experiment was performed by Zervakis *et al.* (2001). A wide range of temperature was selected and found that the optimum temperature for *P. ostreatus* and *P. pulmonarius* is 30°C which depends on the nutrient medium used.

6. PASTEURIZATION *VS.* STERILIZATION

Sterilization is the process of killing of vegetative as well as spore forms of microbes whereas pasteurization is a process of killing vegetative cells of microbes. Pasteurization is done at lower temperature (62-72°C) compared to sterilization (121°C) that is not sufficient to kill all forms of microbes. Mushroom cultivation techniques relies on pasteurization or sterilization process which is required to reducing the microbial population associated with the substrate. These microbes may contaminate the substrate and reduce the spawn running rates and finally affect yield of mushroom. Therefore, it is suggested that every agricultural waste should be properly sterilized or pasteurized before use. The choice of process depends on the microbial load in the substrate. The use of fresh substrates requires only pasteurization. However, the stored and contaminated substrates require autoclaving at 121R°C for 1-2 hrs.

7. EFFECT OF OTHER MICROBES

Pleurotus spp. may be positively or negatively affected by the bacterial population, present in the substrate after pasteurization. *Pleurotus* spp. reacts in various ways in the presence of bacteria. There is scarcity of literature reporting the influence of microbes, present in the substrate, on cultivation of edible fungi. To determine the role of microbes, Velázquez-Cedeño *et al.* (2004) performed an experiment by using wheat straw without heat treatment, sterilized wheat straw and pasteurized wheat straw. Variation in heat treatments of the wheat straw was found to affect not only the bacterial populations but also affect the colonization of the substrate and production of extracellular lignocellulolytic enzymes by *P. ostreatus*. Total number of bacteria in the substrate was the same initially but varies after heat treatment which affected the colonization of the substrate. Finally, these bacteria increase the yield of mushroom when cultivated with them due to fungal antagonistics.

On co-culturing of *Pleurotus spp.* with soil microflora, higher substrate mineralization and colonization was observed (Lang *et al.*, 1997). The presence of *Pseudomonas fluorescens,* in the substrate, induces positive effect on the growth of *P. ostreatus* (Cho *et al.*, 2002). Similarly, on adding *Pseudomonas* sp. to the culture media of *Pleurotus eryngii,* time required for spawn run, pinhead formation and fruiting body formation was reduced (Kim *et al.*, 2008).

8. PRESENCE OF ENZYMES

Pleurotus can be cultivated on a large variety of agricultural or agro-industrial wastes due to the presence of different types of enzymes. The production of these enzymes relies on the type of waste used for the preparation of substrate and environmental conditions. Mushrooms use a combination of enzymes to degrade a variety of wastes which acts as substrate for their growth. The enzyme complex includes lignin degrading oxidative enzymes laccase and manganese peroxidase (MnP) and hemicellulose or cellulose degrading hydrolytic enzymes hemicellulases, xylanases and cellulases (Rodrigues *et al.*, 2012).

The enzyme activities are increased during fruitification because there is need to utilize large amounts of carbon for fruiting bodies formation. The lignocellulolytic enzyme activity is dependent on several factors such as composition of substrate, C/N ratio, availability of oxygen, pH and temperature. Laccase enzyme production is increased (3-4 times) by supplementing the substrate with nitrogenous source (Kahraman and Gurdal, 2002).

Enzymatic activities of *Pleurotus* are also required for nitrogen metabolism. This nitrogen is further required for the growth of mycelium and fruiting bodies. The preferred sources of natural nitrogen for *Pleurotus* sp. is L-glutamate and L-aspartate. The mycelium and fruiting bodies contain both NAD-dependent glutamate dehydrogenase (NAD-GDH) and glutamine synthetase (GS). The highest amount of these enzymes are present in fruiting bodies. NADP-dependent GDH (NADP-GDH) apparently catalyses the synthesis of glutamate whereas the NAD-dependent enzyme (NAD-GDH) catalyses its deamination. NAD-GDH is reported to be derepressed by ammonia and repressed by higher concentrations of L-glutamate. Similarly, GS is also derepressed by ammonia, L-glutamate, L-aspartate and L-alanine. Repression of GS was observed in the presence of L-glutamine, L-arginine, nitrate, urea and low concentrations of ammonia. An increase in the ratio of glucose to nitrogen resulted in a decrease in the level of GS in the mycelium (Mikes *et al.*, 1994).

9. CALCIUM SOURCE SUPPLEMENTATION

Generally, calcium content of mushroom is lesser than that of vegetables (Kawai *et al.*, 1986; Kawai *et al.*, 1990). It is well documented that mushroom can absorb the nutrients from the substrates on which they are cultivated. Tabata and Shinohara (1995) used a culture media supplemented with calcium and reported that *P. ostreatus* has calcium absorption efficiency. It is also reported that calcium salt supplementation in sawdust medium affect the calcium content of *P. eryngii* and its mycelial growth.

Choi *et al*. (2011) reported the positive effect of calcium rich oyster shell powder on the mycelial growth and calcium content of *P. eryngii*. On supplementation of 1% oyster shell powder, calcium content of mushroom was found to increase without affecting the yield spawn running or mature fruiting body formation time.

Similarly, Naraian *et al*. (2014) used the combination of wheat straw and oyster shell powder and reported the significant increase not only in yield but also nutritional content of *Pleurotus florida* PF05. Therefore, increase in calcium content, increase the nutritional content of mushroom.

Calcium not only affects the nutritional content of mushroom but also improves their storage after harvesting. During cultivation, use of 0.75% H_2O_2 and 0.3% $CaCl_2$ containing water for irrigation can reduce the surface bacterial population and improves the storage of mushroom fruiting bodies after harvesting (Chikthimmah *et al*., 2005).

10. EFFECT OF NITROGEN CONTENT

Mushrooms require carbon, nitrogen and inorganic compounds as their nutritional sources. Oyster mushroom can be grown easily on the different types of wastes and agrowastes and convert these residues into protein rich basidiocarps (Banik and Nandi, 2004). These wastes are also suitable for the cultivation of *Pleurotus* spp. which require high C/N ratio. The protein content of the mushrooms depends on the supplementation of nitrogen in the substrate. Nitrogen is an important element for the growth and various metabolic activities and for the synthesis of enzymes and proteins. The straws used for cultivation of *Pleurotus* are the poor source of nitrogen (0.5 to 0.8%) (Upadhyay *et al*., 2002) and the available amount is totally consumed by the mushroom till fruiting body formation. This results in the depletion of nitrogen in the substrate and subsequently affects the mushroom yield. Therefore, the supplementation of substrate with nitrogen source is required to produce satisfactory yield (Carvalho *et al*., 2010). These supplements maintain the carbon/nitrogen ratio of the substrate and provide good yield of mushrooms with rich protein content.

According to Macaya-Lizano (1988) substrate with high N contents and, consequently, low C:N ratio, did not provide the total substratum colonization and did not allow appropriate basidiocarp production for *Pleurotus* spp. Upadhyay *et al*. (2002) evaluated seven different organic nitrogen sources: wheat bran, rice bran, soybean floor, de-oiled soybean meal, mustard cake, cotton seed cake and cotton seed meal for their effect on mushroom yield. It was reported that the addition of 1% de-oiled soybean meal and 2.5% cotton seed cake are the optimum doses for these supplements to enhance the yields of *P. ostreatus var florida*. Their supplementation also gave higher dry matter content of mushroom than unsupplemented bags.

In contrast to this, Fanadzo *et al.* (2002) reported the adverse effect of nitrogen supplement on the growth of *Pleurotus sajor-caju* and *P. ostreatus*. Similar data was presented by Rajarathnam and Bano (1989) according to which high nitrogen contents leads to the increase of toxic substances in the culture medium, limiting the mycelial development and, consequently, fruiting bodies production. Therefore, it is recommended that the C:N ratio must be in between 5 to 30. Any imbalance in the ratio will lead to decrease yield of mushroom.

11. EFFECT OF MINERALS

The nutritional quality of mushroom relies on the quality of substrate. Mineral content of the mushroom also depends on the substrate used for cultivation and it changes with the variation in substrate. Mineral content can be determined by solubilizing the sample using acid digestion. Then, atomic absorption spectrometry (for calcium, magnesium, iron, copper, zinc and manganese), atomic emission (for sodium and potassium) and colorimetry (for phosphorus) techniques can be used for the determination of mineral content of mushrooms (Sales-Campos, 2009).

Pre-cultivation and post-cultivation analysis of mineral content of fruiting bodies and substrate reveals that the mineral contents of the fruiting bodies of *P. sajor-caju* increase when cultivated on substrates with higher mineral contents. A total of eight minerals was determined (calcium, phosphorus, potassium, magnesium, sodium, iron, manganese and zinc) and found that the potassium content was highest followed by phosphorus, magnesium and sodium, depending on the substrate composition (Patrabansh and Madan, 2004). It is also reported that addition of manganese at levels of 150 and 250 microg/g significantly increase the yield of mushroom (Rodriguez and Royse, 2007).

12. CONCLUSIONS

Pleurotus spp. are good candidates for cultivation due to their efficiency to degrade a large variety of lignocellulosic material, easily to grow without cumbersome efforts, complete their life cycle in lesser time compared to other species of mushroom. For the cultivation of this species raw material is abundantly available in each and every country in the form of different types of wastes. Besides waste recycling and remediation, it is a source of income, alternative source of protein, provides employment to the people. It is the product of demand nowadays due to its culinary, nutritional, health benefits. All these benefits focus on the fact that the cultivation of *Pleurotus* is beneficial business and can be helpful in fulfilling the demand of protein sources. Therefore, farmers should get encouragement and some type of financial support to start this at small scale. Furthermore, researchers should try to optimize the conditions for the better yield of mushroom.

13. REFERENCES

Agostini, E.C.D., Mantovani, T.R.A., Valle, J.S., Paccola-Meirelles, L.D., Colauto, N.B. and Linde, G.A. 2011. Low carbon/nitrogen ratio increases laccase production from basidiomycetes in solid substrate cultivation. *Sci. Agric.*, 68(3): 295–300.

Alam, N., Shim, M.J., Lee, M.W., Shin, P.G., Yoo, Y.B. and Lee, T.S. 2009. Vegetative growth and phylogenetic relationship of commercially cultivated strains of *Pleurotus eryngii* based on ITS sequence and RAPD. *Mycobiology*. doi: 10.4489/MYCO.2009.37.4.258.

Alananbeh, K.M., Bouqellah, N.A. and Al Kaff, N.S. 2014. Cultivation of oyster mushroom *Pleurotus ostreatus* on date-palm leaves mixed with other agro-wastes in Saudi Arabia. *Saudi J. Biol. Sci.* doi: 10.1016/j.sjbs.2014.08.001.

Alemu, F. 2014. Cultivation of *Pleurotus ostreatus* on *Grevillea robusta* leaves at Dilla University. *J. Yeast Fungal Res.* doi: 10.5897/JYFR2014.0139.

Badham, E.R. 1991. Growth and competition between *Lentinus edodes* and *Trichoderma harzianum* on sawdust substrates. *Mycologia,* 83: 455–463.

Banik, S. and Nandi, R. 2004. Effect of supplementation of rice straw with biogas residual slurry manure on the yield, protein and mineral contents of oyster mushroom. *Ind. Crop. Prod.* doi: 10.1016/j.indcrop.2003.11.003

Bech, K. and Ra smussen, C.R. 1968. Further investigations on organic and inorganic supplementation of mushroom compost. *Mushroom Sci.,* 7: 329–342.

Bettin, F., Montanari, Q., Calloni, R., Gaio, T.A., Silveira, M.M. and Dillon, A.J. 2009. Production of laccases in submerged process by *Pleurotus sajor-caju* PS-2001 in relation to carbon and organic nitrogen sources, antifoams and Tween 80. *J. Ind. Microbiol. Biotechnol.* doi: 10.1007/s10295-008-0463-1.

Carvalho, C.S.M., Sales-Campos, C. and De Andrade, M.C.N. 2010. Mushrooms of the *Pleurotus* genus: A review of cultivation techniques. *Interciencia,* 35(3): 177–182.

Chang, S.T. 1999. World production of cultivated and medicinal mushrooms in 1997 with particular emphasis on *Lentinula edodes* (Berk.) Sing. in China. *Int. J. Med. Mushrooms.,* 1: 291–300.

Chang, S.T. and Chiu, S.W. 1992. Mushroom production: An economic measure in maintenance of food security. *In*: DaSilva, E.J., Ratledge, C. and Sasson, A. (*eds.*), Microbial Technology: Economic and Social Aspects Cambridge University Press, Cambridge, UK, pp. 110–141.

Chikthimmah, N., La Borde, N.F. and Beelman, R.B. (2005) Hydrogen peroxide and calcium chloride added to irrigation water as a strategy to reduce bacterial populations and improve quality of fresh mushrooms. *J. Food Sci.* doi: 10.1111/j.1365-2621.2005.tb11446.x.

Cho, Y.S., Kim, J.S., Crowley, D.E. and Cho, B.G. 2002. Growth promotion of edible fungus *Pleurotus ostreatus* by fluorescent pseudomonads. *FEMS Microbiol. Lett.* doi: 10.1016/S0378-1097(02)01144-8.

Choi, U.K., Lee, O.H. and Kim, Y.C. (2011). Effect of calcinated Oyster shell powder on growth, yield, spawn run, and primordial formation of king Oyster mushroom (*Pleurotus eryngii*), molecules. doi: 10.3390/molecules16032313.

Cohen, R., Persky, L. and Hadar, Y. 2002. Biotechnological applications and potential of wood-degrading mushrooms of the genus *Pleurotus*. *Appl. Microbiol. Biotechnol.* doi: 10.1007/s00253-002-0930-y.

Eichlerová, I., Homolka, L., Nerud, F., Zadrazil, F., Baldrian, P. and Gabriel, J. 2000. Screening of *Pleurotus ostreatus* isolates for their ligninolytic properties during cultivation on natural substrates. *Biodegradation.* doi: 10.1023/A:1011165919887.

Elisashvili, V. 2012. Submerged cultivation of medicinal mushrooms: Bioprocesses and products (review). *Int. J. Med. Mushrooms. doi*: 10.1615/Int J Med Mushr. v14. i3.10.

Eswaran, A. and Ramabadran, R. 2000. Studies on some physiological, cultural and post-harvest aspects of oyster mushroom, *Pleurotus eous*. *Trop. Agric. Res.,* 12: 360–374.

Fanadzo, M., Zireva, D.T., Dube, E. and Mashingaidze, A.B. 2002. Evaluation of various substrates and supplements for biological efficiency of *Pleurotus sajor-caju* and *Pleurotus ostreatus*. *Afr. J. Biotechnol.,* 9: 2756–2761.

Flores, C., Casasanero, R., Trejo-Hernández, M.R., Galindo, E. and Serrano-Carreón, L. 2010. Production of laccases by *Pleurotus ostreatus* in submerged fermentation in co-culture with *Trichoderma viride*. *J. Appl. Microbiol.* doi: 10.1111/j.1365-2672.2009.04493.x.

Freixo, M.R., Amin, K. and José, A. 2012. Production, purification and characterization of laccase from *Pleurotus ostreatus* grown on tomato pomace. *World J. Microb. Biot.* doi: 10.1007/s11274-011-0813-4.

Gharehaghaji, A.N., Goltapeh, E.M., Masiha, S. and Gordan, H.R. 2007. Hybrid production of Oyster mushroom *Pleurotus ostreatus* (Jacq: Fries) Kummer. *Pak. J. Biol. Sci.* doi: 10.3923/pjbs.2007.2334.2340.

Goltapeh, E.M., Gharehaghaji, A.N., Masiha, S. and Gordan, H.R. 2007. Hybrid production of oyster mushroom *Pleurotus ostreatus* (Jacq: Fries) Kummer. *Pak. J. Biol. Sci. 10*: 2334-2340.

Ha, B.S., Kim, S. and Ro, H.S. 2015. Isolation and characterization of monokaryotic strains of *Lentinula edodes* showing higher fruiting rate and better fruiting body production. *Mycobiology*. doi: 10.5941/MYCO.2015.43.1.24.

Hayes, W.A. 1972. Nutritional factors in relation to mushroom production. *Mushroom Sci.,* 8: 662–674.

Inácio, F.D., Ferreira, R.O., de Araujo, C.A.V., Peralta, R.M. and de Souza, C.G.M. 2015. Production of enzymes and biotransformation of orange waste by Oyster mushroom, *Pleurotus pulmonarius* (Fr.) Quél. *Adv. Microbiol.* Doi:10.4236/aim.2015.51001.

Jandaik, S. and Guleria, D.S. 1999. Yield loss in *Agaricus bisporus* due to *Trichoderma* infection. *Mushroom Res.,* 8: 43–46.

Kahraman, S.S. and Gurdal, I.H. 2002. Effect of synthetic and natural culture media on laccase production by white rot fungi. *Bioresour. Technol.,* 82: 215–217.

Karam, C., Sassine, Y.N., Dima, M. and Ghassan, A.M. 2010. Prospective of local oyster mushroom production in Lebanon. *J. Appl. Sci. Res.,* 6(12): 2139.

Kawai, H., Sugahara, T., Fujishiro, S., Matsuzawa, M., Aoyagi, Y. and Hosogai, Y. 1990. Mineral contents in edible mushrooms growing on wood. Comparison with mineral contents of mushrooms growing in soil. *J. Jpn. Soc. Food Sci. Technol.* doi: 10.3136/nskkk1962.37.6_468.

Kawai, H., Sugahara, T., Matsuzawa, M., Sumiyashiki, K., Aoyagi, Y. and Hosogai, Y. 1986. Mineral contents in edible mushrooms. *J. Jpn. Soc. Food Sci. Technol.,* 33: 250–255.

Kim, M.K., Math, R.K., Cho, K.M., Shin, K.J., Kim, J.O., Ryu, J.S., Lee, Y.H. and Yun, H.D. 2008. Effect of *Pseudomonas* sp. P7014 on the growth of edible mushroom *Pleurotus eryngii* in bottle culture for commercial production. *Bioresource Technol.,* 99(8): 3306–3308.

Kulshreshtha, S., Mathur, N. and Bhatnagar, P. 2011. Pros and cons of *P. florida* cultivation for managing waste of handmade paper and cardboard industries. *The IIOAB Journal Spl. Issue,* 2(1): 45–48.

Kulshreshtha, S., Mathur, N. and Bhatnagar, P. 2014. Mushroom as a product and their role in mycoremediation. *AMB Express. doi:* 10.1186/s13568-014-0029-8.

Kulshreshtha, S., Mathur, N., Bhatnagar, P. and Jain, B.L. 2010. Bioremediation of industrial wastes through mushroom cultivation. *J. Environ. Biol.,* 31: 441–444.

Kulshreshtha, S., Mathur, N., Bhatnagar, P. and Kulshreshtha, S. 2013. Cultivation of *Pleurotus citrinopileatus* on handmade paper and cardboard industrial wastes. *Ind. Crop. Prod.* doi: 10.1016/j.indcrop.2012.04.053.

Kurt, S. and Buyukalaca, S. 2010. Yield performances and changes in enzyme activities of *Pleurotus* spp. (*P. ostreatus* and *P. sajor-caju*) cultivated on different agricultural wastes. *Bioresour. Technol.* doi: 10.1016/j.biortech.2009.12.011.

Lang, E., Kleeberg, I. and Zadrazil, F. 1997. Competition of *Pleurotus* sp. and *Dichomitus squalens* with soil microorganisms during lignocellulose decomposition. *Bioresource Technol.* doi: 10.1016/S0960-8524(97)00016-3.

Lara-Herrera, I., Mata, G. and Gaitán-Hernández, R. 1998. Evaluation of the viability of *Pleurotus* spp. strains after liquid nitrogen cryopreservation. *Revista de Microbiologia Rev. Microbiol.* doi: 10.1590/S0001-37141998000300009.

Lee, C.Y., Park, J.E., Kim, B.B., Kim, S.M. and Ro, H.S. 2009. Determination of mineral components in the cultivation substrates of edible mushrooms and their uptake into fruiting bodies. *Mycobiology.* doi: 10.4489/MYCO.2009.37.2.109.

Lee, H.J., Kim, S.W., Ryu, J.S., Lee, C.Y. and Ro, H.S. 2014. Isolation of a variant strain of *Pleurotus eryngii* and the development of specific DNA markers to identify the variant strain. *Mycobiology.* doi: 10.5941/MYCO.2014.42.1.46.

Levanon, D. 1993. Mushroom cultivation: Environmental problem to environmental benefit. *In*: Chang, S.T., Buswell, J.A. and Chiu, S.W. (*eds.*), Mushroom Biology and Mushroom Products: The Chinese University Press, Hong Kong, pp. 331–338.

Macaya-Lizano, A.V. 1988. Cultivo de *Pleurotus ostreatus* y especies afines (Fungi: Pleurotaceae) sobre medios naturales semi-esteriles. *Revista de Biologia Tropical,* 36(2a): 255–260.

Mamiro, D.P., Mamiro, P.S. and Mwatwala, M.W. 2014. Oyster mushroom (*Pleurotus* spp.) cultivation technique using re-usable substrate containers and comparison of mineral contents with common leafy vegetables. *J. Appl. Biosci.* doi: 104314/jab.v80i1.13.

Mantovani, T.R.A., Tanaka, H.S., Umeo, H.S., Zaghi, Jr. L.L., do Valle, J.S., Paccola-Meirelles, L.D., Linde, G.A. and Colauto, N.B. 2012. Cryopreservation at –20 and –70°C of *Pleurotus ostreatus* on grains. *Indian J. Microbiol.* doi: 10.1007/s12088-012-0289-4.

Mikes, V., Zofall, M., Chytil, M., Fulneèek, J. and Schánì, L. 1994. Ammonia-assimilating enzymes in the basidiomycete fungus *Pleurotus ostreatus. Microbiology*, 140: 977–982.

Morais, H., Forgács, E. and Cserháti, T. 2005. Enzyme production of the edible mushroom *Pleorotus ostreatus* in shaken cultures completed with agro-industrial wastes. *Eng. Life Sci.* doi: 10.1002/elsc.200420065.

Naraian, R., Narayan, O.P. and Srivastava, J. 2014. Differential response of oyster shell powder on enzyme profile and nutritional value of oyster mushroom *Pleurotus florida* PF05. *BioMed. Res. Int.* doi: http://dx.doi.org/10.1155/2014/386265.

Obodai, M., Cleland-Okine, J. and Vowotor, K.A. 2003. Comparative study on the growth and yield of *Pleurotus ostreatus* mushroom on different lignocellulosic by-products. *J. Ind. Microbiol. Biotechnol.,* 30(3): 146–149.

Ogundele, G.F., Abdulazeez, R.O. and Bamidele, O.P. 2014. Effect of pure and mixed substrate on oyster mushroom (*Pleurotus ostreatus*) cultivation. *Journal of Experimental Biology and Agricultural Sciences*, 2(2S): 215–219.

Oseni, T.O., Dube, S.S., Wahome, P.K., Masarirambi, M.T. and Earnshaw, D.M. 2012. Effect of wheat bran supplement on growth and yield of oyster mushroom (*Pleurotus ostreatus*) on fermented pine sawdust substrate. *Experimental Agriculture and Horticulture. http://www.bapress.ca/eah/v1n2/Tajudeen%20O.%20Oseni.pdf. Accessed on 24th Oct, 2014.*

Pathmashini, L., Arulnandhy, V. and Wijeratnam, R.S.W. 2008. Cultivation of oyster mushroom (*Pleurotus ostreatus*) on sawdust. *Cey. J. Sci. (Bio. Sci.).* doi: 10.4038/cjsbs.v37i2.505.

Patrabansh, S. and Madan, M. 2004. Mineral content of the fruiting bodies of *Pleurotus sajor caju* (FR.) Singer cultivated on different kinds of biomass. *Acta Bioth.* doi: 10.1002/abio.370190203.

Rajarathnam, S., Bano, Z. and Keith, H.S. 1989. *Pleurotus* mushrooms. Part III. Biotransformations of natural lignocellulosic wastes: Commercial applications and implications. *Crit. Rev. Food Sci. Nutr.* doi: 10.1080/10408398909527491.

Rani, P., Kalyani, N. and Prathiba, K. 2008. Evaluation of lignocellulosic wastes for production of edible mushrooms. *Appl. Biochem. Biotech.* doi: 10.1007/s12010-008-8162-y.

Ro, H.S., Kim, S.S., Ryu, J.S., Jeon, C.O., Lee, T.S. and Lee, H.S. 2007. Comparative studies on the diversity of the edible mushroom *Pleurotus eryngii*: ITS sequence analysis, RAPD fingerprinting and physiological characteristics. *Mycol. Res.,* 111: 710–715.

Rodrigues, da Luz J.M., Nunes, M.D., Paes, S.A., Torres, D.P., Soares da Silva, M.C. and Kasuya, M.C.M. 2012. Lignocellulolytic enzyme production of *Pleurotus ostreatus* growth in agroindustrial wastes. *Braz. J. Microbiol.* doi: 10.1590/S1517-83822012000400035.

Rodriguez Estrada, A.E. and Royse, D.J. 2007. Yield, size and bacterial blotch resistance of *Pleurotus eryngii* grown on cottonseed hulls/oak sawdust supplemented with manganese, copper and whole ground soybean. *Bioresour. Technol.*, 98(10): 1898–906.

Royse, D.J. 2014. A global perspective on the high five: *Agaricus, Pleurotus, Lentinula, Auricularia*, and *Flammulina*. Proceedings of the 8th International Conference on mushroom biology and mushroom products, pp. 1–6.

Royse, D.J., Rhodes, T.W., Ohga, S. and Sanchez, J.E. 2004. Yield, mushroom size and time to production of *Pleurotus cornucopiae* (oyster mushroom) grown on switch grass substrate spawned and supplemented at various rates. *Bioresour. Technol.*, 91(1): 85–91.

Sales-Campos, C., da Eira, A.F., Minhoni, M.T.A. and de Andrade, M.C.N. 2009. Mineral composition of raw material, substrate and fruiting bodies of *Pleurotus ostreatus* in culture. *Interciencia,* 34(6): 432–436.

Sánchez, C. 2010. Cultivation of *Pleurotus ostreatus* and other edible mushrooms. *Appl. Microbiol. Biotechnol.* doi: 10.1007/s00253-009-2343-7.

Tabata, T. and Shinohara, H. 1995. Absorption of calcium salts added culture media by Hiratake (*Pleurotus ostreatus* (Fr.) Quel.) and Nameko (*Pholiota nameko* J. Ito). *J. Jpn. Soc. Food Sci. Technol.,* 42: 682–686.

Tuckwell, D.S., Nicholson, M.J., Mc-Sweeney, C.S., Theodorou, M.K. and Brookman, J.L. 2005. The rapid assignment of ruminal fungi to presumptive genera using ITS1 and ITS2 RNA secondary structures to produce group specific fingerprints. *Microbiology,* 151: 1557–1567.

Upadhyay, R.C. and Sohi, H.S. 1988. Apple pomace – A good substrate for the cultivation of edible mushrooms. *Curr. Sci.,* 57: 1189–1190.

Upadhyay, R.C. and Verma, R.N. 2000. Non-conventional substrates for growing oyster mushrooms. *Mushroom Res.,* 9(1): 35–38.

Upadhyay, R.C., Verma, R.N., Singh, S.K. and Yada, M.C. 2002. Effect of organic nitrogen supplementation in *Pleurotus* species. *Mushroom Biology and Mushroom Products,* 105(3): 225–232.

Upadhyay, R.C., Vijay, B. and Verma, R.N. 1996. Use of industrial tea leaf waste for cultivation of oyster mushrooms, *In*: Royse, D.J. (*ed.*), *Proceed. II Conf. on Mushroom Biology and Mushroom Products,* pp. 423–428.

Velázquez-Cedeño, M.A., Farnet A.M., Ferré, E. and Savoie, J.M. 2004. Variations of lignocellulosic activities in dual cultures of *Pleurotus ostreatus* and *Trichoderma longibrachiatum* on unsterilized wheat straw. *Mycologia,* 96(4): 712–719.

Zervakis, G.I., Venturella, G. and Papadopoulou, K. 2001. Genetic polymorphism and taxonomic infrastructure of the *Pleurotus eryngii* species-complex as determined by RAPD analysis, isozyme profiles and ecomorphological characters. *Microbiology,* 147: 3183–3194.

Zhang, J.X., Huang, C.Y., Chen, J., Gao, W. and Zheng, S.Y. 2010. Evaluation for wild germplasm of cultivated mushroom species. *Journal of Plant Genetic Resources,* 11(2): 127–131.

2.5

Cultivation of Shiitake

SHWETA KULSHRESHTHA[1]*

ABSTRACT

Shiitake is one of the popular species in China and Japan. It is highly priced mushroom and much more costly than button mushroom. It is cultivated for its medicinal and nutraceutical properties. Shiitake grows normally on rotten wood logs as it is wood decaying fungi. It can be cultivated on logs, sawdust containing bag under the specific conditions of light, temperature and humidity. The beginners can cultivate this mushroom on wooden logs. In this chapter, various method and conditions for the cultivation of shiitake have been discussed which helps the growers in cultivating shiitake mushroom.

Key words: Shiitake cultivation, *Lentinus* cultivation, Log cultivation, Polypropylene bag cultivation.

1. INTRODUCTION

Shiitake (*Lentinus edodes*) is one of the most popular edible mushrooms in the world. This mushroom is highly nutritious due to its higher protein, dietary fibers and important mineral contents (Khan *et al.*, 2009). It also possesses many medicinal properties like anticancer, antidiabetic, hypotensive, hypocholesterolemic, antimicrobial activities (Wasser, 2005). It possesses many bioactive compounds which are important for the pharmacology, food and cosmetology industries (Kues and Liu, 2000; Gaitán-Hernández *et al.*, 2006). Recently, it is reported that the combinations of this mushroom derived polysaccharides and chemotherapeutic drugs are useful in mutagen induced tumour cell growth and therefore, may be used

[1] Amity Institute of Biotechnology, Amity University Rajasthan, Jaipur.
**Corresponding author:* E-mail: shweta_kulshreshtha@rediffmail.com

for cancer treatment (Ren *et al.*, 2014). Similar type of observation is recently reported by Xu *et al.* (2014) which is related to antitumor and immunomodulating potential of *Lentinus edodes* (Xu *et al.*, 2014). The polysaccharide lentinan, eritadenine, and lectins, antibiotics, anticarcinogenic and antiviral compounds have been isolated intracellularly (fruiting body and mycelia) and extracellularly (culture media) from shiitake (Bisen *et al.*, 2010).

Shiitake (*Lentinula edodes*) was initially grown in Japan and China and is now sold in the European market. Shiitake is the most important mushroom among the species cultivated mushroom. Nowadays, it is gaining importance due to several health benefits, excellent flavor and higher shelf life (Moonmoon *et al.*, 2011). Higher shelf life is an important criterion because of its high value in the international market and increase demand of dried mushrooms and their import by other countries. It possesses many bioactive compounds which are important for the pharmacology, food and cosmetology industries (Kues and Liu, 2000). These health promoting properties may act as a tool to promote their sale at high prices (Kimmons *et al.*, 2010).

At present, China is the leading country in the world market for the production of shiitake mushroom. This is because they have developed the technology for cultivating this mushroom on wood shavings instead of logs which results in increased production of shiitake in recent decades (Gaitán-Hernández *et al.*, 2011). Like China, other countries need some modified technology for the cultivation of this mushroom.

The main focus of developing the mushroom cultivation technology is to obtain any pharmaceutical and nutritionally important compound in bulk amount. It may be cultivated in solid state or in liquid state by fermentation process. Efforts are made to reduce the fruiting time and further reduce the production time. Hence, it is necessary to develop new methods and practices of fermentation or cultivation which can increase biological efficiency or yield of mushroom (Gaitán-Hernández *et al.*, 2011).

2. CULTIVATION OF SHIITAKE

2.1. Strains

The main criterion for strain selection is biological efficiency or yield of mushroom. Generally, higher yield providing strains are selected to get profit but occasionally, these high yield strains do not provide expected results after inoculation. This is because there are several other factors that affect the strain yield.

The desired strain can be obtained from various centers which collect and preserve these strains. A number of strains may be isolated and assessed for their cultivation potential on different types of substrates. The following parameters should be considered before selecting mushroom for growth:

Adaptability of strain to the substrate: Recently, new varieties of substrates have been used for the cultivation of shiitake and it is also possible that the conventional strain may not provide comparable yield on the new substrate due to its non-adaptability. Therefore, adaptability of the strain to the new substrate must be assessed prior to cultivate shiitake mushroom.

Adaptability of strain can be assessed by mycelial growth rate on various substrates under various environmental conditions. For this purpose, substrate is ground and filled in petri dishes. These petri dishes are sterilized at 121ºC for 30 min. and allowed to cool down. After this, petri dishes are inoculated with mycelium of shiitake strains by placing the mycelium containing disc (1cm) in the center of petri dish and incubate at 24°C for three to four weeks. Petri dishes should be observed daily and mycelial growth is recorded and compared with one another (Levanon *et al.*, 1993).

2.1.1. *Selection based on temperature*

The selection of shiitake strain depends on the expected range of ambient temperatures throughout the productive life of the log. Shiitake strains are categorized on the basis of fruiting temperature requirements: low temperature (less than 10ºC), mid-temperature (10–18ºC), warm temperature (greater than 20ºC), and wide-range (5–35ºC) (Chen *et al.*, 2000). Three strains of shiitake - wide-range strain WR46, warm-weather strain WW44, and cold-weather strain CW50 (Field and Forest Products, Peshtigo, WI) have been reported by Bruhn *et al.* (2009).

2.1.2. *Genetic variability of strain*

Genetic constitution of strain is an important selection criterion to get high yield. A minor variation in the strain may cause great loss of mushroom crops and therefore, it is necessary to assess it. Levanon *et al.* (1993) reported the use of esterase isoenzymes, to indicate genetic variations.

Among all enzymes, the great variation is found in the esterases profiles of the strain and therefore considered as an indicator of genetic variability (Ohamasa and Furakawa, 1986; Itavaara, 1988). At first, genetic variability of any strain is determined and then it is inoculated on the substrate for observing biological efficiency. This combined strategy is used to select best strain for cultivation (Levanon *et al.*, 1993). The identification of genetic differences among the fungal isolates can be done by isoenzyme electrophoresis. This preliminary screening not only finds out the genetic difference but also reduces the number of strains to be tested for biological efficiency (Royse and May, 1987; Levanon *et al.*, 1993).

Shiitake mycelium is grown on potato dextrose agar which is further used for the preparation of agar discs. Four Agar discs of 1cm size were cut and placed in the 100 ml Potato Dextrose Broth (PDB) medium and incubate the medium in an orbital incubation shaker at 24°C and 120 r/min. After three weeks of incubation, the fungal biomass was separated by filtration and rinsed with distilled water. It was homogenized in 0.1 M phosphate buffer at pH 7.0 (1:1, w/v) and centrifuged at 13000 g for 3 min. The pellet was discarded and supernatant was used for electrophoretic studies of the separation of the esterase isoenzymes. For esterase, polyacrylamide slab gel electrophoresis can be performed according to the method of Dewald *et al.* (1974) at pH 8.9 in 12.5% gel. The esterase enzyme can be assessed after staining the gel with α-naphthyl acetate (Levanon *et al.*, 1993).

2.1.3. *Strain maintenance*

The selected strains are grown on malt extract agar or potato dextrose agar and maintained at 25°C or low temperature in refrigerator. These strains are checked from time to time for their viability. Non-viable or contaminated strains should be discarded.

2.2. Spawn

Millet seeds (*Panicum miliaceum* L.) were soaked in water and then water was drained off and is mixed with coffee pulp powder (8.8%) and calcium sulphate (1.3%) and peat moss (1.3%) according to the dry weight of the material. When spawn reaches 70% moisture then this mixture is placed in plastic bags and sterilizes at 121°C for 90 min. The sterilized mixture was inoculated with 1cm^2 piece cut from malt extract agar medium possess culture of *L. edodes*. After inoculation, these bags were incubated at 25°C for 15 days in complete darkness.

If readymade spawn is available, it can also be used for further preparation of secondary spawn. In this method, plastic bags were filled with seed by the method mentioned above and then inoculated with little amount of readymade spawn and incubated in the same condition (Gaitán-Hernández *et al.*, 2006).

Recently, Gaitán-Hernández *et al.* (2014) have prepared different types of spawn by using millet seeds and their combinations. Control (C) (millet seed *Panicum miliaceum* L., 100%)

(Pearl seed / Canadian origin), Formula 1 (F1) [Millet seed, 88.5%; Wheat bran, 8.8%; Peat moss (*Sphagnum* / Canadian origin), 1.3%; $CaSO_4$, 1.3%], and Formula 2 (F2) (Millet seed, 88.5%; Powdered wheat straw, 8.8%; Peat moss, 1.3%; $CaSO_4$, 1.3%), percentages are based on dry matter. The final moisture content was 65%. Rest of the procedure of preparing spawn is same as mentioned above. All these formulations were reported to affect the biological efficiency of shiitake which was higher for the supplemented spawn compared to that of non-supplemented spawn. Secondary spawn was prepared by filling the new bags with the sterile mixture and inoculated with the first spawn (15 g master spawn) (Gaitán-Hernández *et al.*, 2014).

Spawn can also be prepared from sawdust. The most common forms of spawn are nutritionally supplemented sawdust spawn and wooden dowel spawn (Chang and Miles, 2004), but sawdust spawn plugs capped with a thin layer of styrofoam (thimble spawn) are also available.

2.3. Substrate Used For Cultivation

Hardwoods logs, obtained from oak trees, are used as substrate for the cultivation of shiitake. The main problem associated with the use of log is – its limited supply due to its slow growth rate and overexploitation. This method of cultivation is not an ecofriendly approach and a limiting factor for the shiitake cultivation. Thus, there is a need to develop a more efficient, faster and more reliable production system.

Researchers have also developed a technique to cultivate shiitake on the shavings of *Carpinus*, *Bursera*, *Alnus*, and *Elaeocarpus* and even coffee pulp and sugarcane bagasse. The availability of some of these substrates is limited and therefore, there is a need of doing research on alternative substrate for cultivation.

Many researchers reported the successful production of shiitake on bulk, pasteurized wheat straw (Delpech and Olivier, 1990; Delpech and Olivier, 1991; Levanon *et al.*, 1993). The use of wheat straw as substrate for shiitake cultivation is recommended to cultivate shiitake mushroom. Substrate was prepared by enriching the wheat straw with chicken feather meal. 75 g/g benomyl or carbendazim was added to it in order to control *Trichoderma* spp. infection. In this case, high amount of spawn was added to the substrate to avoid contamination (Delpech and Olivier, 1991). Other researchers adopted hygienic measures despite of adding fungicides. Researchers have improved substrate formulation by mixing with sphagnum peat to suppress green mold (Mata *et al.*, 1998).

Many researchers have focused on the use of sterilized sawdust to enhance the biological efficiency and to reduce the production cycle (Kalberer, 1998). Royse and Sanchez-Vazquez (2001, 2003). They have developed the formulation of substrate by using the mixture of sawdust and wheat straw. The collected sawdust mainly consists of sawdust of red oak (*Quercus rubra* L.) and partially white oak (*Quercus alba* L.) wood. This sawdust was used to prepare substrate by mixing with wheat straw. The general substrate formulation (all ingredients based on oven-dry substrate wt.) consisted of 52% sawdust/wheat straw, 24% millet (*Panicum miliaceum* L.), 12% wheat bran (*Triticum aestivum* L.), and 12% rye (*Secale cereale* L.). Mixed substrate ingredients were pasteurized, cooled, inoculated and filled in polypropylene bags. These bags were incubated for spawn run, log browning. Log soaking were conducted as described by Royse and Sanchez-Vazquez (2001, 2003). The combination of substrate provided better yield of shiitake compared to that of unmixed substrate (wheat straw or oak sawdust) (Philippoussis *et al.*, 2007; Royse and Sanchez, 2007). Royse and Sanchez (2007) again modified the composition of the substrate by developing three formulae containing ground wheat straw and oak sawdust in the mentioned ratio (0%WS + 52%, OWC; 8% WSS + 44% OWC and 16% WSS + 36% OWC). These substrates were pasteurized at 111°C for 20 min. in an autoclave. After sterilization, these substrates were filled in polypropylene bags and subjected to uniform environmental conditions for running of spawn and browning of fruiting bodies. The mature fruiting bodies were harvested and found that mushroom yields were 19% higher (maximum) when 16% wheat straw was substituted for portions of oak sawdust (Royse and Sanchez, 2007).

Lentinula edodes (shiitake) can also be produced by using different mixtures of wheat straw (WS), corncobs (CC), and oak-wood sawdust (OS) as substrates. Oak wood saw dust alone delayed sporophore initiation but increased protein content of mushroom. Wheat straw promoted early fructification and provided good quality mushroom. However, higher productivity was achieved on corncobs. This shows that all these substrates can be used for the production of shiitake. However, its quality and productivity varies according to the substrate (Moonmoon *et al.*, 2011). Moonmoon *et al.* (2011) cultivated shiitake mushroom on saw dust with different levels (10%, 15%, 20%, 25%, 30%, 35% and 40%) of supplements of wheat bran, rice bran, maize powder and their combinations in order to assess their effect on yield. It was reported that 25% supplementation of sawdust with wheat bran is very effective for higher yield. However, on supplementing the sawdust with 40% wheat bran provided better quality of *L. edodes* (shiitake).

The sawdust may be supplemented with various materials prior to spawning for enhancement of the yield of mushrooms (Przybylowicz and Donoghue, 1990; Hadwan *et al.*, 1997). Sawdust may be supplemented with different starch based supplements, such as wheat bran, rice bran, millet, rye and maize powder to enrich the medium and make it suitable for growth (Royse, 1997, 2001). These supplements not only improve the production and biological efficiency but also improve the quality, flavor and shelf life of shiitake (Mau *et al.*, 2002; Chang, 1996; Isikhuemhen *et al.*, 2000; Chiroro, 2004; Okhuoya *et al.*, 2005). These studies reported the suitability of supplemented and non-supplemented sawdust as substrate for shiitake cultivation.

There are many advantages of producing shiitake on sawdust instead natural logs. For *e.g.*, less time is required to complete a growing cycle and mature fruit bodies formation. It provides greater biological efficiency compared to natural logs. However, the major disadvantage is the requirement of high initial investment costs compared to natural logs. Shiitake cultivators are facing the problem of shortage of sawdust due to its increase use in fuel, animal food and feed, bedding, meat smoking, pulping, mulching, soil conditioning, and particle-board production (Royse and Sanchez, 2007). To continue the production of shiitake, cultivators are trying to find out new alternatives substrates for its production. Sanchez *et al.* (2002) reported the potentiality of vineyard pruning for mushroom production and Gaitán-Hernández *et al.* (2006) has utilized this as substrate for the cultivation of shiitake. Shiitake was cultivated on vineyard-pruning and results were compared with that of barley-straw (BS) and wheat-straw (WS) in solid state fermentation. After inoculation, primordia appeared in lesser time period compared to control which initiated the formation of sporophore. Vineyard pruning waste was found to support the formation of fruiting bodies and requires the shortest primordium formation time (28 days), highest biological efficiency (93.25%), highest yield (37.46%), and shortest production cycle (6 days). The fruiting bodies possessed high energy (379.09 to 392.95 kcal) and contents of protein (12.37 to 17.19%), but low contents of fat (1.82 to 2.15%) (Gaitán-Hernández *et al.*, 2011). Thus, vineyard pruning is a potential substrate for shiitake production due to its low cost, short production cycles, and high biological efficiency.

Another alternative for use as substrate is Hazelnut husk which is generated as agricultural waste. Hazelnut husk was dried in air prior to use for the preparation of substrate. Hazelnut husk (3.0-3.5 cm) was mixed with wheat straw (<10 cm size) and beech wood chip (0.5-4.0 cm size) on the basis of their dry weight and a homogenous mixture was produced. These combinations were wetted with water to maintain 60-65% moisture content. This mixture is filled in polypropylene bags sterilized in an autoclave at 121°C for 90 min. The sterilized waste was inoculated with 0.8% (w/w) spawn, and incubated at 22 ± 2°C in darkness. When the mycelium colonization was completed, the bags were exposed to day-light for 10 h photoperiod to promote mushroom formation at 18 ± 2°C room temperature and 80–90% relative humidity in a controlled room. After each flush of mushrooms was harvested, synthetic logs were resoaked to increase log weight to 0.9 kg (Royse, 1985). The yield was found to be low on using the hazelnut husk alone as compared to the controls. When the hazelnut husk content in the mixtures was kept below 50%, the yield was relatively high. When the hazelnut husk content increased to 75% in the mixture, the comparable yield was obtained by adding 10% of wheat straw as nutrients. The results showed that hazelnut husk can be used as a new basal ingredient for substrate preparation in shiitake cultivation (Ozc'elik and Peksen, 2007).

Many conventional (wood logs) and non-conventional substrates (wheat straw, vineyard pruning, sawdust etc.) have proved to be useful for the cultivation of shiitake mushroom. The successful cultivation of mushroom on any substrate with good biological efficiency depends upon the nutritional composition of substrates and their utilization by mushroom mycelium. Another important point to be considered for selection of substrate is local availability of waste (Gaitán-Hernández *et al.*, 2011).

3. METHOD OF CULTIVATION

Traditionally, wood logs are used for the cultivation of shiitake. However, new methodologies have been developed for the cultivation of this mushroom species on new substrates in the last decades. Due to this, it is becoming popular in the other countries of the world. Shiitake cultivation requires two steps in order to support different physiological stages of the fungus. These steps are – (i) Vegetative growth and (ii) Fruit body formation. These steps require different environmental conditions such as temperature, humidity, dark period or light period (Gaitán-Hernández *et al.,* 2011).

3.1. Cultivation on Wood Logs

3.1.1. *Log-grown shiitake production method*

Traditional method of cultivation of shiitake is the use of wood logs. Generally, oak tree logs are used for cultivation which is cut in green stage from white oak tree after leaf drop in the fall but before bud break in the spring (mid-November to late March). It should possess approximately 100 cm in length and 10 -15 cm in diameter (Kimmons *et al.*, 2010).

A method of cultivation of shiitake is previously reported by Chang and Miles (2004). It is suggested that logs can be prepared for inoculation by drilling holes (1–1.5 cm diameter 1.5–2 cm deep) at pre-determined intervals and subsequently filled with spawn (*i.e.,* mycelium-colonized substrate). After inoculation logs are sealed with hot and molten cheese wax. There is no use of cheese wax to seal the inoculation sites if thimble spawn is used because thimble spawn possesses a styrofoam cap (Bruhn *et al.*, 2009). These inoculated logs are placed in a "laying yard" for spawn run. Spawn run refers to the colonization of substrate by the fungal mycelium which occurs due to the enzymatic degradation of substrate and then its utilization by fungi. In the natural conditions, sufficient rainfall can stimulate the fruiting bodies formation. However, in artificial conditions, it can be done by force fruiting technique which is based on the submersion of inoculated logs in water for 1–3 days depending on season and other variables (Chang and Miles, 2004).

A modified method is proposed by (Kimmons *et al.*, 2010). To prepare logs for shiitake cultivation, these logs were initially incubated in a tight lean-to stack position by maintaining approximately 90% shade. In this condition it was maintained for approximately 9 months and hydrated periodically using sprinklers in order to prevent drying of logs. After completing the incubation period, holes (1.2 cm diameter and 2.5 cm depth) were drilled in the logs in a diamond pattern configuration (15-22 cm x 4-5 cm). These logs were inoculated with sawdust spawn and then sealed with food grade cheese wax with heating (approximately 190°C). In the September and October, fruiting bodies of shiitake appears following inoculation and this time logs are transferred to "A-frame" stacks under a densely shaded, deciduous trees. Logs are left in this position for the formation of basidiocarps. Fruiting usually occurred after a heavy rainfall which provide favourable temperature (Kimmons *et al.*, 2010).

In the United States, log-based mushroom production is very popular due to its application with many other land management practices. It not only recycles by-products of forest but also provides profit in the form of salable fruiting bodies of shiitake. Therefore, it is considered as a source of income for the workers, work in the forest (Bruhn *et al.*, 2009).

3.1.2. *Cultivation in polypropylene bags*

Traditionally, shiitake is cultivated on tree logs, which is not ecofriendly approach, may face the problem of shortage. Therefore, this log cultivation method is replaced by bag cultivation. This cultivation technique is applicable when straw based substrates or sawdust are used for cultivation of shiitake. Gaitán-Hernández *et al.* (2006) used this method for the cultivation of shiitake on vineyard pruning (*Vitis vinifera* L.), barley straw (*Hordeum vulgare* L.) and wheat straw (*Triticum aestivum* L.) Gaitán-Hernández *et al.* (2006). All these were used as a substrate for the cultivation of *Lentinus* spp. These substrates are chopped in small pieces (5-8 cm length) by using electric chopper or by mechanical cutting method. The chopped pieces are dipped in water in separate containers for 12 h for hydration purpose. After this, water is drained off to maintain appropriate moisture content (vineyard pruning waste: 60%; barley straw: 73%; wheat straw: 75%). These substrates were filled in polypropylene bags and sterilized for 90 mins at 121°C. The bags were cooled down and then inoculated using 5% of spawn and incubated in a dark room at a controlled temperature of 25°C. When mycelium covers the whole substrate, polypropylene bags are removed. The bundles are transferred to the production room for the growth of fruiting bodies. In the production room, relative humidity is maintained between 85-90% and temperature is 18°C with air circulation facility. Air circulation in the room is helpful in maintaining the level of CO_2. The CO_2 level should be kept below 1000 ppm as recommended for the proper primordium development. A photoperiod of 12 h is also provided during the day to favour fruit bodies formation. For this purpose, light lamps of 350 lx illumination can be used. Fruit bodies appear in the form of pinhead and take few days to mature. Maturity of fruit bodies can be determined by exposed gills with completely extended cap margins. These mature fruiting bodies are harvest for consuming purpose (Gaitán-Hernández *et al.*, 2006).

A modified method of cultivation was used by Philippoussis *et al.* (2007) for *L. edodes* (Berk.) Pegler strain. The oak wood sawdust, wheat straw and corncob were mixed in 2:1 ratio giving 75% of the total substrate weight (on a dry weight basis), with the basal ingredient used in the substrate mixture. This mixture was soaked in water for 12-24h. After this soaking period excessive water is drained off and mixed with supplements. 25% supplements were added to the substrate which includes 10% millet, 10% wheat bran, 4% soybean flour, and 1% $CaCO_3$. 2kg of this mixture was filled in autoclavable polypropylene bags and sterilized for 60 min and 121°C in an autoclave. Sterilization of bags was done twice. The moisture content of sterilized substrate was maintained as 60-65%. The substrates were inoculated with 3% (w/w on fresh weight basis) spawn and incubated at 25°C in dark. When colonization of the substrates took place in growth chambers then these were transferred to fruiting room and subjected to a cold shock for 3 days to induce primordia formation at 14–15°C, 90% relative humidity, and 24-h/day illumination. For maturation and colour development, bundles were illuminated by 200 lux intensity fluorescent lamps for 24h, air exchange rates were controlled to maintain low CO_2 level (<1200ppm), relative air humidity was adjusted to 75–90%, and temperature was set at 17.5±1.0°C. The mature fruiting bodies were harvested. After the first two flushes, blocks were soaked in cold tap water (14–15°C) for 8–12h for soaking and inducing the new fruiting bodies formation.

This method is used for various types of substrates including wood logs, sawdust, wheat straw, vine pruning etc. The yield of shiitake mushroom depends on characteristics and supplementation of substrates with suitable supplements and their amount (Moonmoon *et al.*, 2011). This method is successfully used for the cultivation of shiitake on supplemented substrates. To increase the production of good quality mushroom, Moonmoon *et al.* (2011) cultivated shiitake mushroom on saw dust with different levels (10%, 15%, 20%, 25%, 30%, 35% and 40%) of supplements of wheat bran, rice bran, maize powder and their combinations in order to assess their effect on yield. All this substrate mixture was filled in polypropylene bags and put them for sterilization. These bags were inoculated with spawn and incubate for mycelial run and its maturation. At stage all the packets were opened from the top side and soaked in water for 15-20 mins. These packets were placed on cemented floors in the culture room which had 18–22°C temperature, 60–70% humidity and 10–20 lux intensity light. Proper misting was done to prevent the drying of substrate and proper aeration was maintained to release CO_2 and the supply of O_2 for primordial initiation and fruiting body development. Mushroom fruiting bodies were harvested when the mushroom cap surface was flat to slightly up-rolled at the cap margins (Moonmoon *et al.*, 2011).

3.2. Harvesting

Mushrooms are harvested at one of three development stages: bud, veil break, and open caps. Bud stage refers to the initial basidiocarp (mushroom cap) appearance. Buds are formed when cap is swollen and appear distinct from its stalk. The buds are dome shaped in diameter. When the veil begins to open and separate from stalk these expose the gills or lamellae. This stage of basidiocarp development is known as veil break. At this stage, the cap is still dome shaped. In few days, basidiocarp expands and cap becomes flat with fold margins. After this stage, the cap becomes upfold because outside edges are slightly curled and gills exposed at the fully open stage (Kimmons *et al.*, 2010). This is maturation stage which is suitable for the harvesting of mushrooms. Harvest period generally varies from 28-38 days (Royse and Sanchez, 2007).

Harvesting is a very crucial stage because many characteristics of mushroom depend on their harvesting period. It is reported that mushroom harvested at immature stage (during bud break or before veil break) are rich in high molecular weight polysaccharide, known as lentinans. It is also reported that first harvest of shiitake mushroom possess higher amount of lentinans compared to subsequent harvests (Kimmons *et al.*, 2010).

4. FACTORS AFFECTING THE MUSHROOM CULTIVATION

The mushroom yield, biological efficiency varies according to the variation in substrate because it depends on the degradation capacity of lignocellulosic components, nature of substrate and its chemical composition (Gaitán-Hernández *et al.*, 2006). Strains of mushroom, its genotype, breeding of adaptable mushroom strains, efficient inoculum delivery methods etc. are also found to effect the mushroom yield. Shiitake productivity also depends on environmental factors such as moisture content, temperature and supplementation in the substrate. The length of spawn is also found to affect the mushroom yield (Ito, 1978; Chang and Miles, 2004; Bruhn *et al.*, 2009; Gaitán-Hernández*et al.*, 2011).

4.1. Strains

The selection of strain is considered as the most important criterion for successful cultivation of shiitake. A strain possessing desirable characteristic should be used for successful cultivation because the yield, biological efficiency and characteristics of basidiocarps depend on it. For *e.g.,* higher amount of lentinans are produced by shiitake strains NN-430 and 569-430 compared to other shiitake strains (Kimmons *et al.*, 2010).

It is also reported that strain selection varies with the season. There are cold weather strains (CW-25) and hot weather strains (29-430, 569-430, MY-602 and NN-430) of shiitake. Cold weather strains grow slowly and need more time for spawn run and unable to produce sufficient quantity of mushroom. However, hot weather strains possess fast growth rate, and provide good yield (Kimmons *et al.*, 2010). Recently, hybrid lines have been developed with desirable characteristics and can be used at commercial level. For *e.g.*, hybrid Lel2 line is obtained from cross between single- spore isolates of WC305 and WC380.

4.2. Spawn

The selection and preparation of spawn is critically important to ensure the highest biological efficiency in the short time period. The spawn should be able to adapt to the substrate and environmental conditions. Generally, shiitake spawn is prepared by millet seeds as the basic material which is sometimes supplemented with some other material for increasing yield (Royse,1996; Gaitán-Hernández *et al.*, 2014). The supplemented spawn is nutritionally rich which not only induces mycelium run, fruiting bodies formation but also reduces the contamination by competing microbes. The ability of shiitake to grow on lignocellulosic substrates is related to the vigor of its mycelium to cover the substrate in shortest possible time so that contaminating fungi cannot access the substrate and compete for nutrients and space (Gaitán-Hernández *et al.*, 2014).

4.3. Characteristics of Wood Logs

Wood logs must possess some characteristics required for successful cultivation of shiitake. Log must possess following characteristics:

(i) ***Intact bark:*** Barks of the wood logs must be intact otherwise cannot be used for the cultivation of shiitake.

(ii) ***Size:*** The suitable length of the bark is 1m to use it for shiitake cultivation because environmental conditions can be maintained uniformly in these logs. Larger logs can provide fruiting bodies for up to at least six years after inoculation. Many logs will be most productive during the second to fifth years after inoculation (Bruhn *et al.*, 2009).

(iii) ***Layers of log:*** Ideal shiitake logs must have a thick sapwood layer, little heartwood without rotten heart. The stem wood region, made up of living cells and reserve carbohydrates is known as sapwood. This discolored wood lies between the bark and central, pigmented heartwood core. The inner stem-wood region which does not possess

living cells or reserve material is known as heart wood. This region may be coloured due to fungal infection or any injury. Due to thick sapwood layer, sugar maple is proved as a superior substrate to white/ northern red oak. The discolored wood is the cylinder of sapwood between the bark and any central, pigmented heartwood core (Bruhn *et al.*, 2009).

(iv) ***Physical shock:*** The suitable barks are, at first, selected, inoculated with spawn and then incubated. During incubation, wood logs are forced to fruit by applying physical shock (Bruhn *et al.*, 2009).

4.4. Effect of Season on Log Characteristics

Season of collection of log wood affects the yield of shiitake mushroom. Therefore, it is necessary to harvest the wood logs in appropriate season. (Bruhn *et al.*, 2009) suggested that sugar maple can be harvested in spring season and stored under shade however, oak logs can be harvested shortly before spring season (Bruhn *et al.*, 2009).

4.5. Effect of Moisture Content

Moisture content of logs is an important factor to be considered for shiitake productivity and biological efficiency. The final substrate moisture should be 55% to get maximum yield of shiitake (Shen *et al.*, 2008).

4.6. Effect of Soaking on Mushroom Yield

After browning of mycelium, substrate logs need soaking in order to stimulate shiitake mushroom production. Soaking is required to replenish the loss of water which occurs due to evaporation and mycelial respiration. Logs are dipped in water tank for soaking purpose for 3-24 h. Soaking time varies and depend upon the stage of crop, age of logs, flush of mushroom harvest, soak times may average 3-4 h on the first soak, 8-16 h on the second soak and 14-24 h on the third soak. Soaking time increases with the increase in age of mushroom due to their hydrophobic nature. After each flush of mushroom soaking time need to be increased.

The problem associated with this is rate of absorption of water varies and depending on the log characteristics. To solve this problem, it is suggested that dried logs must be placed at the bottom of soaking tank so that these can absorb more water compared to logs that are at the surface. This arrangement helps in maintaining the uniform moisture content in the logs. However, it is difficult to apply this technique on large scale farms. Therefore, vacuum soaking technique is developed for uniform soaking of logs (Royse *et al.*, 2002). For vacuum soaking, two logs were placed in metal container [35×25×25 cm (L×W×H); 25 l capacity] and fixed in its place by locking with perforated lid. Water was added until the water level reached 8 cm above the submerged logs. This container was placed in oven and sealed it. A vacuum pump was connected to the oven to remove air present inside. This created a partial vacuum (54kPa) within 60s which reached the target log. Now vacuum is released in approximately 60-70 s. These logs were again exposed to vacuum for second soak. This partial vacuum (74.5 kPa) was given for 80 s and then released. After this, steel container is taken out from the oven and wood logs were removed from it and put back to the production room for shiitake production. This vacuum soaking logs provides better yield compared to regular soaking process. Moreover, time required for regular soaking is less (only 3 min) compared to regular soaking time (3-15 h) (Royse *et al.*, 2002).

4.7. Effect of Particle Size

Particle size is an important parameter to consider because fine particles allow the faster breakdown of lignocellulosic material than that of coarse particles. Sawdust was collected, ground and then filtered with sieve of standard pore sizes. Four sawdust particle size ranges (2.8±4.0; 1.7±2.8; 0.85±1.7, and <0.85 mm) were screened, collected and used for inoculation. Biological efficiency was found to be increased on decreasing the particle size of substrate. Substrate with the finest particles (<0.85 mm) size did not provide good yield. The possible reason for this is over-compaction of substrate logs which restricts the gaseous exchange between the environment and the substrate. Over-compaction makes small air spaces in the substrate which further results in the slower gaseous exchange and lower productivity of shiitake.

4.8. Carbon Nitrogen Ratio

Carbon and nitrogen ratio is an important criterion to be considered for substrate formulation because it affects the biological efficiency. Mushroom production in bags is correlated with nitrogen content of the substrates. Increase in nitrogen content poses positive effect on fruiting bodies formation. However, lower C/N ratio induces earlier sporophore induction (Philippoussis *et al.*, 2007).

4.9. Enzymes

Shiitake mushrooms produce hydrolytic and oxidative enzymes that are required for the degradation of substrate. The complex components of the substrates are converted into simple components by the combined activities of cellulases, hemicellulases and ligninases which further utilize it for their vegetative growth and fruiting bodies formation. Gaitán-Hernández *et al.* (2011) reported that the formation of enzymes is positively correlated with the substrate composition such as presence of lignin, cellulose and total carbohydrate in the substrate. The enzyme production by shiitake depends on substrate composition. Besides this, various environmental factors affect the production of enzymes (Gaitán-Hernández *et al.*, 2006).

4.10. Calcium Carbonate Supplementation

Calcium carbonate ($CaCO_3$) is an important constituent for the preparation of substrate which must be added in the artificial log preparation. Royse *et al.* (2003) determine the effect of calcium carbonate supplementation on mushroom yield and size. The yield and biological efficiency of shiitake was found to higher in supplemented log compared to non-supplemented artificial logs. It is also reported that productivity of mushroom is more consistent on adding 0.6% calcium carbonate to the substrate. However, total weight of produced mushroom was found to be low in supplemented straw than the non-supplemented straw (Royse *et al.*, 2003).

4.11. Effect of *Trichoderma viride* Infection

Nowadays, different types of material are used as substrate (artificial logs) for the production of shiitake. This is better approach to continue the production of shiitake due to the short supply of oak wood logs. However, this approach has major limitation *i.e.,* infection of substrate by *Trichoderma viride*, which is an antagonistic fungi. This fungus competes with the shiitake for space and access to nutrients. Its infection may result in reduced yield and biological efficiency.

Mata *et al.* (1998) made two modifications in the substrate: (i) Addition of peat (ii) Use of new activated inoculum. These changes in the cultivation method increased the overall yield of shiitake due to increase in its competitive ability.

5. HARMFUL EFFECTS OF SHIITAKE CONSUMPTION

Shiitake (*Lentinula edodes*) is an edible mushroom which possesses many medicinal properties. Recently, it is reported that ingestion of raw or undercooked shiitake is responsible for many type of allergies. Shiitake consumption may cause Flagellate erythema disease which is first reported in Japan in 1974. Many researchers have reported the recent occurrence of this disease (Wang *et al.*, 2013; Boels *et al.*, 2014). The symptoms of this disease includes eruption of pruritic, erythematous multiple linear streaks on the trunk and extremities. Skin biopsy did not show any specific findings, including a sparse perivascular and interstitial dermatitis. On administrating topical steroids and oral antihistamines helps in resolving the lesions (Wang *et al.*, 2013).

Shiitake mushroom-induced toxicoderma, or shiitake dermatitis, is a very popular disease in Japan, China and Korea. Recently, it is reported in the European countries. The symptoms of the disease appear within 1 to 2 days after consumption of raw or cooked shiitake mushrooms. It is characterized by specific pattern of whiplike, linear, erythematous wheals, papules and erythematous-violaceous linear streaks (Adriano *et al.*, 2013; Chu *et al.*, 2013). It is reported that lentinan, a polysaccharide of shiitake, is thought to instigate a toxic reaction which results in the appearance of a rash. This disease is self-limited and resolves automatically within days to weeks of its appearance (Chu *et al.*, 2013).

6. CONCLUSIONS

Shiitake mushroom is rich in nutrients and can be consumed for its nutritional and medicinal values. There are many methods for the cultivation of Shiitake like cultivation of wood logs and on wheat straw. Many factors like selection of strain, cultivation on log, different seasons, particle size of straw, etc. affects the production of shiitake. A few reports on the allergies due to consumption of shiitake mushroom are discussed revealing its allergic potential. Therefore, it is suggested to check its toxic and allergic ability before consumption.

7. REFERENCES

Bisen, P.S., Baghel, R.K., Sanodiya, B.S., Thakur, G.S. and Prasad, G.B. 2010. *Lentinus edodes*: Macrofungus with pharmacological activities. *Curr. Med. Chem.*, 17(22): 2419–2430.

Bruhn, J.N., Mihail, J.D. and Pickens, J.B. 2009. Forest farming of shiitake mushrooms: An integrated evaluation of management practices. *Bioresour. Technol.*, 100: 6472–6480.

Chang, S.T. 1996. Mushroom research and development equality and mutual benefits. *In*: Royse, D.J. *(ed.)*, Mushroom Biology and Mushroom Products. Pew State University Press; Philadelphia, p. 283.

Chang, S.T. and Miles, P.G. 2004. Mushrooms: Cultivation, Nutritional Value, Medicinal Effect, and Environmental Impact, Second Ed. CRC Press, New York.

Chen, A.W., Arrold, N. and Stamets, P. 2000. Shiitake cultivation systems. *In*: Griensven Van *(ed.)*, Science and Cultivation of Edible Fungi. Balkema, Rotterdam, pp. 771– 778.

Delpech, P. and Olivier, J.M. 1990. Champignon parfumØ (ou shii-take).UnemØthode francaise pour sa culture. *Rev. Hort.*, 305: 25–30.

Delpech, P. and Olivier, J.M. 1991. Cultivation of shiitake on straw based pasteurized substrates. *Mush. Sci.,* 13: 523–528.

Dewald, B., Dulaney, J.T. and Toustee, O. 1974. Solubilization and polyacrylamide gel electrophoresis of membrane enzyme with detergents. *Methods in Enzymol.,* 32: 82–91.

Gaitán-Hernández, R., Cortés, N. and Mata, G. 2014. Improvement of yield of the edible and medicinal mushroom *Lentinula edodes* on wheat straw by use of supplemented spawn. *Braz. J. Microbiol.,* 45(2): 467–474.

Gaitán-Hernández, R., Esqueda, M., Gutiérrez, A. and Beltrán-García, M. 2011. Quantitative changes in the biochemical composition of lignocellulosic residues during the vegetative growth of *Lentinula edodes*. *Braz. J. Microbiol.,* 42(1): 30–40. doi: 10.1590/S1517-83822011000100004.

Gaitán-Hernández, R., Esqueda, M., Gutiérrez, A., Sánchez, A., Beltrán-García, M. and Mata, G. 2006. Bioconversion of agrowastes by *Lentinula edodes*: The high potential of viticulture residues. *Appl. Microbiol. Biotechnol.,* 71(4): 432–9.

Hadwan, H.A., Al-Jaboury, M.H. and Hassan, A.O. 1997. Suitability of different substrates and amendments on the cultivation of oyster mushroom, Collection of Thesis Materials, S & T, Development, Environment and Resources. Proc. 96 FUZHOU International, Symposium on the Development of Juncau Industry, pp. 215–221.

Itavaara, M. 1988. Identification of Shiitake strains and some other basidiomycetes: Protein profile, esterases and acid phosphatase zymograms as an aid in taxonomy. *Trans. Brit. Mycol. Soc.,* 91: 295–304.

Kalberer, P. 1998. Influence of the substrate component on the crop yield of shiitake (*Lentinus edodes* (Berk) singer). *Gartenbauwissenschaft*, 63: 15–19.

Khan, M.A., Khan, L.A., Hossain, M.S., Tania, M. and Uddin, M.N. 2009. Investigation on the nutritional composition of the common edible and medicinal mushrooms cultivated in Bangladesh. *Bangladesh J. Mushroom,* 3: 21–28.

Kimmons, T.E., Phillips, M. and Brauer, D. 2010. Effects of management factors on the concentration of a high molecular weight polysaccharide fraction from log-grown shiitake mushrooms (*Lentinula edodes* (Berk.) Pegler). *J. Agric. Food Chem.,* 58: 4331–4335. doi:10.1021/jf904268m.

Kües, U. and Liu, Y. 2000. Fruiting body production in basidiomycetes. *Appl. Microbiol. Biot.,* 54: 141–152.

Levanon, D., Rothschild, N., Danai, O. and Masaphy, S. 1993. Strain selection for cultivation of shiitake mushroom (*Lentinula edodes*) on straw. *Bioresour. Technol.,* 45: 9–12.

Mata, G., Savoie, J.M., Delpech, P. and Olivier, J.M. 1998. Reduction in the incidence of *Trichoderma* spp. using substrate supplementation with peat and an alternative spawn during cultivation of *Lentinula edodes* on pasteurised wheat straw. *Agronomie,* 18: 515–520.

Mau, J.L., Lin, H.C. and Chen, C.C. 2002. Antioxidant properties of several medicinal mushrooms. *J. Agric. Food Chem.,* 50: 6072–6077.

Moonmoon, M., Shelly, N.J., Khan, M.A., Uddin, M.N., Hossain, K., Tania, M. and Ahmed, S. 2011. Effects of different levels of wheat bran, rice bran and maize powder supplementation with sawdust on the production of shiitake mushroom (*Lentinus edodes* (Berk.) Singer). *Saudi J. Biol. Sci.,* 18(4): 323–8. doi: 10.1016/j.sjbs.2010.12.008.

Ohamasa, M. and Furakawa, H. 1986. Analysis of esterase and malate dehydrogenase isozymes on *Lentinus edodes* by isoelectric focusing for the identification and discrimination of stocks. *Trans. Mycol. Soc. Japan,* 27: 79–90.

Philippoussis, A., Diamantopoulou, P. and Israilides, C. 2007. Productivity of agricultural residues used for the cultivation of the medicinal fungus *Lentinula edodes. Int. Biodeterior. Biodegradation,* 59(3): 216–219. doi:10.1016/j.ibiod.2006.10.007.

Przybylowicz, P. and Donoghue, J. 1990. Shiitake growers handbook. The Art and Science of Mushroom Cultivation, Kendall Hunt, Dubuque.

Ren, M., Ye, L., Hao, X., Ren, Z., Ren, S., Xu, K. and Li, J. 2014. Polysaccharides from *Tricholoma matsutake* and *Lentinus edodes* enhance 5-fluorouracil-mediated H22 cell growth inhibition. *J. Tradit. Chin. Med.,* 34(3): 309–16.

Royse, D.J. and May, B. 1987. Identification of genotypes zby multilocus enzyme electrophoresis: Catalog of lines. *Biochem. Genetics,* 25: 716–19.

Royse, D.J. and Sanchez, J.E. 2007. Ground wheat straw as a substitute for portions of oak wood chips used in shiitake (*Lentinula edodes*) substrate formulae. *Bioresour. Technol.,* 98(11): 2137–41.

Royse, D.J. and Sanchez-Vazquez, J.E. 2001. Influence of substrate wood-chip particle size on shiitake (*Lentinula edodes*) yield. *Bioresour. Technol.,* 76: 229–233.

Royse, D.J. and Sanchez-Vazquez, J.E. 2003. Influence of precipitated calcium carbonate ($CaCO_3$) on shiitake (*Lentinula edodes*) yield and mushroom size. *Bioresour. Technol.,* 90(2): 225–228.

Wasser, S.P. 2005. Shiitake (*Lentinus edodes*). *Encyclopedia of Dietary Supplements. In*: Coates, P., Blackman, M.R., Cragg, G., Levine, M., Moss, J. and White, J. (*eds.)*, Informa World Online Publication, pp. 653–664.

Xu, X., Yan, H., Tang, J., Chen, J. and Zhang, X. 2014. Polysaccharides in *Lentinus edodes*: Isolation, structure, immunomodulating activity and future prospective. *Crit. Rev. Food Sci. Nutr.,* 54(4): 474–87. doi: 10.1080/10408398.2011.587616.

2.6

Cultivation of *Volvariella volvacea*

Shweta Sulshreshtha[1]*

ABSTRACT

Volvariella volvacea is commonly known as paddy straw mushroom due to its efficiency to grow on paddy straw. It can be cultivated under artificial conditions on a variety of substrates such as cotton waste, sugarcane bagasse, banana leaves, sorghum straw etc. The main problem with its cultivation is related to its uncertainty in the growth of mycelium and fruit body which ultimately affects its yield and production. The interesting feature of this mushroom is its requirement of high temperature for growth and therefore, it can be cultivated in the summer season when the cultivation of other mushroom like Agaricus and Pleurotus is not possible. In this chapter, a detailed method of cultivation of Volvariella is explained which is helpful for the beginners.

Key words: *Volvariella* cultivation, Nanhua mushroom, Tropical mushroom, Paddy straw mushroom.

1. INTRODUCTION

Volvariella volvacea commonly known as paddy straw mushroom, straw mushroom or Chinese mushroom, is able to grow at high temperature and therefore, considered as an edible mushroom of tropics and subtropics. In China, it is also known as "Nanhua mushroom" which was named after Nanhua temple of Chaohsi in northern Guangdong Province, China. The Buddhist monks of this temple cultivated this mushroom for their own consumption. It is also called as "tropical mushroom" or "warm mushroom"

[1] Amity Institute of Biotechnology, Amity University Rajasthan, Jaipur.
**Corresponding author:* E-mail: shweta_kulshreshtha@rediffmail.com

due to its ability to grow at relatively high temperature, *i.e.*, vegetative growth at 32-34°C.

It belongs to the family Pluteaceae Kotl. and Pouz of Basidiomycetes. The cultivation of *V. volvacea* has begun around 300 years ago in China. Then, it is introduced into the other countries such as Malaysia, Philippines, South East Asia *etc.* by migrated Chinese people. *Volvariella volvacea* contains good nutrient profile due to the presence of proteins, crude fibres and ash.

V. volvacea is more effective in utilizing cellulosic material than any other mushroom. The optimum carbon nitrogen (C:N) ratio required for *Volvariella volvacea* is about 40:60. *Volvariella volvacea* is the mushroom which can be cultivated in summer season when the cultivation of most popular mushroom species such as *Agaricus* and *Pleurotus* is not possible due to the high temperature. It is easy to cultivate *Volvariella volvacea* but consistent yield is not assured (Chang *et al.*, 1993).

2. NUTRITIVE VALUE OF MUSHROOM

On fresh weight basis, *Volvariella volvacea* contains 90% water, 30-43% crude protein, 1-6% fat, 12-48% carbohydrates, 4-10% crude fibre and 5.13% ash. It also possesses mineral content which are mentioned here as per 100 g of fresh mushroom such as phosphorus 0.10 g, potassium 0.32 g, iron 1.70 g, calcium 5.60 mg, riboflavin 0.61 mg, niacin 2.40 mg, ascorbic acid 18 mg (Verma, 2002).

The fat content increases with the maturation stage. Mature fruit bodies possess highest fat content (5%). In the button stage, N-free carbohydrates are present which increases up to egg stage and then start decreasing till maturation stage. This reveals that the nutrient content of mushroom depends upon the stages of maturation. Nutrient content also depends upon the method of cultivation.

3. DIFFERENT STAGES OF *VOLVARIELLA VOLVACEA*

3.1. Mycelium

Mycelium is meshwork of fungal cell which is made up of small units of fungal cells called hyphae. It is of two types: (i) Primary mycelium – when mycelium is directly derived from basidiospore, it is uninucleate and haploid. Therefore, it is called as primary mycelium. (ii) Secondary mycelium - when plasmogamy occurs between two mycelia then it form dikaryotic mycelia which bear clamp connection. It is known secondary mycelium.

In the case of *Volvariella volvacea*, mycelium cannot be identified as primary or secondary mycelium due to some special properties like absence of clamp connection. It possesses multinucleate mycelium and number of nuclei varies in each cell. It ranges from 3 to 105 and average nuclei per cell are 22 which are randomly distributed in the cell.

3.2. Morphology of Fruit Body

The fruit body of mushroom is known as basidiocarp or carpophore which is mainly divided into six different stages: pinhead, tiny button, button, egg, elongated fruit body, mature carpophores. The small fruit body appear on the substrate possesses very tiny stalk with round head which is called as pinhead. At this stage, pileus and stipe are not visible in vertical section.

After this, it forms tiny button stage. At this stage the top of the veil is brown while the rest is white. On vertical section, a thin narrow band on the lower surface of pileus, can be seen which is called as lamellae. In button stage, stipe of fruiting body and pileus is very small and covered by a universal veil. Stipe is visible only in longitudinal section of fruiting body.

Round fruiting bodies become elongated and looks like egg. In egg stage, stipe can be observed only after longitudinal dissection of fruit body. In the egg stage, pileus grows and pushed out of the veil and takes the shape of volva. Gills do not bear basidiospores when observed under microscope. However, the formation of sterigmata, a structure that bear basidiospore, may occur at this staeg. This stage is also marketable and mushroom harvested at this stage can also be sold in the market.

The stipe of these fruiting bodies elongated and reached its full length and therefore, this stage is known as elongation stage. Maximum growth in the stipe is found at the attachment site of pileus. When pileus of fruit body open up and take the shape of umbrella, it is called as mature fruiting body. Initially, pileus is covered with thin layer of cells called veil which attaches to the stipe. As the fruit body grows, veil breaks down which allow the pileus to take umbrella like shape and expose the gills. Gills possess basidia that bear oval shaped basidiospores.

At mature stage, the structure is divided into three regions: (i) the pileus or cap, (ii) stipe or stalk and (iii) the volva or cup (Fig. 1).

3.3. Histology of Gills

In the case of *Volvariella volvacea*, cross-section of gills reveals three layers – trama (innermost), sub-hymenium (middle layer) and hymenium (outermost layer). Hymenium gives rise to basidium which possesses basidiospores. One basidium bears four basidiospores that are attached to basidium by the help of sterigmata. The tetrad of basidiospores can be observed under microscope. The basidiospores of *Volvariella volvacea* have asymmetric morphology and these can be seen in egg-shaped, spherical and ellipsoidal spores. These spores are smooth but have thick covering. A short triangular spine protrudes from the narrow end of spores. When it is viewed microscopically the colour of spores changes from transparent and light yellow to pink and dark brown depending on the maturity status of mushroom.

4. SPAWN PREPARATION

Spawn is a seed material for mushroom cultivation.There are many types of spawn which is used for *Volvariella volvacea* cultivation.

4.1. Grain Spawn

Rye, sorghum, wheat seeds can be used for the preparation of spawn. The grains are washed twice with fresh water and soaked overnight. After soaking, grains are boiled till they become soft. Care should be taken that it should not be over boiled and broken. Excess water is drained off and boiled grains are spread over clean blotting paper for removing excessive moisture. This will help in breaking lumps and maintain 80% moisture. Then, 6% chalk powder (calcium carbonate) and 2% gypsum are mixed with boiled grains and filled in sterile bottles. Two-third bottle should be filled with glucose which act as carbon source. The mouth of the bottle should be plugged with non-absorbent cotton and then, wrapped with paper. These grains filled bottles are sterilized at 15 lbs/inch^2 pressure in autoclave for 2 hours. After cooling, these bottles are transferred to sterile inocultaion chamber or laminar air flow. For sterilizing the surface in laminar air flow or inoculation chamber 2% formaldehyde is used. 2% formaldehyde solution is poured in petri dish and left it open inside the chamber. Prior to inocultaion, UV tube is switched on for half an hour. UV light should be switched off before using laminar air flow or inoculation chamber. The bottles can be inoculated with the culture of *Volvariella volvacea* under sterile conditions. For this purpose, mycelium is picked up with the help of inoculating needle and then, transferred in the bottle. The culture is placed on the grains in the central position. After inoculation bottles are transferred into the growing chamber where 34°C temperature is maintained. The optimum temperature for this mushroom spawn is 33 ± 2°C, however, the *V. volvacea* spawn can be grown upto 40°C.

4.2. Straw Spawn

Rice straw is considered as good substrate for the cultivation of *V. volvacea*. This is also act as a good substrate for the preparation of spawn. This rice based spawn is called as straw spawn. For the preparation of this spawn, the paddy straw is screened for the presence of contaminants. It should be free from fungal infection. It should be bright and light brown in colour.

After screening, straw is chopped into 1 inch bits and soaked in water for 2 to 4 hours. After this, the straw is taken out from water and spread on a clean surface or on blotting paper to drain off excess moisture. Care should be taken that the straw must possess 85% moisture. Now, straw is mixed with 4% gypsum by weight and compact in glucose bottles. The mouth of the bottle is plugged with non-absorbant cotton and autoclaved for 2 hours. Then, it is left over night for cooling, inoculted by a small bit of culture under sterilized conditions.

4.3. Spawn Prepared by Using Waste Substrates

Spawn is prepared from used tea leaves which is discarded as waste. For this purpose, used tea leaves can be collected and washed thoroughly with water to remove any type contaminants. This is already wet and therefore, there is no need of wetting this waste before use. In the tea leaves waste, 2% calcium carbonate is added to the substrate. This will help in maintaining the range of pH *i.e.,* 6.8 to 7.8. Then, the substrate should be mixed thoroughly. This substrate is filled in glass bottle. After this, bottle is sterilized and inoculated in the same way as mentioned earlier.

Cotton waste can also be used for the preparation of spawn. It is processed in the same way as tea leaves are processed. In this spawn preparation, calcium carbonate is used for maintaining the pH.

4.4. Substrate

V. volvacea uses lignocellulosic material for growth. Lignocellulosic material is composed of lignin and cellulose which can be utilized as carbon source by mushroom. During cultivation, *V. volvacea* produces lignocellulolytic enzyme for utilizing the substrate. Many scientists have used various waste substrates in composted and uncomposted form for the cultivation of this mushroom. A variety of waste material such as paddy straw (Chang, 1965), cotton waste (Hamlyn, 1989; Yau, Chang, Hu *et al.*, 1973), water hyacinth (Cheng and Mok, 1971), oil palm bunch (Naidu, 1971), oil palm pericarp waste (Graham and Yong, 1974), sugarcane bagasse (Hu *et al.*, 1973, 1976a, 1976b), (Chang, 1993; Reyes *et al.,* 1998) pineapple waste, sawdust 32, banana leaves, agricultural waste (Obodai *et al.*, 2003; Onuoha *et al.*, 2009) *etc*. All these substrates contain the combination of lignin and cellulose but the composition of waste varies. The main component used by *V. volvacea* is cellulose and therefore, it produces cellulolytic enzymes such as endoglucanases (Zheng and Ding, 2013), cellobiohydrolases and two glycosidases (BGL I and BGL II) (Buswell *et al.*, 1996; Cai *et al.*, 1998) for the degradation of cellulosic component of waste substrate. It also produces Beta-glucosidase enzyme which is involved in cellulose degradation (Li *et al.*, 2005). However, it does not produce lignin degrading enzymes in high amount. Therefore, it is not efficient in utilizing lignin containing substrate which affects its yield.

It is also important to know about cellulolytic enzyme system of *V. volvacea* in order to increase the efficiency of cellulose utilization and improved yield. Jia *et al.* (1999) have done the cloning of cbhI and cbhII genes involved in cellulose utilization by this mushroom. Fruit body production is correlated with alterations in cellulase activity (Wood *et al.*, 1988). Knowledge of genes helps in manipulation of cloned cellulase genes. The productivity of this mushroom depends on the potential of hydrolytic enzyme production, substrate quality and method of its preparation, and the growing conditions (Ahlawat *et al.*, 2011).

Analysis of *Volvariella volvacea* cbhI gene revealed that it possesses a typical C-terminal family I cellulose binding domain (CBD), a linker and catalytic domain regions. Certain glutamic acid and aspartic acid residues play an active role in catalytic mechanism. Analysis of *V. volvacea* cbhII gene revealed architecture very similar to that of other cbhII genes. It possesses a typical N-terminal family I cellulose binding domain, linker and catalytic domain regions. These genes were cloned and amplified by randomly amplification of cDNA Ends (RACE)-PCR was performed to obtain full-length cDNA clones. The results of this study revealed that these genes are involved in utilization of cellulose (Jia *et al.*, 1999).

V. volvacea is also known as paddy straw mushroom because paddy straw is mainly used as a substrate for its cultivation. The paddy straw used should be uncrumpled fresh, not more than one year old, and not very leafy. All leaves must be removed before using these straws for cultivation. It should be stored at a proper place so that it does not get wet during rain because moistened straw harbours unwanted microbes (Fig. 2). These microbes compete with the mushrooms and thus the yield of mushroom is reduced considerably. The half dried or decaying straw may harbour fungi which adversely affects the crop yield.

The other substrate can also be used for its cultivation like combination of rice straw-cotton with waste compost or dried banana leaves. Farm wastes were also used as substrate by Ukoima *et al.* (2009) and maximum yield was reported for palm fibre which is followed by rice husk and sawdust.

4.5. Cultivation Techniques for *V. volvacea*

4.5.1. *Outdoor cultivation technique*

Traditionally, *V. volvacea* is cultivated on beds of banana leaves of rice straw. These substrates were used in raw form without composting or pasteurization. These straws were tied at both ends and laid on the floor to make a layer. A second layer is placed on the top of first layer and in this way; bed of substrate is prepared for mushroom cultivation. These beds must be covered with plastic sheet in order to avoid drying of beds by direct sunlight. The yield of mushroom bed fluctuates as this method does not provide protection from microbes and insects. Environmental conditions cannot be controlled by this type of method. This method is not used by mushroom farmers nowadays.

4.5.2. *Indoor cultivation techniques*

It is done in a room where set-up of equipment is established in order to control environmental conditions. This room has proper arrangement of light. Temperature is controlled by air-conditioners and heating coils. Beds of substrates are prepared on trays or in racks or in polythene bags for mushroom cultivation. Indoor cultivation of mushroom relies on three points which must be considered before cultivation: (1) Composting (2) Pasteurization and (3) Cultivation in the mushroom house. This method is better than outdoor cultivation technique because environmental conditions can be controlled.

4.5.2.1. *Substrate preparation by unfermented paddy straw*

The paddy straw is cut into small pieces and soaked overnight. To prevent the bundle from floating, place some heavy object such as bricks or stones. The paddy straw is taken out from the tank after 10-15 hours. Wet paddy straw is kept on cemented floor to drain off excessive moisture and then bundles are made. Each bundle should possess equal weight *i.e.*, 1 kg. The unequal protruding parts of the paddy straw are cut off from the bundle. Then, bundles are pasteurized for two hours by steam. The paddy straw can be used without fermentation.

In contrast to above, cotton waste ferments before use. Cotton waste is collected and pile up in the field. It is wetted and left open for sometime to ferment. It is turned up once or twice a day in order to produce uniform compost. It takes 3 to 7 days to ferment. Care should be taken that cotton pile must be covered with polythene. The temperature of the centre of cotton pile is about 60°C during composting which must be reduced to 40°C before use it for bedding purpose. After reaching this temperature, substrate must be spread in shelves or trays or in plastic bags. Composted cotton is distributed evenly in shelves in the form of layer. This layer is sterilized by steam sterilization process.

4.5.2.2. *Pasteurization*

Pasteurization is a practice to eliminate harmful microbes, nematodes and insect-pests and help to get contamination free substrate. Any type of contamination can decrease the yield of mushroom. Hot water treatment, steam treatment or chemical treatment can be used to make the substrate germ free. However, chemical treatment is not well adapted procedure.

4.5.2.3. *Steam sterilization*

Steam has good penetration power and therefore, it is considered as good technique for making the substrate germ free. For this purpose, bundles are kept in autoclave in free flowing steam condition. Substrate is kept for 2 hrs in autoclave for sterilization. After pasteurization bundles are allowed to cool.

4.5.2.4. *Hot water pasteurization*

In hot water pasteurization, the paddy straw bundles are dipped in hot water which is at 80-85°C for 60 minutes. After 60 minutes the bundled are removed from hot water and kept on cleaned cemented floor or raised support to drain off excess moisture and allowed it to cool down at room temperature. The same process is followed for any type of substrate use for bed preparation. In the bed preparation of paddy straw mushroom gram dal powder, cotton seed husk or chicken manure is used as supplement. These constituents are spread over the spawn in the form of layer. These supplement should also be pasteurized by hot water treatment. For this purpose, these supplements are filed in polythene bags and dipped in hot water at 80-85°C for one hour.

4.5.2.5. *Chemical pasteurization*

Chemical pasteurization is done by treating bundles with 5% formaldehyde solution for 24 hours. Then, the bundles are soaked in water for 10-15 hours. After 15 hours of soaking, bundles are taken out and kept on some support to drain off excessive water.

4.5.3. *Preparation of beds or inoculation of spawn*

Bed preparation is done by stacking 1m×1m bed either on cemented floor or on a bamboo frame. Bamboo frame is made by placing and tie-up the sticks in such a way that it will give frame like structure. A raised platform is contructed on this bamboo frame. For this purpose, bamboo frame is placed on the ground and then, bricks are arranged on it in order to make a raised platform like structure. This structure will not allow the contact of paddy straw layer with the ground. The bed is prepared with 32 bundles, having four layer of eight bundles each. For preparation of bed four bundles of paddy straw are placed over platform side by side in one direction. Another four bundles are placed on the top in opposite direction so the level get equalised. The number of the bundles should be placed in such a way that it approximately covers the length of the straw. To make a squrare bed trim the sides of the bundles. The width of bundle should be roughly 1m×1m. Small bits of spawn are placed on the straw 10 to 12 cm from the edges and about 15 cm apart and dal powder is sprinkled on spawn. Again, a layer of rice straw is made over it. A third layer is placed across to this layer and spawned with mycelial bits. However, fourth layer is spread all over the bed and is covered with a thin layer of soaked paddy straw. Two spawn bottles of 100 g are used for the bed. Finally, the bed is slightly pressed down to set it in a proper form. Sprinkle water gently over the bed. Cover the bed with polythene sheet to maintain the optimum temperature and moisture content. It is advisable to remove the polythene sheet folding from the side of the bed as temperature rises beyond 40°C. The optimum temperature for good spawn run and fruit body formation is 30-38°C .

Care should be taken during preparation of bed. The packing of bed should not be too tight or too loose because too tight packing will restrict the aeration whereas loose packing will cause quick drying of beds due to the loss of moisture. Both factors adversely affect the yield of mushroom. The size of bed should not be very small because both the conditions were found to adversely affect mushroom yield (Stamets, 1993).

4.5.4. *Harvesting and packaging*

Mushrooms are harvested at button stage. Harvesting is done early in the morning or late in the afternoon. This prevents opening of mushroom caps and prevents the excessive loss of moisture. Harvesting is done by holding the fruit body by hand and twist and pull it. This action must not disturbe the other fruit bodies which are in developing stage.

Harvested mushrooms must be cleaned by removing the straws attached at their base. These can be packed in a perforated plastic bags. These mushrooms cannot be refrigerated as they are highly perishable and cannot withstand at low temperature.

Thiribhuvanamala *et al*. (2012) suggested some modification in order to increase the yield and commercialize its cultivation. They suggested the use of circular compact bed instead of rectangular bed. This change provides good yield with biological efficiency of 23.8%. If 15-20 days old spawn with well developed chlamydospores is used, higher yield was recorded as compared to new spawn. When rice straw was used in combination with oil palm bunch waste, higher yield (876.3 g/ bed) was recorded as compared to using it alone (832 g/bed). Similarly, the combination of rice straw with cotton waste provided significantly higher yield (876.3) as compared to using it alone (778 g/bed). Strain Vvb1 required $CaCO_3$ 400 ppm + $CaCl_2$ 50 ppm + KH_2PO_4 50 ppm + NaCl 50 ppm + Na_2HPO_4 50 ppm while Vvb2 required

$CaCl_2$ 400 ppm + Na_2HPO_4 100 ppm as micronutrient based boosters which supported the formation of mycelium with aerial silky strands, early chlamydospore formation and increased chlamydospore density (7.8×10^2 for Vvb1 and 7.2×10^2 for Vvb2 spores/ml) (Thiribhuvanamala *et al.*, 2012).

4.5.5. *Indoor cultivation in wooden crates*

In this method, rice straw is soaked overnight in clean tap water. After soaking, straw become soft. Soaked rice straw is supplemented with 1% molasses and 0.5% fertilizer. This is piled up for composting which takes 14 days. Initially, this pile is covered with plastic sheets so that thermophilic decomposers can grow and proliferate. The partially composted rice straw is shredded after 3 days to release ammonia and other toxic gases. Then, it is mixed with 1% agricultural lime thoroughly, and continued the process for completing the 14 days period. After this, the composted substrate is shredded for aeration. This is followed by crating process. In this process, a wooden open crate 15 x 45 x 60 cm is used. The composted substrate is packed in the crates and leave it for sometime to release toxic gases like ammonia. These crates are shifted to mushroom growing chamber. Care should be taken that there must be sufficient space between the crates. These should be kept in such a way that these should not block uniform circulation of steam. These crates must receive uniform light during incubation and fruiting. Steaming is done in such a way that the temperature of compost must be 60°C for 4-6 hrs. These crates allowed to cool till their substrate reach 30°C temperature. Now, these crates are used for spawning purpose. The crates are inoculated with 10-20% spawn as per spawning rate of the substrate. The crates are incubated for the growth of *V. volvacea* at 35°C and 85% relative humidity. Humidity is maintained by sprinkling water on the floor. In 7-10 days, mycelium colonizes the substrate. At first, tiny fruit body appears which is called as fruiting primordia which reach button shape in 3-4 days. These button shape fruit body can be harvested and sold in the market (Reyes, 2000).

4.5.6. *Indoor cultivation in wooden framed trapezoid mould*

This type of mould is opened at the top and bottom. The dimensions of the mould are 36 cm base, 30 cm top, 36 cm height and 91 cm length. It can be used to make the bed of 5 kg substrate. When substrate is kept in this mould, it is pressed by hand to remove trapped air. Ten gram of spawn is spreaded on the periphery over the first layer. Similarly, second layer of substrate is made and spreaded with spawn. Then, third layer is made by sprinkling spawn evenly on the entire surface of bed. After spawning, bed are covered with plastic sheets and required environmental conditions are maintained. Mycelia are allowed to grow and form fruit bodies that can be harvested at the button stage (Obodai *et al.*, 2003).

5. FACTORS AFFECTING MUSHROOM CULTIVATION

Successful cultivation of mushroom depends on many factors such as, temperature, relative humidity, moisture content of substrate, ventilation in mushroom house and aeration, quality of substrate and spawn etc. These factors play an important role for the mycelial growth and fruit body formation of mushroom. Factors affecting mushroom production must be studied before cultivating *V. volvacea* because the productivity of *V. volvacea* is very low (about 10–15% on rice straw) compared to other cultivable species of mushroom.

However, other characteristics of superior yielding strains *viz.*, mycelial growth rate, mycelial density, and formation of aerial hyphae and chlamydospores on growing medium vary under varied growth conditions.

5.1. Temperature

The temperature required for *V. volvacea* (paddy straw mushroom) is around 30°C. Therefore, it can be cultivated during summer season and tropical region of the country. The paddy straw mushroom develops well at 30 to 36°C but it can tolerate temperature upto 45°C. However, there is slight reduction in yield in paddy straw mushroom when temperature is above 36°C.

5.2. Relative Humidity

Generally, mushrooms require high relative humidity during spawn running. The relative humidity should be 90 to 95% inside the spawn room. However, during cropping of mushroom a bit low relative humidity is required, *i.e.*, 80-90%. During cropping period, high humidity results in the incidence of bacterial contamination. However, lower relative humidity (compared to the optimum value) results in reduced yield. During dry season, fresh air should be supplied in the cultivation room in order to maintain temperature. Water must be sprayed frequently on walls, floor and also on gunny bags to wet them for maintaining relative humidity.

5.3. Moisture Content

Mushroom crops should be protected from dry atmosphere. Moisture content is an important factor to be considered for mushroom cultivation. It is essential to maintain moisture especially during vegetative and reproductive stage of mushroom cultivation. The moisture content of the substrate should be 68-70% for *V. volvacea* cultivation. The surface of the substrate should be kept moist by covering the substrate with paper and spraying water on the paper twice a day. Direct application of water on bed is injurious to growing crops. In case of polythene bags, moisture content should be 65-

68%. Normally, substrate in polythene bags does not require additional spray of water because rate of moisture loss is very slow. In beds, frequency of spraying water depends on the atmospheric condition of mushroom house. At this stage, no ventilation is required to maintain moist atmosphere, but the walls and floor should kept wet which reduces water loss from the substrate.

5.4. Ventilation

Ventilation is one of the important factors for paddy straw mushroom cultivation. Ventilation is required for maintaining the removal of toxic gases by introduction of adequate fresh air. During spawn running, little ventilation is required. Carbon dioxide (4%) is required for good spawn growth. The good spawn run rate can be achieved even without supplying fresh air if the rise in temperature can be checked. Therefore, the fresh air should be introduced only when it requires to control the rise in temperature. Ventilation is required after spawn run.

The carbon dioxide concentration should be 0.07 to 0.09% for good and healthy crop of mushroom. Low concentration of carbon dioxide during cropping produces abnormalities in mushroom crop. If carbon dioxide concentration is too high, mushroom appears with too thick stalk and too small cap. The optimum carbon dioxide concentration and temperature can be maintained by complete air exchange with fresh air. This is done 2-4 times per day. The air speed should not increases 0.2 to 0.4m/ sec. The high speed of air and low humidity causes scale formation and cracks in caps particularly.

5.5. Effect of Phenolics

A study conducted by Cai *et al.* (1993), showed that *V. volvacea* exhibits different sensitivity profiles to lignin-related phenols and tannin derivatives. It is also reported that these differences are completely associated with their different abilities to grow and producing fruit bodies.

5.6. Effect of Temperature and Lipids

Supplementation of substrate with vegetable oils can affect the vegetative and reproductive growth of mushrooms. Therefore, a number of lipids are used commercially for mushroom cultivation. The degree of unsaturation in the fatty acid of the mushroom grown at varied temperature is different. Many mushroom *L. edodes, Pleurotus* spp. possess ability to modify their membrane lipid composition in order to maintain membrane fluidity. In some mushrooms, lowering the growth temperature leads to increase in the degree of unsaturation which is associated with the membrane fluidity. In contrast to this, *V. volvacea* showed decrease in the degree of unsaturation at 0°C temperature. This is observed in polar lipid fraction *i.e.,* phospholipid, which is an important constituent of cell membrane structure. This shows that *V. volvacea* is not capable of adjusting its fatty acid composition at low temperature. This causes chilling injury and autolysis of the mushroom fruit body. This is also the reason of short shelf life of this mushroom in cold storage.

5.7. Nutrient Requirement

Nutrient requirement varies from strain to strain. Reyes *et al.* (1998) evaluated nutrient requirement for the efficient mycelial colonization of *V. volvacea* strains Vvc1 and Vvc2 (commercial strains) and EAAC-001 and EAAC-002 (wild strains). These strains vary in their requirement for carbon sources. Strain Vvc1 requires starch or cellulose while Vvc2 requires sorbitol. Wild strains require starch as a preferred carbon source. Soytone, peptone and glycine were found to support the mycelial growth of all strains. Ascorbic acid, calcium pantothenate and biotin are good sources of vitamin.

5.8. Disturbance in The Substrate

It is fast growing mushroom because it requires only 8-10 days time to produce mature fruiting bodies after spawning. One crop takes 20 days of time for producing two flushes. It is important to note that after spawning, substrate should not be disturbed. Ignoring this will lead to the breaking off mycelia network in the substrate which further prevent the colonization of the substrate. This is done because *Volvariella volvacea* has poor colonizing ability. The minor disturbances in the substrate inhibit its ability to run on the substrate. Spawn run also depends on the suitable environmental conditions. If there is any disturbance in the spawn run, then yield is reduced drastically even after having favourable environmental conditions. On undisturbed substrates and under favourable environmental conditions pinhead formation occurs in 4-5 days after spawning. Therefore, it is the fastest growing species and produce fruiting bodies in very less time. However, it has lower biological efficiency compared to other cultivable mushroom species.

6. CONCLUSIONS

Volvariella volvacea is the mushroom which needs high temperature to grow (above 30°C). The best substrate for the growth of this mushroom is rice straw. Unlike other mushrooms, this mushroom can be easily cultivated in tropical regions due to requirement of high temperature to grow. These cannot be preserved at low temperature.

7. REFERENCES

Ahlawat, O.P., Singh, R. and Kumar, S. 2011. Evaluation of *Volvariella volvacea* strains for yield and diseases/insect-pests resistance using composted substrate of paddy straw and cotton mill wastes. *Indian J. Microbiol.,* 51(2): 200–205. doi: 10.1007/s12088-011-0126-1.

Buswell, J.A., Cai, Y.J., Chang, S.T., Peberdy, J.F., Fu, S.Y. and Yu, H.S. 1996. Lignocellulolytic enzyme profiles of edible mushroom fungi. *World J. Microbiol. Biotechnol.,* 12: 537–542.

Cai, Y.J., Buswell, J.A. and Chang, S.T. 1993. Effect of lignin-derived phenolic monomers on the growth of the edible mushrooms *Lentinus edodes, Pleurotus sajor-caju* and *Volvariella volvacea. World J. Microbiol. Biotechnol.,* 9: 503–507.

Cai, Y.J., Buswell, J.A. and Chang, S.T. 1998. Beta-glucosidase components of the cellulolytic system of the edible straw mushroom *Volvariella volvacea. Enzyme Microbiol. Technol.,* 22: 122–129.

Chang, S.T. 1965. Cultivation of the straw mushroom in S.E. China. *World Crops,* 17: 47–49.

Chang, S.T., John, A.B. and Chin, S. 1993. Mushroom biology and mushroom products. Hong Kong: Chinese University Press, pp. 1–73.

Cheng, S. and Mok, S.H. 1971. Preliminary experiment of water hyacinth used as a medium for the cultivation of paddy straw mushroom. *J. Hortic. Soc. China (Taiwan),* 17: 194–197.

Graham, K.M. and Yong, Y.C. 1974. Studies on the paddy mushroom (*Volvariella volvacea*) II. Effect of bed depth on yield in oil palm pericarp waste. *Malay. Agric. Res.,* 3: 1–6.

Hamlyn, P.F. 1989. Cultivation of edible mushrooms on cotton waste. *The Mycologist,* 3(4): 171–173.

Hu, K.J., Song, S.F. and Liu, P. 1973. Experiments on Chinese mushroom cultivation, I. The comparison of cultivating materials. *J. Taiwan Agric. Res.,* 22: 145–148.

Hu, K.J., Song, S.F. and Liu, P. 1976a. The comparison of composts made of different raw materials for *Volvariella volvacea. Mushroom Sci.*, 9(1): 691–700.

Hu, K.J., Song, S.F., Liu, P. and Peng, J.T. 1976b. Studies on sugarcane rubbish for Chinese mushroom culture and its growth factor. *Mushroom Sci.,* 9(1): 691–700.

Jia, J., Dyer, P.S., Buswell, J.A. and Peberdy, J.F. 1999. Cloning of the cbhI and cbhII genes involved in cellulose utilization by the straw mushroom *Volvariella volvacea. Mol. Gen. Genet.,* 261: 985–993.

Li, X., Pei, J., Wu, G. and Shao, W. 2005. Expression, purification and characterization of a recombinant glucosidase from *V. volvacea. Biotechnol. Lett.,* 27: 1369–1373. Doi: 10.1007/s10529-005-3683-8.

Naidu, N.R. 1971. Cultivation of paddy straw mushroom *Volvariella volvacea* (Fr.) Sing. using soil palm bunch as a medium. *Planter,* 47: 190–193.

Obodai, M., Cleland-Okine, J. and Johnson, P.N.T. 2003. Use of agricultural wastes as substrate for the mushroom *Volvariella volvacea. Trop. Sci.,* 43: 121–124.

Onuoha, C.I., Oyibo, G. and Judith, E. 2009. Cultivation of straw mushroom (*Volvariella volvacea*) using some agro-waste material. *J. Am. Sci.,* 5(5): 135–138.

Reyes, R.G. 2000. Indoor cultivation of paddy straw mushroom, *Volvariella volvacea,* in crates. *Mycologist,* 14(4): 174–176.

Reyes, R.G., Eguchi, F., Iijima, T. and Higaki, M. 1998. Physiological considerations for efficient mycelial colonization of Philippine strains of *Volvariella volvacea*. *J. Wood. Sci.,* 44: 408–413.

Stamets, P. 1993. Paul Stamets, Ten Speed Press, Berkeley.

Thiribhuvanamala, G., Krishnamoorthy, S., Manoranjitham, K., Praksasm, V. and Krishnan, S. 2012. Improved techniques to enhance the yield of paddy straw mushroom (*Volvariella volvacea*) for commercial cultivation. *Afri. J. Biotechnol.,* 11: 12740–748.

Ukoima, H.N., Ogbonnaya, L.O., Arikpo, G.E. and Ikpe, F.N. 2009. Cultivation of mushroom (*Volvariella volvacea*) on various farm wastes in Obubra Local Government of Cross River State, Nigeria. *Pakistan J. Nutrition,* 8(7): 1059–1061.

Verma, R.N. 2002. Cultivation of paddy straw mushroom (*Volvariella* spp.). *In*: Verma, R.N. and Vijay, B. (*eds.*), Recent advances in the cultivation technology of edible mushrooms. National Research Centre for Mushroom, Solan (HP), India, pp. 221–220.

Wood, D.A., Claydon, N., Dudley, K.J., Stephens, S.K. and Allan, M. 1988. Cellulase production in the life cycle of the cultivated mushroom *Agaricus bisporus*. *In*: Aubert, J.P., Beguin, P. and Millet, J. (*eds.),* Biochemistry and genetics of cellulose degradation. Academic Press, London, pp. 53–70.

Yau, C.K. and Chang, S.T. 1972. Cotton waste for indoor cultivation of straw mush. *World Crops*, 24: 302–303.

Zheng, F. and Ding, S. 2013. Processivity and enzymatic mode of a glycoside hydrolase family 5 endoglucanase from *Volvariella volvacea*. *Appl. Environ. Microbiol.,* 79(3): 989–96. doi: 10.1128/AEM.02725-12.

Coloured Figures of All Chapters

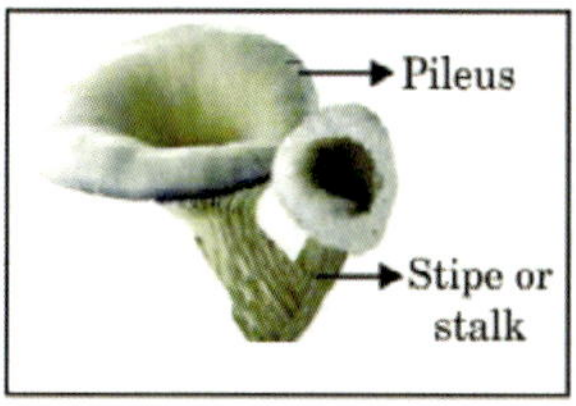

Ch-1—Fig. 1
MFB

Ch-1—Fig. 2
Gills

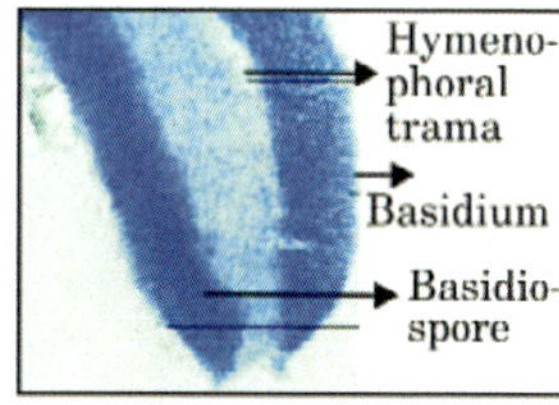

Ch-2—Fig. 2
Histology of gills

Ch-2.1—Fig. 1
Apparatus SC

Ch-2.1—Fig. 2
Basidiospores *P. florida*

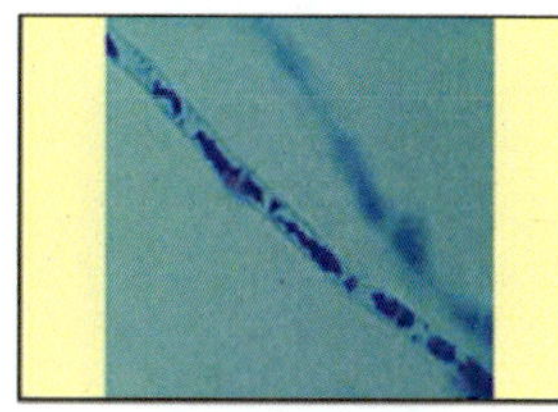

Ch-2.1—Fig. 3
Hypha *P. florida*

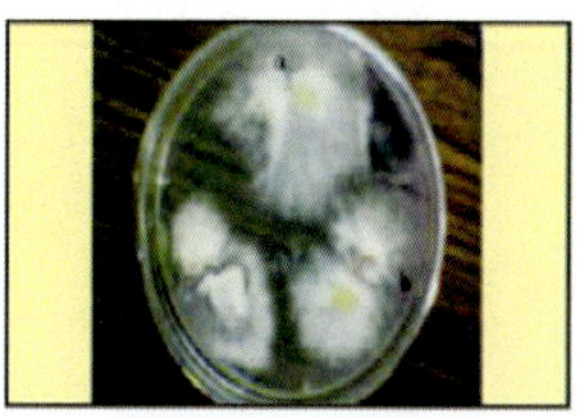

Ch. 2.1—Fig. 4
Monokaryons C or IC mating

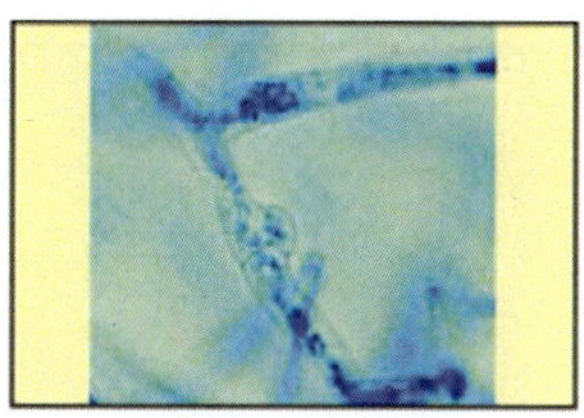

Ch. 2.1—Fig. 5
Dikaryon of *P. florida* (CC)

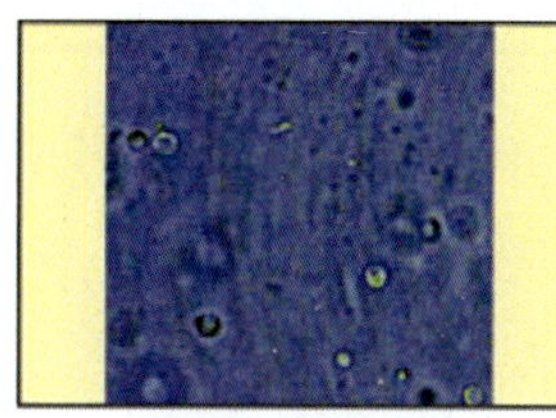

Ch. 2.1—Fig. 6
Protoplasts Ci-3 strain

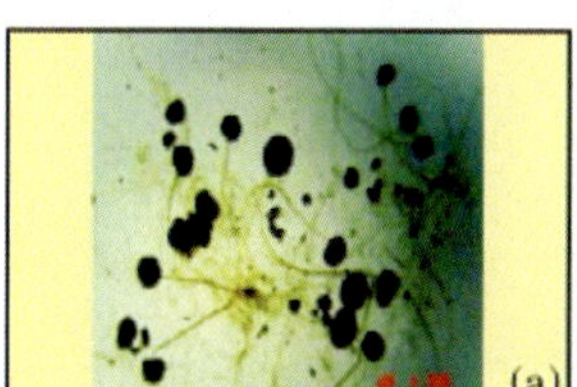

R. stolonifer

C. cinereus

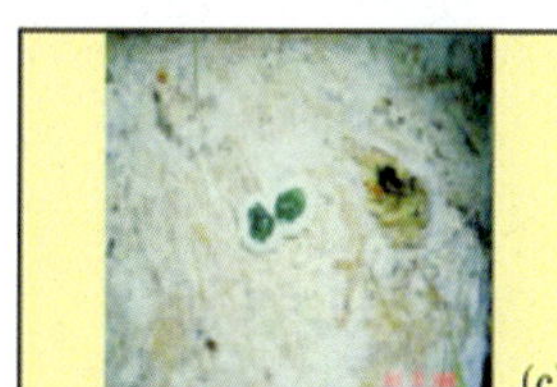

A. fumigates

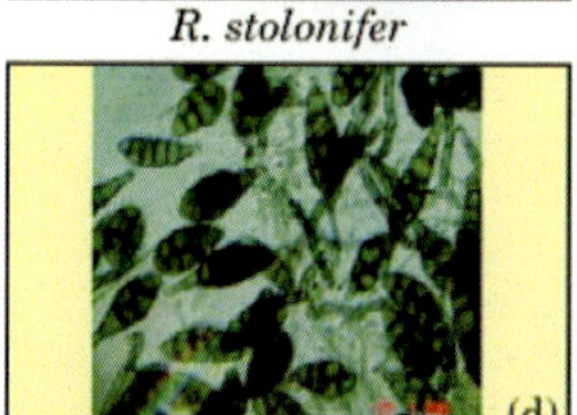

A. alternate

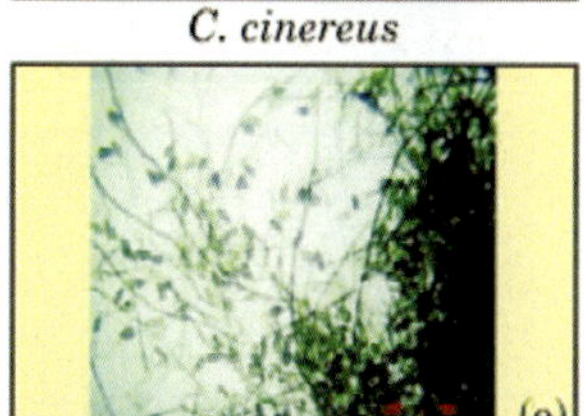

C. lunata

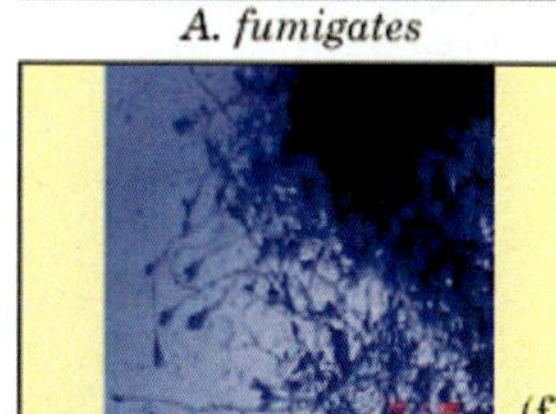

Penicillium spp.

Ch. 2.2—Fig. 1: Fungal competitors (a–f)

Ch. 2.4—Fig. 1
Gills of *P. florida*

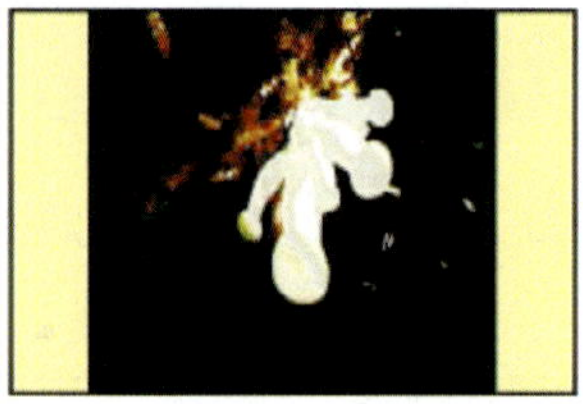

Ch. 2.4—Fig. 2
Pinhead of *Pleurotus*

Ch. 2.4—Fig. 3
Immature FB (Intial)

Ch. 2.4—Fig. 4
Immature FB

Ch. 2.4—Fig. 5
Mature FB

Ch. 2.4—Fig. 6
Highly mature FB

Ch. 2.4—Fig. 7
P. florida in plastic bottles

Ch. 2.6—Fig. 1
FB *V. volvacea*

Ch. 2.6—Fig. 2
Paddy straw UI   Paddy straw I

Ch. 2.9—Fig. 1 (a–h)
Different varieties of milky mushroom

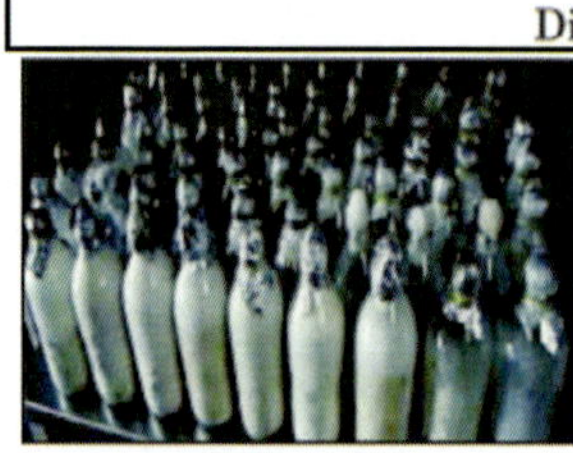

Ch. 2.9—Fig. 2
Mother spawn filled bottles

Ch. 2.9—Fig. 3
Grains filled polypropylene bags

Ch. 2.9—Fig. 4
Chemical sterilization of straw

Ch. 2.9—Fig. 5
Spawing bags

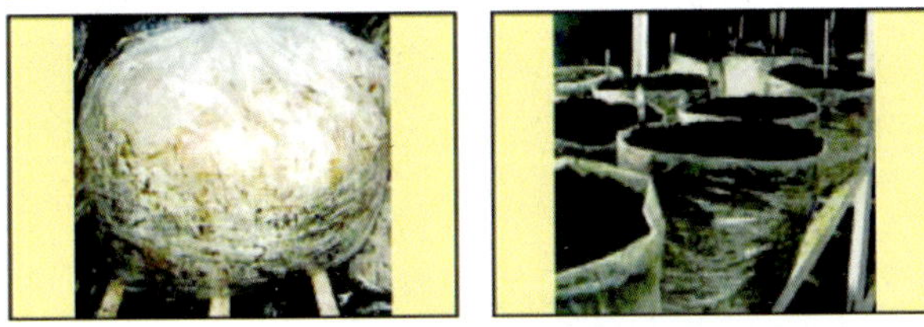

Ch. 2.9—Fig. 6
Casing covered with polythene sheet

Ch. 2.9—Fig. 7
Mushroom initiation after incubation

Ch. 2.9—Fig. 8
Milky mushrooms ready for harvesting

Ch. 2.9—Fig. 9
Harvested milky mushrooms

Ch. 2.9—Fig. 10
Contamination by different microbes are shown in this figure. Contamination in substrate containing bottle (A); *Coprinus* fruit body (B); *Stemonitis* (C); (*Contd*...)

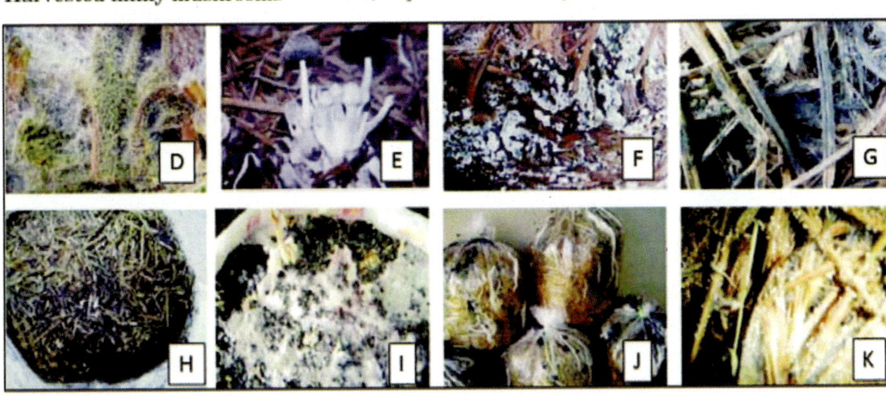

Ch. 2.9—Fig. 10 (*Contd*...)
Penicillium (D); *Coprinus* (E); *Trochoderma* (F); *Aspergillus* (G); *Rhizoctonia* (H); Unidentified (I); *Coprinus* (J); *Sclerotium rolfsii* (K)

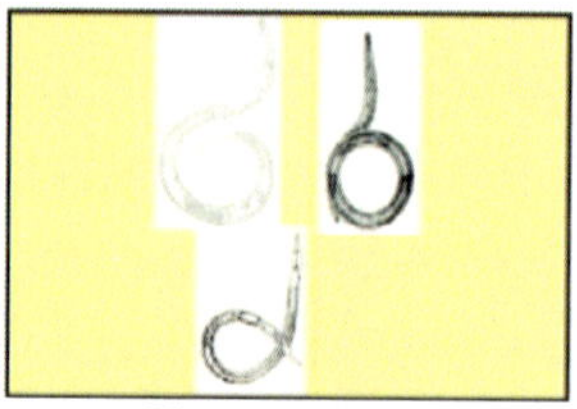

Ch. 3.2—Fig. 1
Nematode pests of *Agaricus*

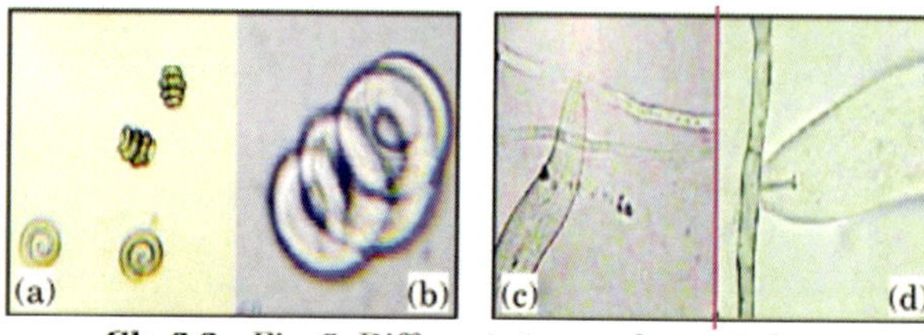

Ch. 3.2—Fig. 2: Different stages of nematodes
(a) A. swarupi (b) A. bisporus (c) A. swarupi (d) D. mycellophagus

Ch. 3.2—Fig. 3
Nematode infected

Ch. 3.2—Fig. 3
Nematode uninfected

Ch. 3.2—Fig. 4
Crop failure

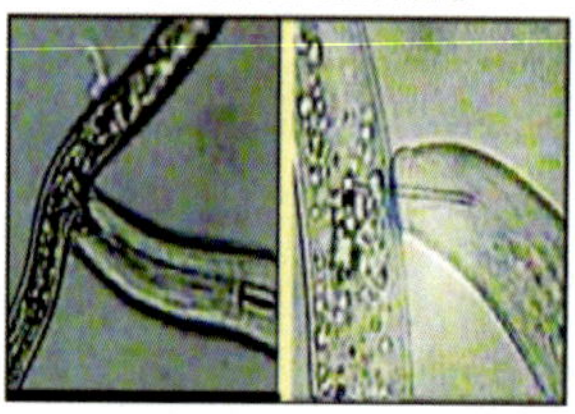

Ch. 3.2—Fig. 5
Fictor sp. *Seinura* sp. MN

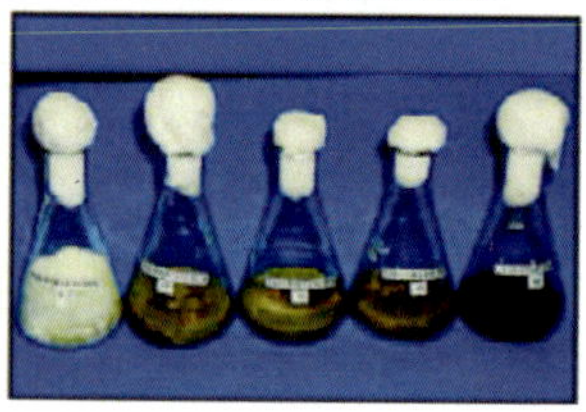

Ch. 3.2—Fig. 6: Effect
Paraphelenchus on *C. indica*

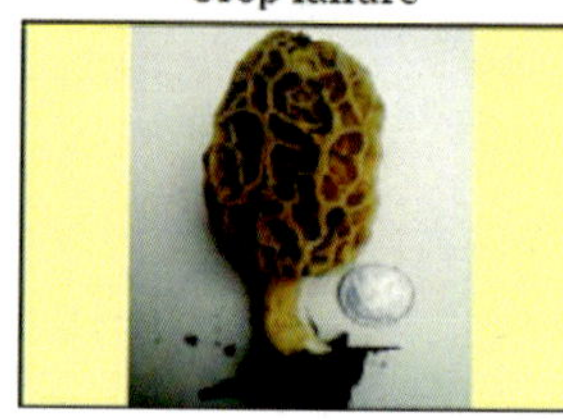

Ch. 4.3—Fig. 1
Morchella ascocarp

Ch. 4.3—Fig. 2
Morchella continuous cavity

Ch. 4.3—Fig. 3
Morel mushroom

Ch. 4.3—Fig. 4
Morchella as ornament

2.7

Cultivation of *Grifola* (Maitake) Mushroom

SHWETA KULSHRESHTHA[1]*

ABSTRACT

Grifola frondosa (hen of the wood) is one of the most important medicinal mushrooms. It has a crispy texture and nice flavour. It has immune boosting power with tumour modulating activities. It produces one large cluster of mushroom and therefore, also known as dancing mushroom. Although its yield is low 0.5-2 lb / bag, it is cultivated for its medicinal benefits.

Key words: Dancing mushroom, *Grifola frondosa*, Maitake cultivation.

1. INTRODUCTION

Maitake mushroom (*Kingdom: Fungi; Phylum: Basidiomycota; Class: Agaricomycetes; Order: Polyporales; Family: Meripilaceae; Genus: Grifola; Species: Frondosa*) is known as the 'King of Edible Mushrooms' due to its large size (more than 20 kgs). It is also known hen of woods, sheep's head, cloud mushroom *etc.* due to its amazing structure. It is known as maitake (dancing mushroom) and kumotake (cloud mushroom) in Japan because Japanese people started dancing and celebrating when they found its beneficial effects. One story is that the mushroom's appearance - small overlapping fan-shaped caps with stalks that often fuse together in a heavy mass - resembles a cluster of dancing butterflies. However, it came by its name; maitake's medicinal properties are giving us something to dance about today. It is referred to as the 'signorina' by Italians, 'Laubporling' or ram's head by Germans and 'Huishuhua' by Chinese. In France it is called Poule de Bois. It is also known as monkey's bench, shelf fungus *etc.* Its scientific name is *Grifola frondosa.*

[1] Amity Institute of Biotechnology, Amity University Rajasthan, Jaipur.
**Corresponding author:* E-mail: shweta_kulshreshtha@rediffmail.com

The first report of maitake was recorded in a tale from the 11th century Japanese text Konjaku Monogatari (gourmet book), *Grifola frondosa* (Dicks: Fr.) S.F. It was previously described as *Polyporus frondosus* (Dicks.: Fr.) (Fries, 1821). Prior to this, it was known as *Boletus frondosus*

(Dicks.) (Dicks, 1785). At species level Gilbertson and Ryvarden (1987), Farr *et al.* (1989) and Zhao and Zhang (1992) recognize *G. frondosa* as the only species of *Grifola* whilst *G. frondosa* is generally accepted as being in the family Polyporaceae (Corner, 1989; Shen, 2001).

Maitake is well known edible fungus, found in Asia, which has both nutritional value and medicinal value. Maitake is saprophytic fungus that grows on fallen trees and stumps of Fagaceae tree. This mushroom has no characteristic flavour and therefore, it can be the part of any dishes. In this way without affecting the taste and aroma it increases the nutritional value of dish. It is sold in many forms: fresh, dried, capsules and tea (Stamets, 1993a,b; Royse, 1997a,b).

Maitake has been artificially cultured and eaten as a food in Japan since long time. It is famous in Korea, China, North America, Canada and Europe. The mushroom has been used as food supplement, soups, tonic, nutritional drinks *etc.* It is also used for medicinal purposes and hence, comes under medicinal mushroom category. Maitake is a perennial mushroom and can live on the same tree for decades. Generally, it grows on hardwood trees and occasionally on coniferous trees. It is found on the bases of oak and beech, produced fruit body at the extreme base of the trunk showing its parasitic nature and on dead buried trees as well as rotten roots showing its saprophytic nature *etc.* It is a white-rot saprophyte, decomposing (hemi) cellulose and lignin of dead wood and is generally found at the base of deciduous hardwoods that are dying or dead due to other causes it is found in association of roots of many plant species like *Pyrus, sorbus, Fraxinus, Quercus serrate* and *Q. crispula* (oak), *Fagus crenata* (beech), *Acer* (maple), *Corylus, Ulmus* (elm), *Nyssa* (black gum), *Cedrus, Castanea sativa, Rhododendron ponticum, Ilex aquifolium* and broad leaved trees (Stamets and Chilton, 1983; Bon, 1987; Cortecuisse and Duhem, 1995; Mayuzumi and Mizuno, 1997; Stamets, 2000). It has been reported to grow on Douglas fir (Gilbertson and Ryvarden, 1987).

2. MORPHOLOGY

Maitake mushroom is having soft and large body which can weigh around 20-23 kilograms. It is fleshy, polypore fungi. Polypore fungi lack gills on underside of cap. Instead it releases its spores through small pores. Spores are the "seeds" that helps in propagation of mushroom, edible mushroom in which pilei or caps are generally dark grey and brown at initial stages and with aging, becomes lighter in color, light grey to light yellow but edges are dark colored. Fruit body of maitake is made up of many overlapping, curved, fan or spoon shaped and finely velvety or bald caps with wavy margins. Fruit body is 4-10 cm in diameter and 0.5-1 cm in thickness. These sprout from branching stem like tough structure, creamish or pale grayish in color. These are eccentrically attached which means that they arise from a common base and extend radially due to which they share a large fist-sized, underground located common base system. Young fruit bodies are adorned with fine grey fibrils. Caps are organized in large clusters and are having multiple grayish-brown caps which are occasionally, curled. The margins of caps are mostly wavy and also the cap's surface contains whitish to yellowish pores or hymenophores. The surface, beneath the cap, contains 2-4 angular pores per mm and tubes to 5 mm deep. Spore prints of this mushroom shows that the colour of spore is white. The stipe is milky-white.

When fruit body undergoes maturation process, the branching in stem or stipes becomes hard and stiff. The mycelia of the maitake are longitudinally lincar and white in color. The odour and taste of maitake mushroom is mild and pleasant when it is young. It fruits in the season of September and October. Hyphal growth is fluffy, and non-rhizomorphic, extending in uneven circles when grown on agar. When maitake is grown on sawdust or woodchips, a dense matty structure of mycelium is formed which discharges orange or rusty fluids (Stamets, 2000). When suspended in liquid culture, growth is a "feather-like mycelial clump" (Lee *et al.*, 2004). The microscopic features of maitake include the size of spores which is generally 5-7 x 3.5-5μ and having characteristic smoothness. The spores are white, smooth, and elliptical. Cystidia are absent whereas hyphal system is dimitic. The clamp connections present on generative hyphae but absent on skeletal hyphae.

3. NUTRITIONAL PROFILE

Regardless of medical benefits, maitake is one of the healthiest foods. It contains high levels of immune supporting beta-glucans, natural fibers. It helps in controlling blood pressure and good cholesterol level to maintain the health of the person. Moreover, maitake maintains the nutritional profile of the original fungus. Studies show that organically-grown maitake has 377 calories per 100 g dry weight, 19-25 percent protein, 3-4 percent fats (1 percent poly-unsaturated fat; 2 percent total unsaturated fat; 0.3 percent saturated fat, 0 percent monosaturated fats), approximately 60 percent carbohydrates (41 percent are complex carbohydrates), approximately 29 percent fiber, 19-20 g of sugar, 0 percent cholesterol, 2 percent magnesium, 0 percent calcium and 0 percent vitamin A. It also possesses vitamin B. The composition of vitamin B (mg/100 g) is: niacin (64.8 mg); riboflavin (2.6 mg); and pantothenic acid (4.4 mg). It is high in vitamin D, potassium, and niacin. It contains high concentration of potassium: 2,300 mg/100 g (or 2.3 percent of dry mass), 1 mg sodium. Moreover, maitake also possesses L-glutamate, a natural flavor-enhancer compound (Chan *et al.*, 2009). Sugar components link together to form polysaccharide beta-1,6-glucan, a complex carbohydrate substance, which is the main ingredient of maitake. One fraction of glucan was isolated and purified. This extracted form of glucan is called the maitake D-fraction which was later patented. Maitake glucan and maitake D-fraction terms can be used interchangeably. *Fraction X* and *fraction ES* are the two other fractions, extracted from maitake were by Harry Preuss, a medicine and pathology professor at Georgetown University Medical Center in Washington, DC.

3.1. Chemical Constituents of *Grifola frondosa*

Amino acids

The fruit body of maitake possesses copious amount of Glutamate, a flavour enhancer amino acid, provides delicious flavor to this mushroom. It also contains alanine, threonine aspartic acid, valine, lysine, and arginine. During the maturation of fruit body, the amount of free amino acid decreases (Mizuno and Zhuang, 1995).

Lectins

Recently, N-acetylgalactosamine- specific lectin was also isolated from *Grifola frondosa* by affinity chromatography. This fraction can agglutinate erythrocyte and shows the property of hemagglutination. Lectin, isolated from maitake, is known as maitake lectin. This possesses a saccharide content of 3.3%. Isoelectric point of this lectin is pH 5.9. Although this lectin was cytotoxic for the HeLa cells, yet it can be used after pre-incubating lectin with the haptenic sugar N-acetylgalactosamine (Mizuno and Zhuang, 1995; Kawagishi *et al.*, 1990).

Free sugars

Maitake contains mainly three sugars *i.e.,* glucose, trehalose and mannitol. Glucose content decreases with the growth of fruit bodies. It is present in low amount in the mature fruit body. With the growth of fruit bodies, the amount of mannitol decreases while trehalose increases (Mizuno and Zhuang, 1995).

Polysaccharide fractions

Two polysaccharide fractions have been extracted from maitake mushrooms: (i) β-D-glucan (ii) α-D-glucan. β-D-glucan was extracted using hot water and chemical extraction procedure followed by ethanol precipitation, ion-exchange chromatography, gel filtration and affinity chromatography. A fraction of β -D-glucan was isolated which was having a *Beta*-(1→6) bond in the main chain and Beta-(1→3) bond in the branched chains (Mizuno and Zhuang, 1995). This polysaccharide is known for its antitumor activity (Alonso *et al.*, 2017).

Polysaccharide fraction of maitake has immunomodulatory power, antioxidant activity, anti-hyperglycemic activity. This polysaccharide is also effective on the skin and hematopoietic stem cells (He *et al.*, 2017).

The α-D-glucan fraction was separated from β-D-glucan fraction using hot water extraction procedure. This fraction possesses alpha-(1→4) glucan with alpha-(1→6)-branched chain. This fraction do not exhibited antitumor activity.

Enzymes

Maitake possesses a metal protease which is an aminoendopeptidase, specific to lysine. This enzyme digests albumin, elastin, and casein. This enzyme is unable to decompose synthetic polypeptide composed of leucine, tyrosine and phenylalanine (Mizuno and Zhuang, 1995).

Organic acids

A number of free organic acids were reported to be present in maitake during the fruit body formation period. These include-lactic acid, acetic acid, formic acid, malic acid, citric acid and succinic acid, oxalic acid, fumaric acid. During the growth of maitake, a very few changes occur in the fruit body (Mizuno and Zhuang, 1995).

Maitake fiber content

Maitake possesses good fibre content which is having many medicinal values. These are useful in lowering the serum cholesterol level by increasing the excretion of cholesterol in the feces. These contents also induce adipocytokine signaling pathway to promote fatty acid oxidation. These contents can suppress SAA which encodes serum amyloid A and prevent arteriosclerosis (Mizuno and Zhuang, 1995; Fukushima *et al.*, 2001; Sato *et al.*, 2013).

Lipids

Maitake contains three types of lipids (i) Neutral lipids (ii) Sterol lipids (iii) Sphingolipids. These lipids are extracted out in the chloroform and purified by chromatographic methods. A neutral lipid fraction of maitake is extracted and then eluted with chloroform using silicic acid column (Mizuno and Zhuang, 1995).

Maitake mushroom possess sterol lipids. Acylsterol and free sterol are reported to be present in *Grifola frondosa* var. *tokachiana* WMX 12-8. This mushroom also possesses steryl glycoside. The most predominant lipid is ergosterol whereas subsidiary lipid is fungi sterol. The relative ratios of polar and non-polar lipids vary in mushroom. The main two types of polar lipids in the maitake are glucosyl-tetraenic C_{28} sterol and glucosyl-dienic C_{29} sterol (Ohnishi *et al.*, 1985). Maitake contains sphingolipids in the form of ceramide and cerebroside. N-2'-hydroxy lignoseroyl-4-hydroxy-sphingosine is present as ceramide whereas 1-O-glucosyl-A'-2'-hydroxypalmitoyl-9-methyl-trans-4-trans-8-sphingosine is present as cerebroside (Mizuno and Zhuang, 1995).

Maitake also possesses phospholipids such as phosphatidylcholine, phosphatidylserine, phosphatidylinositol, and phosphatidic acid. In mycelium, phosphatidylethanolamine is present.

Other constituents

Maitake are rich in vitamins like B1, B2 and provitamin D. These are found to possess nucleotides. The amount of nucleotides, present in maitake, is approximately 106 to 366 mg per 100 g of mushroom fruit body. A number of inorganic compounds are found to present in maitake. The fruit body of maitake is rich in potassium (K) and phosphorous (P) but possess very low amount of Mg, Ca, Na, Zn.

Maitake D-fraction is the bioactive extract of maitake mushroom (*Grifola frondosa*) and its active constituent is the protein-bound polysaccharide (proteoglucan), or more specifically known as β-glucan.

4. MEDICINAL PROPERTIES

Chinese and Japanese herbal medications have an ancient history of maitake, that being used as a medicine against diverse health severities. Both of them refer to it as the *medicinal mushroom*. However, scientific investigation had begun on maitake only in the 80's in Japan. European nations have developed interest in it very recently. Maitake comes under medicinal mushroom due to having antiviral, antitumor, antimalarial, anticancerous, hypoglycemic, anti-inflammatory, antihypersensitive properties.

Maitake exhibits multiple medicinal properties including i) Spleen improvement, ii) Assuagement of stomach ailments, ii) Antihemorrhagic, and mind and nerve calming ability (Jones, 1998). Though studies on humans are very rare, a single survey of non-human studies concluded maitake as being anti-diabetic, antihypertensive, antineoplastic, antiviral, immuno-suppressant, and also showing interferon like activity (Ulbricht *et al.*, 2009).

5. CULTIVATION OF MAITAKE

Maitake is giant mushroom which possesses short, chunky, highly branched stems and cluster of many overlapping caps. It requires temperate climate and aerobic conditions to grow. Maitake requires a separate room for cultivation due to its specific requirement and therefore, it cannot be cultivated with other mushrooms. *Artificial culture techniques were adopted in Japan for the cultivation of maitake after the elucidation of cultural biology.*

5.1. Strain Selection

The production of maitake depends on strain selection because strains of maitake do not fruit well like the other mushrooms such as *Agaricus* and *Pleurotus*. There are considerable differences among the strains whether these are isolated or procured from culture collection centers. Every strain is different in terms of growth rate and temperature requirement which also affects the yield of mushroom. The shape and colour of fruiting bodies also varies which focuses on the fact that every strain must be tested before use for cultivation. The strains are compatible for mating irrespective of their genotype and phenotype.

5.2. Spawn

Spawn is called as seed of mushroom which is prepared by inoculating pure culture of maitake on grains, sawdust, wooden sticks, liquid media *etc.* under sterile conditions. Maitake is aerobic species and therefore, it needs a microfilter-windows containing polyethylene bag. The purpose of these windows is to provide air for breathing of culture.

Spawn in colonized synthetic logs are sold as ready-to-fruit blocks which can be used by growers for the cultivation of species of maitake. Trained personnel are able to prepare such type of logs in laboratory by inoculating desired strain on the synthetic logs. Spawn is prepared in 4 different ways:

5.2.1. *Sawdust spawn*

Sawdust spawn can be prepared by two methods: Firstly, the hardwood sawdust is sterilized and filled in special filter containing polyethene bags. Secondly, the sterilized bags are filled with sawdust and then sawdust is sterilized. After sterilization, the sawdust was inoculated with mycelium under the sterile conditions for the preparation of spawn. The bags can be incubated at 21°C by maintaining 70% humidity conditions. When mycelium covers the surface of sawdust, spawn is ready to use. It can be preserved at low temperature.

5.2.2. *Plug spawn*

This type of spawn is prepared by using hardwood dowel pins. At first, hardwood is collected and cut into small pieces of equal size. These are known as hardwood dowel pins. Generally, 3/8"x2" sized hardwood dowel pins are used. These dowel pins are soaked for overnight for moistening and then, boil for 5 min. This is followed by drying to maintain the moisture content. These hardwood dowel pins are inoculated with maitake mycelium by alternate inoculation method. In this method, 4-5 dowel pins are filled in sterilized glass bottles and then, inoculate with mycelium which is again filled with dowel pins (Oei, 1996). A comparison of plug spawn and sawdust spawn method is shown in Table 1.

Table 1: Comparison of spawn plugs and sawdust spawn.

Sl. no.	*Spawn plugs*	*Sawdust spawn*
1	It is costlier than sawdust spawn.	It is more economic than plugs especially when it is used to inoculate large number of logs.
2	It dries out slowly compared to sawdust spawn.	It dries out faster than spawn plugs.
3	It is inoculated by drill and fill method or stick method.	It is used to inoculate by the totem method only.

5.2.3. *Liquid spawn*

It is prepared by inoculating the desired strain to 4% ELME (Extra Light Malt Extract) medium. This is liquid medium which supports the rapid growth of maitake. In this preparation, a specially designed syringe is used which makes it a labor-saving approach. Liquid-surface still culture and submerged fermentation have been used to produce liquid spawn. 1 cc of liquid spawn has potential to produce gallons of more liquid culture within weeks. This can be done at home by using common household items or at an industrial scale by submerged fermentation. It is used to colonize different types of spawn media such as grain, sawdust, seeds *etc.*

5.2.4. *Grain spawn*

A variety of grains, such as millet, rye, wheat, sorghum and milo can be used for the spawn preparation. Grains are sterilized and filled in a container and then inoculated by alternate inoculation method.

5.3. Substrates

The use of economical and readily available material is suggested for expanding maitake cultivation business in India. It becomes easy to cultivate maitake on artificial media and controlled conditions when hardwood sawdust is used. This sawdust can be obtained from oak, popular, cottonwood, elm, willow, *Eucalyptus* and alder. In Australia, *Eucalyptus* sawdust is readily available, economic substrate and therefore, used for the cultivation of maitake. These sawdust containing substrates are supplemented with brans such as wheat bran, rice bran, oat bran and maize meal (Mayuzumi and Mizuno, 1997).

Many researchers have assessed the response of maitake to *Eucalyptus* sawdust and its combination with bran of different types of grains such as wheat, rice, maize. The Japanese bag method was used for this assessment. Data obtained from the study indicated that *Eucalyptus* sawdust supplemented with 10% rice bran, 20% wheat bran and 20% maize meal, have the highest efficiency for colonizing substrate. Maitake mycelium took only 33-79 days for colonization of substrate. Similar results were found by Kirchoff (1996) who reported that the primordia formation is initiated by supplementation of 20% corn meal or 10% wheat bran to beechwood sawdust.

Hardwood was used as substrate for the cultivation of *Grifola frondosa* and *Grifola gargal*. Harada *et al.* (2015) used various combinations of substrates for the cultivation of *G. gargal* in order to find out the best substrate. The suitability of substrate is decided on the basis of increment in the yield, biological efficiency, and period required for cultivation; and quality of mushroom in terms of shape and colour. Wood substrates such as hardwoods and sugi are commercially available for mushroom cultivation from the wood industry.

There are many other substrates used for cultivation of *Grifola frondosa* like hardwood sawdust which is the most preferred substrate and also suitable for the cultivation of mushroom at commercial scale. Sawdust derived from oak (*Quercus rubra* L.) is the most popular choice in the United States and Japan. In China, cottonseed hulls are used as a substitute for sawdust for the cultivation of maitake. A mixture of hardwood sawdust and wood chips, mixture of hardwood sawdust and wheat bran, poplar stumps, elm trees, willow, alder, beech, larch trees wood can also be used for the cultivation of maitake. Wheat bran, rye, millet and corn meal supplements can be used alone or in various combinations with mixed oak sawdust to increase the biological efficiency and chemical quality of mushroom.

Recently, a lot of research has been conducted on corncob stalks. This is considered as the best agricultural residue due to its easy and abundant availability and economic prices. It is generated as by-product after the processing of maize. Corn cob can replace sawdust as main substrate of maitake. Corn cob increases the yield of maitake however; sawdust shortens the growth cycle of maitake (Song *et al.*, 2018). This shows the importance of two substrates. The optimized high-yield formula was 73.125% corn cob, 1.875% rice straw, 23% wheat bran, and 2% light $CaCO_3$ (C/N = 48.40); the average yield of the first flush was 134.72 ± 4.24 g/bag, and the BE was 44.91 ± 1.41% (Naraian *et al.*, 2009; Niu *et al.*, 2016; Song *et al.*, 2018). This reveals the importance of the use of corn cob and its combination with sawdust as substrate.

5.4. General Three Methods of Cultivation

Mushrooms have existed on earth even before the civilization of man. However, these are never cultivated and harvested for edible purposes. Early civilization of Egyptians, Greeks, Roman, Chinese, started the use of this mushroom for food, medicines and religious ceremonies. Generally, maitake cultivation has begun in 1980's in Japan, and thereafter, it became the mushroom species of medicinal and edible importance. Now, *Grifola frondosa* is one of the cultivable mushroom species. Nowadays, China and Japan produced this mushroom in high amount due to its popularity. Naturally, this mushroom grows at the base of oak trees, emanating from the center of the trunk or from large rotting roots, rather than from logs. This mushroom cannot be cultivated easily in artificial condition which is the main reason for the delay in domestication of this mushroom species.

In 1990's, several methods of cultivation of maitake are developed which ensure the availability of mushroom throughout the year. These are as follows:

5.4.1. *Bottle culture method*

This method is suitable for cultivation of maitake for throughout the year. The cultivation process can be mechanized to make the operation easy and less laborious. The size of the crop depends on the size of the bottle. Generally, the bottles (800-1000 ml capacity) are filled with sawdust powder obtained from deciduous broad leaf trees. This is called as sawdust bed. This culture bed is prepared by mixing sawdust powder with 15% (v/w) rice bran and 5% (v/w) wheat bran. The culture bed is sprayed with water to adjust its moisture level to 65% and then, packed in polypropylene bottle. These bottles must not fill completely. If the capacity of bottle is 1 litre then, it is filled with 500 g of moisture adjusted bed. After this, the mouth of the bottle is corked or closed by using screw cap or cork. This is followed by sterilization of mushroom bed at 120°C for 6 h and then cooled. The sterilized bed is inoculated with cultured mycelium or 15 ml of spore preparation. After this, maitake culture is incubated for 40 days at 23°C along with maintaining 70% humidity. For the maintenance of bottle culture, temperature of the room is adjusted at 17°C and at 85% humidity. The room must be properly ventilated because a large amount of oxygen is required for the development of cap of fruiting bodies. The yield is generally 100-130 g/bottle in well maintained conditions.

Bottle culture is preferable method over plastic bag culture because it is easy to do and economic. Small size of bottles are used for cultivation which produced retail sized mushroom cluster for sale.

5.4.2. *Bag culture*

In this technique, plastic bags are used which possess special filters. Research is also going on for searching or modifying the strains in such a way that it becomes suitable for cultivation of maitake. Bag cultivation technique for maitake cultivation is similar to shitake cultivation. This method is used most frequently in Japan for the cultivation of maitake. Sawdust is mixed with 15% rice and 5% wheat brans. After this, the moisture of the mixture is adjusted to 65%, the mixture is packed in a polypropylene bag and is molded into a square-shaped culture bed. One polythene bag contains approximately 2.5 kg of the sawdust substrate. Polythene bags are punched for making holes of 1.5 cm in order to provide aeration. Urethane wool is spread as top layer on the culture bed. This bag is sterilized by steam in an autoclave at 110°C for 7h which is followed by cooling till the temperature of the culture bed reaches 15°C. These cooled culture beds are sprayed with a 300 times diluted soap solution. Prior to spraying, this soap solution is exposed to germicidal lamp to kill contaminating microbes. Sterilized culture beds are inoculated with 30 ml spore suspension. Then, the upper portion of bag is folded and stapled. Maitake culture bags are now incubated for 70 days in a culture room at controlled temperature of 23°C and humidity 70%. Primordial formation begins when the culture bed is illuminated by a plant raising lamp. After reaching the primordial formation stage, culture beds are transferred to a developing room and the upper portion of bag is cut and removed. In developing room, these culture beds are kept for about 15 days at 18°C and 85% humidity. This leads to the good yield of mushroom *i.e.,* 400 g/bag.

5.4.3. *Log cultivation*

Maitake is easy to cultivate on wooden logs. Once these logs are prepared, maitake can be produced for many years in low maintenance. Hardwood logs of oaks, elms, maples, honey, locust, black gum, beech *etc.* can be used for maitake cultivation. These logs need to be soaked in water for 2 days prior to use while this can be ignored in the case of freshly cut hardwoods. Generally, fresh hardwoods are recommended for cultivation because old logs need more inoculum compared to fresh log. These inocula are placed on logs at inoculation sites which are prepared by drilling holes (5/16" x $1\frac{1}{2}$" holes) in spiral pattern at 4-6 inches distance. Old logs are prepared by drilling the holes at closer distance compared to standards mentioned so that higher amount of inoculum can be inoculated. These holes are inoculated by colonized plugs that are pressed and pushed inside the hole to sink the top of the plug. This is followed by sealing the inoculation sites by using molten cheese wax. This wax is dripped onto each hole by using brush, large dropper or 10 cc inoculation syringes. These logs are put in shaded area and buried underground slightly below the surface level. To avoid drying these logs are moisten by sprinkler. In the area with frequent rainfall, moistening of logs is not required. Logs, inoculated in spring season, produce fruiting bodies by rainfall season. However, it takes one year time to become mature. These logs produce mushroom for the period of 2-4 weeks during the season. Properly maintained logs produce fruiting bodies for many seasons.

5.4.4. *Outdoor bed culture*

This method was attempted first time in Japan under natural climatic conditions for cultivation of maitake. This method requires about 6 months of time from inoculation to fruiting body formation. The culture period is long, nevertheless, this method has advantages of low cost and easy handling. Mushroom is growing in natural conditions and therefore, air conditioned facilities are not required. There is some minor variation in the preparation of mushroom bed as compared to bag or bottle culture. In this method, sawdust is mixed with 15% corn bran instead of rice bran which is followed by the addition of soil. Corn bran not only improves the taste of maitake but also helps in producing hard fruiting bodies which increases their durability and shelf life. Usually, silty loam soil with pH 6-7 is preferable over any other type because it makes the culture bead hard and compact. It also prevents the breakage of the culture bed when it is taken out from its bag after the completion of mushroom cultivation process.

In this method, culture beds are inoculated in spring season when temperature is around 15°C in the same manner as bag culture. Temperature and humidity in the room is maintained by spraying water. Occasionally, if temperature is low, to maintain temperature heating can be done. Cultivation of maitake requires proper aeration which affects the yield of maitake; therefore, cultures must be aerated properly. The rooms must be properly ventilated. All other parameters are same as mentioned for bag culture.

After the completion of culturing, culture beds are removed from their bags. These culture beds are transferred to well-drained field and covered with about 3 cm of soil layer where 85% shading is made by spreading a black mesh sheet (victoria lawn) at a height of 180 cm. Soil is covered with the fallen leaves in order to prevent loss of covering soil by rainfall. Fruiting bodies starts in early autumn when atmospheric temperature falls to 23°C. When 2-3 beds are placed adjacent to one another, fruiting bodies growing in the beds are fused into one bed and develop into large mushroom bed. This bed results in higher yield 1-1.5 kg compared to when placed alone.

Maitake mushrooms grown in outdoor beds provide higher yields compared to indoor culture methods. Besides, fruit bodies of maitake possess shelf life similar to the mushroom growing in natural conditions.

5.5. Cultivation Conditions

Different parameters of cultivation of *Grifola frondosa* include temperature, carbon dioxide concentration, humidity, light *etc*. The temperature required for spawn run is 21-24°C, humidity is 95-100% and duration is 14-30 days with 30 days dormancy, CO_2 required is 20,000-40,000 ppm, whereas light is not required for spawn. Primordia formation requires the temperature in between 10 to15.6°C and humidity must be 95%. The duration is mostly 5-10 days and CO_2 concentration is 2000-5000 ppm. The light required is 100-500 lux (Chen, 1999).

5.6. Spawn Run, Primordial Formation and Maturation

Spawn inoculation in the bag culture/ log culture/ bottle culture method leads to the spawn run on the substrate under controlled environmental conditions. Spawn run begin in 9 days and during the spawn run period undifferentiated mycelia colonize the substrate. After spawn run, substrate bundles appear white due to the white mycelial growth on the substrate. This stage completes in 14 to 17 days. Maitake has prolonged spawn run period even after the colonization of substrate by mycelia. Light is not required by the maitake mycelium during the spawn run period. However, supply of low level of light (50 lux) in the room throughout the spawn run period induces the primordial formation branches (Wu *et al*., 1997). This stage is followed by primordial formation stage.

The primordia appear on the surface as small antlers after colonization of substrate. At this stage, light of increased intensity (200 lux) is supplied which induces primordial development. It is also supplied with high amount of oxygen and humidity. The bags remained sealed at this stage. Metabolic activities increased at this stage which leads to the secretion of orange brown exudate. The surface of the substrate becomes grayish with amorphous mass which leads to the primordial development. This stage completes in 40-45 days when primordia acquire 1 to 2 inches diameter. The substrate bags remains closed till this stage. Round to globular yellow or brown primordial dotted with exudates are developed. Care must be taken so that exudate should not wipe off. After 2 to 3 days, these bags are transferred to mushroom fruit body development room and slit is made on the top of the bag to open it.

The development of fruit body includes brain stage, cauliflower stage and cluster flower stage. At brain stage, primordial grow and fold on the surface of the substrate. These folds are so convoluted that they look like brain. These convoluted folds develop into young pilei formed in a cluster. The stalk elongates and gives rise to branches (Wu *et al.*, 1997). However, base looks very short due to extensive branching and colour of fruit body disappears which makes it white. All these characteristics give it cauliflower like appearance.

The fruit body of mushroom continues to grow and produce cluster of fan-shaped pilei. The colour of mushroom becomes lighter during the development of fruit body. All these characteristics gives it a flower like morphology and therefore, this stage is called as cluster flower stage. The colour of mature fruit body depends upon the strain of maitake which may be light gray, grayish white, light brownish yellow (Chauhan and Varma, 2009).

5.7. Time of Harvest

When maitake fruit bodies become mature, these are harvested. Otherwise, over-matured fruit body become soft and gives foul odour (fishy smell). Mature fruit body can be identified by observing their morphological characteristics. The mushroom cluster increase in size and acquire fan shaped pilei and stalks. Colour of fruit body gradually lighter from dark gray brown to light brownish yellow, whitish, or grayish white. At the time of harvesting, margins of pilei should be curled inward. The hymenial tubes forms minute pores on the back of pileus but not on the margins of pilei. The fruit body gives a pleasant aroma. After observing all these morphological characteristics, fruit body of mushroom can be harvested.

Prior to harvesting, misting of water on the crop must be stopped one day before. For harvesting, fruit bodies must be hold at the base and then cut off the thick base by using a sharp knife. In other method, cluster of mushroom can be rotated and pull upward carefully and then cut off the thick base using knife. Harvested fruit bodies can be trimmed to remove debris. Generally, the efficiency of maitake is low *i.e.,* 0.5-2 lb per bag.

The harvested fruit bodies can be sold in the market or preserved at low temperature. Stamets (1993a, b) developed a method of preservation for preserving maitake mushroom. The harvested fruit bodies of mushroom can be cleaned properly and wrapped in rice paper. These can be put in refrigerator for the period of 2 weeks. Thusly, their shelf life is increased.

5.8. Management of Growth Parameters

Size of the bottle is the most important factor which needs to be considered for the cultivation of maitake. Large crop size can be obtained in polythene bags compared to bottles due to the variation in the amount of substrate used for cultivation.

Spawn run, primordia initiation and fruit body development can be carried out in the separate rooms under the maintained temperature, light, humidity and aeration conditions. As there is difference in the requirement of humidity, temperature and other environmental conditions, at different stages of development, it should be cultivated in separate rooms. Bags, used for cultivation, initially remains sealed in order to maintain moisture conditions. After primordial formation, top part is cut down to open the bag or slit can be made for proper aeration of the culture. To maintain moisture content, holes are punched at the bottom of polythene bag.

In the case of log culture method, logs are shifted to other room after primordial formation for 2 to 3 days. In the other room, conditions are maintained to support fruiting of maitake.

The cultivation of maitake requires controlled growth condition at each stage of its development. Therefore, all growth parameters such as temperature, relative humidity, light exposure, proper ventilation for air exchange and maintaining O_2 and CO_2 gases. Spawn run duration requires high temperature as compared to fruit body development. The temperature must be controlled properly otherwise, the variation in temperature led to the condensation of water. Some strains of maitake require high intensity of light and relative humidity during primordial initiation stage.

Management of growth parameters is based on the strain type of maitake. The strains varied in their requirement for environmental conditions like temperature, humidity, aeration of bundles and exposure to light. The quality of maitake is affected by the composition of substrate and size. The design of bag, its size and quality of fitted microfilter also affects the yield of mushroom. The quality and quantity of spawn also affects the production and yield of maitake. The soil used for casing also affects the yield. The soil must have 0.5-0.8 particle size. Garden soil, humus rich soil, soil from paddy field can be used. The use of sandy or clay soil must be avoided. Soil must be sterilized before inoculation on the maitake mushroom bed (Wu *et al.*, 1997).

6. LIMITATIONS OF USING MAITAKE

The king of edible mushrooms, *Grifola frondosa*, is not considered as toxic in several studies however, many people have allergic reaction after consuming this mushroom. It is safe if used as food and medicine. Some studies shows that it causes low blood pressure. It also lowers blood sugar levels and therefore, it is advised to take maitake doses after concerning with healthcare professional. Another drawback of using maitake is that it increases bleeding when taken with certain drugs which include aspirin, anticoagulants such as warfarin (Coumadin) or heparin, antiplatelet drugs such as clopidogrel and nonsteroidal anti-inflammatory drugs such as ibuprofen or naproxen.

Some side effects of maitake as a drug include its interaction with certain components of immune system. For example, it may interact with herbs affecting light sensitivity, immunity, anticancerous agents, clomiphene citrate, antivirals agents, cholesterol-lowering agents, fertility herbs and supplements and interferons. Therefore, proper guidance and medication adjustments are required. The dosing precautions are necessary in people having stomach, immune, skin, liver, kidney and musculoskeletal disorders as maitake can cause itching, rashes, nausea, abdominal pain, joint swelling *etc*. Maitake supplement may also affect pregnant and breast feeding women. Avoid in people who are allergic or sensitive to maitake mushroom, its parts, or members of the *Meripilaceae* family. There have been reports of lung inflammation in mushroom farm workers and in people who were exposed to maitake mushroom spores. There is currently a lack of scientific evidence on the medicinal use of maitake during pregnancy or breast feeding.

7. CONCLUSIONS

Maitake mushroom is used for treating cancer. It is cultivated throughout the world for its medicinal values due to having antitumor, immunomodulating power. Nevertheless, it is considered as toxic in several studies and generates allergic reactions after consumption. In the chapter, liquid, sawdust and plug spawn methods are explained which can be used for the preparation of spawn. Log, bottle and bag cultivation techniques can be used for the cultivation of maitake.

8. REFERENCES

Alonso, E.N., Ferronato, M.J., Gandini, N.A., Fermento, M.E., Obiol, D.J., López Romero, A., Arévalo, J., Villegas, M.E., Facchinetti, M.M. and Curino, A.C. 2017. Antitumoral effects of D-fraction from *Grifola frondosa* (maitake) mushroom in breast cancer. *Nutr. Cancer,* 69(1): 29–43. doi: 10.1080/01635581.2017.1247891.

Bon, M. 1987. The Mushrooms and Toadstools of Britain and North-western Europe. Hodder & Stoughton, London.

Chan, G.C., Chan, W.K. and Sze, D.M. 2009. The effects of β-glucan on human immune and cancer cells. *J. Hematol. Oncol.,* 2: 25. doi: 10.1186/1756-8722-2-25.

Chauhan, A.K. and Varma, A. 2009. *A Textbook of Molecular Biotechnology*, Kindle Ed. I.K. International Publishing House, pp. 1–1354.

Chen, A.W. 1999. A practical guide for synthetic-log cultivation of medicinal mushroom *Grifola frondosa* (Dick.: Fr.) S.F. Gray (maitake). *Int. J. Med. Mush.,* 1: 153–167

Corner, E.J.H. 1989. Beihefte zur Nova Hedwigia, Heft 96: Ad Polyporaceas V. *In. J. Cramer*, Berlin, pp. 62–68.

Cortecuisse, R. and Duhem, B. 1995. Collins Field Guide: Mushrooms and Toadstools of Britain and Europe. Harper Collins, London.

Farr, D.F., Bills, G.F., Chamuris, G.P. and Rossman, A.Y. 1989. Fungi on Plants and Plant Products in the United States. APS Press, St. Pal, Minnesota, USA.

Fries, E. 1821. *In*: Systema Mycologicum: Sistens Fungorum Ordines, Genera et Species, Huc U.S. que Cognitas, Quas ad Normam Methodi Naturalis Determinavit, Disposuit Atque Descripsit., Lundae: ex officina Berlingiana, Vol. 1–4.

Fukushima, M., Ohashi, T., Fujiwara, Y., Sonoyama, K. and Nakano, M. 2001. Cholesterol-lowering effects of maitake (*Grifola frondosa*) fiber, shiitake (*Lentinus edodes*) fiber, and enokitake (*Flammulina velutipes*) fiber in rats. *Exp. Bio. Med.*, 226(8): 758–765. doi:10.1177/153537020222600808.

Gilbertson, R.L. and Ryvarden, L. 1987. *North American Polypores: Vol. I and II.* Fungiflora, Oslo, Norway.

Harada, E., Morizono, T., Sumiya, T. and Kawagishi, H. 2016. Effect of the medicinal mushroom, *Grifola gargal* (Agaricomycetes), on bone turnover markers and serum lipids in middle-aged and elderly Japanese women. *Int. J. Med. Mushrooms,* 18(1): 1–7. doi: 10.1615/IntJMedMushrooms.v18.i1.10.

He, X., Wang, X., Fang, J., Chang, Y., Ning, N., Guo, H., Huang, L., Huang, X. and Zhao, Z. 2017. Polysaccharides in *Grifola frondosa* mushroom and their health promoting properties: A review. *Int. J. Biol. Macromol.*, 101: 910–921. doi: 10.1016/j.ijbiomac.2017.03.177.

Jones, K. 1998. Maitake: A potent medicinal food. *Altern. Complement. Ther.*, 4(6): 420–429.

Kawagishi, H., Nomura, A., Mizuno, T., Kimura, A. and Chiba, S. 1990. Isolation and characterization of a lectin from *Grifola frondosa* fruiting bodies. *Biochim. Biophys Acta*, 1034(3): 247–52.

Lee, B.C., Bae, J.T., Pyo, H.B., Choe, T.B., Kim, S.W., Hwang, H.J. and Yun, J.W. 2004. Submerged culture conditions for the production of mycelial biomass and exopolysaccharides in edible basidiomycetes *Grifola frondosa*. *Enzyme Microb. Technol.,* 35: 369–376.

Mayuzumi, Y. and Mizuno, T. 1997. Cultivation methods of maitake (*Grifola frondosa*). *Food Reviews International*, 13: 357–364.

Mizuno, T. and Zhuang, C. 1995. Maitake, *Grifola frondosa*: Pharmacological effects. *Food Rev. Int.,* 11: 135–149.

Naraian, R., Sahu, R., Kumar, S., Garg, S., Singh, C. and Kanaujia, R. 2009. Influence of different nitrogen rich supplements during cultivation of *Pleurotus florida* on corncob substrate. *Environmentalist,* 29: 1–7. doi: 10.1007/s10669-008-9174-4.

Niu, W.J., Han, L.J., Liu, X., Huang, G.Q., Chen, L.J., Xiao, W.H. and Yang, Z.L. 2016. Twenty-two compositional characterizations and theoretical energy potentials of extensively diversified China's crop residues. *Energy,* 100(1): 238–250. doi: 10.1016/j.energy.2016.01.093.

Ohnishi, M., Mori, K., ITO, S. and Fujino, Y. 1985. Chemical composition of steryl lipid in maitake mushroom. *Nippon Nôgeikagaku Kaishi,* 59(10): 1053–1059. doi: 10.1271/nogeikagaku1924.59.1053.

Royse, D.J. 1997a. Specialty mushrooms and their cultivation. *Horticultural Reviews*, 19: 59–97.

Royse, D.J. 1997b. Specialty mushrooms: Consumption, production and cultivation. *Revista Mexicana de Micologia*, 13: 1–11.

Sato, M., Tokuji, Y., Yoneyama, S., Fujii-Akiyama, K., Kinoshita, M., Chiji, H. and Ohnishi, M. 2013. Effect of dietary maitake (*Grifola frondosa*) mushrooms on plasma cholesterol and hepatic gene expression in cholesterol-fed mice. *J. Oleo Sci.,* 62(12): 1049–58.

Shen, Q. and Royse, D.J. 2001. Effects of nutrient supplements on biological efficiency, quality and crop cycle time of maitake (*Grifola frondosa*). *Appl. Microbiol. Biotechnol.,* 57: 74–78.

Song, B., Ye, J., Sossah, F.L., Li, C., Li, D., Meng, L., Xu, S., Fu, Y. and Li, Y. 2018. Assessing the effects of different agro-residue as substrates on growth cycle and yield of *Grifola frondosa* and statistical optimization of substrate components using simplex-lattice design. *AMB Express,* 8: 46. doi: 10.1186/s13568-018-0565-8.

Stamets, P. 1993a. Growing gourmet and medicinal fungi. Ten Speed Press and MycoMedia, Olympia, USA.

Stamets, P. 1993b. Growth parameters. *In*: Growing gourmet and medicinal mushroom. Ten Speed Press and MycoMedia, Olympia, USA, pp. 370–379.

Stamets, P. 2000. *Growing Gourmet and Medicinal Mushrooms.* 3rd. Ed. Ten Speed Press, Toronto.

Stamets, P. and Chilton, J.S. 1983. Mushroom cultivator: A practical guide to growing mushrooms at home. Agarikon Press, Washington.

Ulbricht, C., Weissner, W., Basch, E., Giese, N., Hammerness, P., Rusie-Seamon, E., Varghese, M. and Woods, J. Maitake mushroom (*Grifola frondosa*): Systematic review by the natural standard research collaboration. *Journal of the Society for Integrative Oncology,* 7(2): 66–72.

Wu, J.L., Lin, G.Z., Fu, M.Z. and Wei, Z.C. 1997. *Grifola frondosa*. Huang, N.L. *(ed.),* Cultivation of eighteen precious and delicious Chinese edible fungi. Beijing, China: Chinese Agricultural Press, pp. 130–145.

Zhao, J. and Zhang, J. 1992. The Polypores of China. J. Cramer, Berlin.

2.8

Cultivation of Medicinal Mushrooms

ANILA DOSHI[1]*

ABSTRACT

Medicinal mushrooms have been utilized as traditional medicine for a long time. Naturally, medicinal mushroom produce metabolites of medicinal importance. To increase the production biotechnological approaches can be used. Medicinally important mushroom species can be cultivated to procure medicines and active metabolites. Here, in this chapter different method of medicinal mushroom is presented which will be helpful in cultivating mushroom species.

1. INTRODUCTION

Cultivation of medicinal mushroom is oldest in China about one thousand years ago. *Auricularia* was grown after this, *Lentinus edodes* was also cultivated in China 900 years ago. The knowledge about medicinal value of mushroom of Chinese, Japanese, Korean and Indians are not new. They used mushroom traditionally as medicine and tonic. The medicinal value of *Auricularia auricula-judae, Lentinula edodes, Ganoderma lucidum, Coriolus versicolor, Tremella fuciformis, Flammulina velutipes, Calocybe indica, Tricholoma, Hericium erinaceus, Podaxis pistillaris* and *Phellorinia inquinans* are well known.

Yang and Jong (1989) from China compiled the medicinal uses of *Ganoderma lucidum, Coriolus versicolor, Hericium erinaceus, Armillaria mellea, Armillaria tabescens, Merasmus, Tremella fuciformis* and *Lentinus edodes* in Chinese medicine. Out of these, some mushrooms are cultivated on commercial scale and many species are cultivated commercially.

[1] Department of Plant Pathology, RCA, MPUAT, Udaipur, Rajasthan.
Corresponding author: E-mail: anila_doshi@yahoo.co.in

2. CULTIVATION OF *GANODERMA LUCIDUM*

Ganoderma has been regarded by the Chinese and Japanese to be a high quality herbal medicine also known as longevity drug in Korea (King, 1993). This mushroom either in mycelium or fruiting form are used in various ailments such as diabetes, ulcers, liver, lung diseases, asthma, high blood pressure and high cholesterol level. This mushroom is also used by Russian cosmonauts. Moreover, it is best known for managing cancer in combination with chemotherapy and also for its anti-HIV effects (Dharmanand, 1988). This mushroom contains a treasure chest of immunoregulating compounds. It is also known as mushroom of immortality in Japan (William, 1991). It is taken as powder in hot water 2-3 gms/day or by boiling the fruiting body with water and mixed with honey also known as *Ganoderma* tea.

2.1. Cultivation Technique

2.1.1. *Spawn for Ganoderma spp.*

(a) Preparation of sawdust spawn

Spawn for *Ganoderma* can be prepared by using sawdust mixed with grains. Sawdust = 600 grams, Wet weight of wheat grains = 6 kg, $CaCO_3$ = 1.0%. Spawn prepared using above formula is filled in glass bottles or polypropylene bags and inoculated with master spawn and kept for incubation. After completion of mycelial growth this can be used as spawn.

(b) Grain spawn

Grain spawn is prepared as per method of Sinden and Schisler (1962). After inoculation bottles are kept at 25°C for incubation for 2-3 weeks.

2.1.2. *Preparation of substrate*

Several substrate mixtures were found to support the mycelial growth of *Ganoderma*. Two best formulations are (i) Mixture of sawdust and rice bran (2:1 ratio), (ii) Rice straw, sawdust and rice bran (2:2:1) can be substituted by corn meal or ipil leaves. This mixture is moistened with water and filled in 12"x16" polypropylene bags and one end of this bag is plugged with the help of PVC pipe and cotton. Cover the plug with paper. The bags were placed in the sterilization chamber having 95-100°C temperature for five hours. These bags are allowed to cool and then inoculated with grain spawn under aseptic conditions. The inoculated bags are kept in a dark room having good ventilation. Complete spawn run takes place in thirty days.

2.1.3. *Fruiting*

The bags are then shifted into mushroom house for initiation of fruiting. The air temperature of the growing room is maintained at 28-30°C with relative humidity 85-90 percent. Inoculated bags are placed horizontally on the vertical racks. Watering in the form of mist spray is carried out at least twice a day. The mushroom primordia appeared in about 7-15 days after completion of spawn run.

The mushrooms are ready for harvest after 2-3 months when the fruiting bodies become red having no whitish margins around the edge.

2.1.4. *Harvesting*

Harvesting is done by cutting the base of mushroom stalk with a knife. The bags continued to yield another one or two harvest. Average yield of each bag was around 28 g dry weight. The harvested fruiting bodies can be dried under the sun for a few days before being packed in plastic bags and stored.

3. CULTIVATION OF SHIITAKE MUSHROOM (*LENTINULA EDODES*)

Shiitake has been grown for thousands of years in Asia on small diameter hard wood logs. In Japan, this industry continues to flourish and is producing 826,200 metric tonnes of shiitake every year (Sharma, 1997). The details of shiitake cultivation are mentioned in chapter 2.5 of this book.

This mushroom contains a chemical which reduces plasma-cholesterol level. This chemical also potentiates physiological constitution of the host results in the activation of many kind of cells which vitally important for the maintenance of homeostasis. This is particularly important in cancer patients in an immunosuppressed state. The leading principle of the function of Lentinan resides in the fact that it can cure the patients by restoring their homeostasis and through enhancement of their intrinsic resistance against such diseases. This implies restoration to a normal from an abnormal stage in a relevant area of immunology, endocrinology, neurology and nutrition. This concept may also play an important role as a bridge between modern immunology, biology and oriental medicine.

3.1. Cultivation of Shiitake

3.1.1. *Spawn preparation*

(a) *Sawdust spawn:* Spawn used for inoculation in log cultivation of shiitake may be produced in the form of sawdust spawn or in the form of small wedge shaped or rod-shaped piece of wood impregnated with active mycelium or log cultivation. Plugs are considered superior to sawdust spawn due to the fact that they are easy to use.

(b) *Grain spawn:* Grain spawn is prepared as per method of Sinden and Schisler (1962). After inoculation bottles are kept at 25°C for incubation for 3-4 weeks.

3.2. Cultivation Methods

3.2.1. Log cultivation

The mycelium mainly grows on dried wood logs absorbing nutrients from the cambium. Logs having 10 cm diameter are suitable to use for mushroom cultivation. These can be collected from 15-20 years old trees which are considered as the most suitable substrate. The logs are cut in autumn before the leaf fall. At this time logs contain highest amount of carbohydrate and other organic substances. Moreover, the outer protective layer bark is also tightly attached to the woody portion. The log should contain a moisture content of 40-50 percent at the time of filling. If the moisture content is more than 60 percent and pH 7 to 8, it will be easily contaminated with other competitor moulds. Normally, the optimum pH range requirement of the log is in between 4.5 to 5.5.

Holes are drilled in the wood at intervals of 5 to 35 cm depending on the species of *Lentinula* used. Plugs of inocula are then inserted into the holes and sawdust spawn filled holes are packed with spawn and then, the inoculum points are sealed with wax.

3.2.1.1. *Spawn run and cropping*

Inoculated logs are placed in a position which is favourable for the development of mycelium at temperature of 14-20°C. At this stage, no watering should be carried out and logs can be arranged in flat position in order to prevent water loss. Spawn run takes place in a log within one or two years.

3.2.1.2. *Fruiting*

Fruiting requires a fall in temperature 12-20°C, relative humidity 70-75 percent, low light intensity and aeration. The primordia of the fruit bodies would take 24 hours to develop and fruit bodies are harvested after 7 days. In another method logs are dipped in a tank of water at about 16°C for about 1-3 nights. After dipping, the logs may be arranged in indoors to fruit.

Fruit bodies are normally harvested upto 3 times from a batch of log and after this, logs are rested before being dipped again. This can be repeated upto 4 times per year with a good harvest. It is possible to obtain mushroom from logs upto 6 years. But the yield by the fifth year is generally very poor. Good yield would be 200 g shiitake per kg wood substrate over a period of three years.

3.2.1.3. *Harvesting*

Harvesting is carried out by hand and this mushroom can be eaten fresh, dried and or pickled in vinegar.

3.2.1.4. *Drying*

Drying of shiitake is carried out in oven and harvested fruit bodies are arranged in single layer on shelves at relatively low temperature for first three years followed by 40-60°C for 18 hours. The dried shiitake should have 10-13 percent moisture.

3.2.2. *Commercial bag cultivation*

Commercial cultivation of shiitake may be carried out in polypropylene bags. Some common shiitake formulation for bag cultivation area as follows:

(i) Sawdust (80%), rice bran -20% and water content 65%.

(ii) Sawdust (80%), millet-10%, wheat bran-10% and water content- 60% (Royse *et al.,* 1985).

(iii) Hardwood saw dust -24.2%, rice bran 5.2 % corn powder 2.6%, $CaCO_3$ -0.3% and water 62%.

(iv) Hardwood saw dust - 89.8%, rice Bran -10%, $CaCO_3$ - 0.2% and water content -60% (Han *et al.,* 1981).

(v) Corn cob -40%, saw dust -10%, wheat bran-12.5%, cane sugar 1.0%, pectin-0.15% and urea 0.2%.

(vi) Sugarcane bagasses-50%; rice bran-12.5%, urea 15% and $MgSO_4$-10%.

All the ingredients are thoroughly mixed, the substrate mixture is filled in polypropylene bags and a spike is introduced in the centre. Sterilization is achieved by staining or by autoclaving. After sterilization and cooling, the bags are inoculated with 2% of saw dust spawn aseptically.

3.2.2.1. *Spawn run and cropping*

The inoculated bags are staked in growing houses for approximately two months for spawn run. Any bag showing presence of contamination is immediately removed and discarded. During spawn run bag should be kept under low light intensity and high humidity with little ventilation. At this stage watering is not required. After spawn run bags are arranged in single layer up right on the floor of growing house for one month.

The top of bags are then removed and vigorous watering is applied by fine mist to induce fruiting. The bags are left inverted overnight and this is repeated upto 3 times. At this stage, watering, light intensity and ventilation should be increased. Under optimum conditions 200-500 g shiitake may be obtained/kg sawdust.

4. CULTIVATION OF *AURICULARIA* OR BLACK EAR MUSHROOM

This mushroom is believed to cure sore throat, anemia, certain digestive disorder especially piles on regular consumption. At present, its cultivation has become a major occupation in some of the South-East Asian countries where farms have been exclusively set up for home consumption and export. Its production has steadily increased over last 4-5 years with a total world production 4201000 metric tonnes and ranking fourth among different mushroom species (Sharma, 1997). Thailand is a major importer of this mushroom for local use and 90 percent of dried produce of Taiwan is exported to Hong Kong, Japan and USA. In India, this mushroom is collected and consumed in North-Eastern hills (Singh and Devi, 1986) while its artificial cultivation has been done by Bhandal and Mehta (1989).

4.1. Cultivation of *Lentinus*

4.1.1. Preparation of spawn

The grain spawn is prepared as per standard method of Sinden and Schisler (1962).

4.1.1.1. *Sawdust spawn*

It is prepared by using (i) Sawdust – 70%, wheat bran -15%, used tea leaves - 15% and water content - 65%, (ii) Sawdust - 78%, sucrose - 1%, wheat bran - 20%, $CaCO_3$ - 1% and water content - 65%, (iii) Sawdust – 880 g, rice bran – 320 g, sucrose – 30 g, KNO_3 – 4 g, $CaCO_3$ – 6 g and water 2 litre.

All the substrates are thoroughly sieved to remove bigger particles and mixed with water. The substrate mixture is filled either in empty spawn bottles or in polypropylene bags and plugged with non-absorbent cotton and covered with waste paper or with aluminium foil. It is then autoclaved at 20 lbs psi for 2 hours. After cooling they are inoculated for 30 days at 24±2°C. Contaminated bottles are discarded.

4.1.2. *Cultivation methods*

4.1.2.1. *Log cultivation*

The log cultivation is done similar to that of *Lentinus* species. Generally, 1 metre log of 3-6 cm diameter are preferred for ease of handling. For inoculation, small holes of 1 x 1 cm and 1.5-2.0 cm deep are made on the logs with the help of drilling machine. The holes are made at a distance of 20-30 cm long axis and 6 cm between each rows. The holes between two rows are alternate in position. Spawn is filled in the holes and are sealed with paraffin wax.

4.1.2.2. *Bag cultivation*

Due to deforestation wood scarcity is common, therefore log cultivation is shifted to bag cultivation. Either wheat/paddy straw supplemented with 4% of rice bran is used for cultivation. The substrate is autoclaved after filling in polypropylene bags. After cooling spawning is done with 2% grain sawdust spawn on wet weight basis of the substrate.

4.1.3. *Spawn run and cropping*

The logs/bags can be placed in cropping room in close proximity to each other. A temperature range of 22-28°C, relative humidity 80±10 % is maintained for spawn run and cropping. Spawn run is completed in 2-3 weeks. Then either logs are is removed or slits are made.

Humidity of the cropping room is maintained by sprinkling water with a fine sprayer once or twice a day depending upon humidity of cropping room. After 30 days of opening mushroom appear and continue till the favourable temperature prevails. At this time 2-3 hours of aeration and light is necessary. If the logs are kept properly, mushroom can be harvested up to 10-12 years.

4.1.4. Harvesting

Mushrooms are easily harvested by hand. Mushroom produced from wood logs have longer down hair, more tough texture plus attractive colour and longer production period than from substrate bags. This mushroom can be easily sun dried with a dry matter of 8-12 percent. The total yield after several years of cropping will be 10-20% of the original weight of logs.

4.1.5. *Management*

Proper hygienic conditions are necessary because addition of rice bran may attract moulds in absence of proper autoclaving. Presence of high level of

CO_2 and absence of light, results in development of abnormal fruit bodies of long stalks and small pileus.

5. CULTIVATION OF *TRICHOLOMA* SPECIES (MATSUTAKE MUSHROOM)

Tricholoma species has immense potential in medicinal world as it has been found to have antibiotic substances. Singer (1961) and is beneficial to stomach and intestinal ailments (Yang and Jong, 1984). This mushroom has excellent nutrition and edible quality. It contains protein 14.06-18.43%, carbohydrate 70.99-78.52 % and ash 6.88-10.58 % on percent dry weight basis (Rantavara, 1947).

In Japan, consumption of *Tricholoma* spp. was between 10,000 – 12,000 tonnes per year in mid 1800's. Now, consumption is of 3000 tonnes per year of which Japan produces 1000 tonnes in a good year and imported from China, Morocco, North America and South America. However, British Columbia exported 600 tonnes to Japan.

Moreover, in Japan price of a matsutake producing forest is worth between 5 and 10 times then the price of forest without matsutake. According to Lakhan Pal (1995) in the USA, especially in the Pacific North-West, a year ago, American matsutake (*Tricholoma magnivelare* (Perk) Read head) grow abundantly with *Pinus contorta*. During the past several years, there has been a tremendous increase in the commercial harvest of this and other wild mushrooms.

5.1. Cultivation of Matsutake

5.1.1. *Spawn preparation*

Grain spawn is prepared as per standard method.

5.1.2. *Substrate preparation*

Tominaga (1978) described in detail the cultivation technology of this mycorrhizal mushroom. This mushroom is associated with a number of conifers *Pinus* species, *Abies* species, *Picea* species and *Tsuga* species. It is reported to be distributed to a latitude of 25 ft to 50 ft north. This mushroom is ectomycorrhizal in nature, therefore, its cultivation is carried out in forests. The growth of this mushroom in the forest is dependent on the location of forest, degree of shade (60-70%), thickness of the leaf litter layer (2-3 cm), optimum ground temperature (around 24°C), soil water content (greater the rainfall during August-September, greater the harvest) and growth of pine roots. A large number of mushrooms are produced around the plants whose age is 30-40 years. For inoculation of pine trees such sites where mycorrhizal roots are in abundance is the best. Density of pine trees also affects the production of mushrooms, usually the most productive woods have 800-1200 trees per acre (20 years old) and 600 trees per acre (30 years old). It has been suggested that woods where pine trees are too dense should be thinned and where the pine trees are too sparse should be planted to serve as a host plant.

5.1.3. *Spawn and spawning*

Different types of spawn and method of spawning have been used:

1. One type of spawn consists of basidiospore suspension. The spore suspension poured into holes made in the woods with a rod 1-3 cm diameter and 10-15 cm depth. The holes are covered with soil so that the basidiospores attached to the fine roots of the pine.
2. The second type of inoculum consists of mycelial strands of the fruiting body by which it is attached to the fine pine roots. Holes equivalent in size to the stipe diameter are made in a forest of optimum age and the mycelial connection is inserted into these holes as inoculum.
3. In the third type of inoculum a part of the fairy ring is used as inoculum. The place where the fairy ring has appeared earlier is marked and a part of this ring is removed and transplanted to an appropriate site having a transplanting hole of the same size. The mycelium connects with the fine feeder roots and produce new fairy rings.

5.1.4. *Hiroshima method of tunnels*

Tominaga (1975) cultivated this mushroom by another method "Hiroshima Method of Tunnel." In this method, the tunnel consists of 60 cm high and 1 metre wide wall covered by vinyl sheet. This structure is constructed over a place where fruit bodies are expected to develop. Humidity and stimulative temperature are maintained and the primordia of pine mushroom develop in this artificial environment. Since, the tunnel protects the sporocarps from inclement weather and other factors, a greater production of fruit bodies is ensured.

5.1.5. *Cultivation in fairy ring near the mycorrhizal sites*

Sites of fairy ring appearance are marked after trees are felt. The sites are then rejuvenated by planting 2-3 years old saplings of the pine at the marked saplings of the pine at the marked fairy ring site. The seedling becomes mycorrhizal with matsutake inoculum and afterwards pine mushroom production begins.

In another method the pine seedlings are inoculated with a pure culture of pine mushroom and raised in pots of sterilized soil. When mycorrhization is achieved the seedlings are transplanted among pine trees of optimum age.

In the USA, especially in pacific North-West this mushroom grows abundantly with *Pinus contorta*. Presently, permits are required for harvesting this mushroom on national forest land. Mushroom picking in general is prohibited in national parks, national forests, wilderness and mushroom study research areas. To protect the "Matsutake" mushroom habitats, guidelines are issued in the form of posters which stress for safe picking and permits for "Matsutake" mushroom.

5.2. Cultivation of *Tricholoma crassa*

For the first time Roy and Samajpati (1981) from India were able to cultivate *Tricholoma crassa* artificially on wheat straw. This mushroom cultivated successfully on chemically sterilized wheat and paddy straw. Later on Zacharia and Doshi (1999) increases its biological efficiency 23.10 percent by supplementing base substrate with cotton linter and termitarium soil.

Casing: This mushroom also requires casing. The best casing material is spent compost followed by FYM and spent compost having pH 6.5-7.5.

6. CULTIVATION OF *FLAMMULINA VELUTIPES*

Flammulina velutipes is one of the cosmopolitan temperate species that grows on woods. In nature, it usually exists as a saprophyte on logs of *Celtis trees*. In Japan, it is commonly known as "Enokitake". Following the introduction of sawdust as a substrate for its cultivation, this mushroom is now only second to *Lentinus edodes* in economic importance in the Japanese mushroom industry. It is sold fresh or canned and is also cultivated commercially in Taiwan.

In India this mushroom has been cultivated on experimental scale on oak or poplar sawdust in bottles or polybags (Rangad and Jandaik, 1980). Dhar (1976) also tried Japanese method of cultivation.

6.1. Cultivation of *Flammulina velutipes*

6.1.1. *Preparation of spawn*

Sawdust or grain spawn is used for its cultivation. The method of preparation of spawn is similar as in other mushroom. After inoculation, bottles/ polypropylene bags are incubated at room temperature 25±2°C. After 15-20 days of inoculation spawn is ready for use. When sawdust is used for preparation of spawn, it is first wetted and steamed for 1 h 30 min and then rice bran 10:1 wet/weight is mixed thoroughly filled in bottles or polypropylene bags.

6.1.2. *Substrate preparation*

Substrate is mainly composed of willow sawdust 1760 g; paddy straw- 452 g and wheat bran – 452 g in the ratio of 4:1:1 and water 3060 ml. All these ingredients are mixed together and filled in either polypropylene bags or in bottles and sterilized in an autoclave at 121°C for 1 hour.

6.1.3. *Spawning and cropping*

The bags/bottles are then inoculated with grain/sawdust spawn and incubated in dark at 20±1°C. Spawn run is completed in 25-30 days in case of bottle culture and 6 days in polypropylene bags. Thielke (1989) cultivated this mushroom on sterilized coffee ground moistened with 0.5 percent yeast extract. Sawdust supplemented with straw and seed bean can increase the yield of *Flammulina velutipes.*

6.1.4. *Fruiting*

After the completion of spawn run the caps of bottles and upper layers of sawdust spawn are removed. The bottles are kept in darkness at temperature between 10 to 12°C and a relative humidity of 80-85 percent. After 10-14 days when the stipes have reached a length of 12 cm, the neck of bottle is wrapped with waxed paper in a plastic film to hold the mushroom upright.

The plastic film is removed when stipes are 13-14 cm long the harvesting is done. Bottle culture is not recommended for commercial cultivation. Kachroo (1991) obtained increased yield by dipping the bags in a solution of copper- 0.4 mg, iron- 0.2 mg and Ca- 5.0 mg after spawn run.

6.1.5. *Harvesting*

When stipe length reaches 13-14 cm, then they are cut in bundles and packed. It is sold fresh or canned, the optimum yield of this mushroom is 142-215 g/kg by block of sawdust.

6.1.6. *Management*

The initiation of fruit bodies start in dark but light is necessary for its further development. At the time of fruit body formation, temperature of cropping room should be lowered to 8-12°C with relative humidity from 80-85 percent.

7. CONCLUSIONS

This chapter describes the methods of cultivation of medically important mushroom species like- *Ganoderma lucidum, Auricularia, Tricholoma, Flamulina velutipes* and shiitake. This will be helpful increasing the cultivation of mushroom for the isolation of active ingredients for producing the drugs or herbal medicines to cure diseases.

8. REFERENCES

Bhandal, M.S. and Mehta, K.B. 1989. Cultivation of *Auricularia polytricha* in India (Mont.) Sacc. On wheat straw. *Mushroom Sci.,* 12: 387–393.

Doshi Anila, Sharma Shyam Sunder and Trivedi Amit 1993. A promising edible mushroom for the tropics *Calocybe indica* P&C. *Mushroom Information,* 10(5/14): 14–22.

Han, Y.H., Veng, W.T., Chen, L.C. and Chan, S. 1981. Physiology and ecology of *Lentinus edodes* (Berk.) Sing. *Mushroom Sci.,* 11: 623–658.

Kachroo, J.L. 1991. The effect of micronutrients on the cultivation of *Flammulina velutipes* (Curt. ex. Fr.) Sing. *Indian J. Mycol. Pl. Pathol.,* 21: 287–288.

King, T.A. 1993. Mushrooms, the ultimate health food but little research in US to prove it. *Mushroom News,* 41(2): 26–29.

Rangad, C.O. and Jandaik, C.L. 1980. Cultivation of *Flammulina velutipes* and *Agrocybe aegerita* in India. *Indian J. Mushroom,* 6(4): 8–52.

Roy, A. and Samajpathi, S.N. 1981. *Tricholoma crassum* (Berk.) Sacc. A new Indian edible mushroom. Edible Mushrooms of West Bengal, VIII Mush. *Newsletter for the Tropics,* 1(3): 18–20.

Royse, D.J., Schisler, L.C. and Diehle, D.A. 1985. Consumption, production, cultivation of shiitake mushroom. *Interdisciplinary Sci. Rev.*

Sharma, S.R. 1997. Scope of specialty mushrooms in India. *In*: Advances in mushroom biology and production. *Proceedings of Indian Mushroom Conference, 1997. Published by Mush. Soc. of India, Solan*, p. 193.

Sinden, J.W. and Schisler, L.C. 1962. Nutrient supplementation of mushroom compost at casing. *Mushroom Science,* 5: 267–280.

Singer, R. 1961. Mushroom and truffles for human consumption. Mushrooms and truffles. Interscience Pub. Inc. New York, pp. 219, 222–223.

Singh, R.I. and Devi, S.M. 1986. Mushroom cultivation in Manipur. *In*: *Souvenir on Mushrooms* (Sohi, H.S., Bhandal, M.S. and Mehta, K.B. (*eds.*)), NCMRT, Solan, India, pp. 60–61.

Tominaga, Y. 1975. Studies on the tunnel cultivation of the so called Hiroshima method of *Tricholoma matsutake. Bull. Hiroshima Ag. College,* 5: 165–180.

Tominaga, Y. 1978. *Tricholoma matsutake. In: The Biology and Cultivation of Edible Mushrooms* (Chang, S.T. and Hayes, W.A. (*eds.*)), Academic Press.

Yang, Q. and Jong, S.C. 1984. Medicinal mushrooms in China. *Mush. Sci.,* 12(1): 631–643.

Yang, Q.Y. and Jong, S.C. 1989. Medicinal mushroom in China. *Mushroom Sci.*, 12(1): 631–43.

Zacharia Sunil and Doshi Anila 1999. Studies on the artificial cultivation of *Tricholoma crassa* (Berk.) Sacc. An ectomycorrhizal mushroom, *Mycologia Italiana*.

2.9

Biology and Cultivation of Milky Mushroom (*Calocybe indica*)

P. Deepthi[1], Siva Sankar Reddy E.[2], Pushpanjali Wagh[3], C.S. Shukla[4] and Harvinder Kumar Singh[5*]

ABSTRACT

Mushrooms were so far considered as luxury food especially among the rich community because of their unique flavor and excitingly different taste but now they have grown to a common man's food. Mushrooms are traced as a special kind of food, since ancient times. The Greeks believed that mushrooms provided strength for warriors in battle and Romans regarded them as "Food of Gods" or "Gods Flesh", which were served only on festive occasions. Mushroom cultivation is an ecofriendly activity where agricultural / industrial wastes are utilized and recycled. Till now more than 80 edible mushrooms are considered for commercial exploitation, among these mushrooms, milky mushroom has become the focal point of exploitation in India. Milky mushroom is robust, fleshy and milky white in color even after flattening. Hence, it has greater consumer attraction. As it grows in hot humid climate the mushroom is highly suitable for cultivation in the most of the plains of India almost throughout the year. The cultivation technology is very simple, involves less cost and no special compost is needed for the cultivation. The cultivation process resembles that of oyster mushroom but for the additional process of casing. The mushroom can be harvested from 24-28 days after spawning and the total crop cycle is only 45-50 days. Most importantly, the milky mushroom has an extended shelf life of 3-5 days compared to other cultivated species, making it more amenable to handling, transportation

[1] Plant Pathology, Horticultural Research Station, Anantapuramu, Dr. YSRHU, Andhra Pradesh.
[2] Genetics and Plant Breeding, College of Agriculture, IGKVV, Raipur.
[3,4,5] Plant Pathology, College of Agriculture, IGKVV, Raipur.
[*] *Corresponding author*: E-mail: harry.0452@gmail.com

and storage. Milky mushroom can be cultivated on a wide range of substrates like, paddy straw, maize stalks, sorghum stalks, pearl millet stalks, palmarosa grass, vetiver grass, sugarcane bagasse, soybean hay, groundnut haulms etc. However, for commercial production paddy straw is the best substrate. Milky mushroom is a rich source of protein with protein content of 32.3% and fetches a high market price compared to oyster mushrooms. It is highly suitable for drying, canning, soup powder preparation and pickle making. Thus, there is a growing interest among the farmers towards cultivation of milky mushroom.

***Key words*:** Mushroom, *Calocybe indica,* Milky mushroom cultivation, Substrate, Casing.

---

1. INTRODUCTION

Mushrooms are the fruiting bodies of macrofungi. They are also called as "white vegetables" or "boneless vegetarian meat" and contain ample amounts of proteins, vitamins and fibre apart from having immense medicinally bioactive molecules properties and thus their commercial cultivation is being done in more than 100 countries of the world today (Thakur and Singh, 2013). Mushrooms contain about 85-95% water, 3% protein, 4% carbohydrates, 0.1% fats, 1% minerals and vitamins (Thakur, 1998). Mushroom cultivation is the most economical and relatively short biological process for the biotransformation of organic materials into protein (20-35% dry weight) rich food (Chiu and Moore, 2001; Martinez-Carrera *et al.*, 2000). At present, world production is estimated to be around 27 million tonnes including 21.54 million tonnes by China alone (Li, 2012) and is ever increasing. *Agaricus bisporus* is the most popular mushroom commercially cultivated all over the world. Till now more than 80 edible mushrooms are considered for commercial exploitation The Indian subcontinent is known worldwide for its varied agro climatic zones with a variety of habitats that favour rich mushroom biodiversity (Verma *et al.,* 2003). In India, at present, four mushroom genera *viz.* (Button mushroom) *Agaricus bisporus,* (oyster mushroom) *Pleurotus* spp. (paddy straw mushroom) *Volvariella* spp. and (milky mushroom) *Calocybe indica* have been recommended for the year round cultivation. Among, tropical mushrooms, oyster mushroom, paddy straw and milky mushrooms are very much common in cultivation in India. This mushroom was first reported from India by Purkayastha and Chandra in the year 1974. The production of tropical mushrooms *viz. Pleurotus* spp., *Volvariella volvacea* and *Calocybe indica* utilizing locally available substrates *viz.* paddy straw, wheat, soybean, cotton wastes, coffee waste, water hyacinth, tree saw dust, sugar cane bagasse, wild grasses and various categories of refuse and lignocellulosic wastes have great potential to exploit and convert it into a highly nutritious food within less period of time per unit area (Singh, 2011).

2. INDIAN SCENARIO OF MUSHROOM PRODUCTION

In India, the total production of mushroom is about 1,29,782 metric tonnes (Sharma *et al.*, 2017) of which major share (73%) goes to button mushroom and that too by a single unit *i.e.,* Agro Dutch, Lalru (Patiyala), followed by oyster mushroom which contributes about 16% Punjab producing around 18,000 metric tonnes/annum (Sharma *et al.*, 2017).

Table 1. Tropical mushroom production in India in 2016 (metric tonnes).

Mushroom species	***Yield (t)***
Oyster mushroom	21,272
Paddy straw mushroom	10,000
Milky mushroom	3,184

Source: Sharma *et al.* (2017).

About two decades ago, *Calocybe indica* P. and C. was identified as a wild edible mushroom in India. This fungus was first time reported in India from West Bengal by Purkayastha and Chandra (1974) and it was for the first time artificially cultivated on a mixture of soil, sand and maize meal (12:6:1) in soil jars (Purkayastha and Chandra, 1976). The production technology further refined with improve production technology at IIHR, Bangalore; MPAU, Udaipur and NRCM, Solan.

However, complete commercial production techniques were evolved for the first time in Tamil Nadu (Krishnamoorthy, 1995). The milky mushroom (*Calocybe indica*) is a potentially new species to the world mushroom growers. It has milky white sporophore and large size fruit body with delicious flavor and resembles button mushroom (Pani, 2012). Its robust size, sustainable yield, attractive colour, delicacy, long shelf-life and lucrative market value have attracted the attention of both mushroom consumers and prospective growers (Chakraborty and Sikdar, 2010). The mushroom is rich in protein, lipids, fibres, minerals, carbohydrate and contains an abundant amount of essential amino acids (Alam *et al.*, 2008; Mallavadhani *et al.*, 2006). It is an excellent source of thiamine, riboflavin, nicotinic acid, pyridoxine and ascorbic acid (Breene, 1990). It is suitable for hot humid climate and can be cultivated indoor in high temperature and high humidity areas like Chhattisgarh (Shukla, 2004), Tamil Nadu, Odisha etc. It grows well at a temperature range of 25-35°C and relative humidity more than 80%. Milky mushroom is gaining more popularity in tropical and sub-tropical regions due to its easy cultivation technology, higher biological efficiency and better shelf life. The nutritive value of milky mushroom is at par with other cultivated mushrooms. Dried sporophore contain approximately 17% protein, 4.1% fat, 3.7% crude fibre, 60% carbohydrate. In addition to this, it has vitamins and mineral salt *i.e.,* sodium, phosphorus, iron, zinc, calcium, potassium, copper and manganese. It has milky white sporophore and large size fruit body with delicious flavour. This mushroom can be used as relay cropping when other mushroom cannot be grown at higher temperature with more than 60% biological efficiency. The cultivation technology is very simple, involves less cost and no special compost is needed for the cultivation. As it grows in hot humid climate the mushroom is highly suitable for cultivation in the most of the plains of India almost throughout the year. The choice of mushroom growers of south India always was towards oyster mushroom which shows wider adaptability for ambivalent climatic conditions. However, during summer season the productivity of oyster mushroom is found to be low with only 24 hours of shelf life. Therefore, milky mushroom is considered as a better proxy for oyster mushroom notably in tropical regions with longer shelf life of 3-4 days. Commercial milky white mushroom growers are mostly confined to Tamil Nadu, India, particularly in Erode, Salem, Coimbatore, Trichy, Madurai and other districts (Lakshmipathy *et al.*, 2012). During last 2-3 years its cultivation is also becoming popular in North, particularly in Haryana. Among vital growth requirements, environmental factors play a major role in the growth and reproduction of edible fungi. Cultivation of *C. indica* is influenced by temperature and relative humidity for its yield. However, optimum conditions favorable for the growth of *C. indica* have not been clearly defined under controlled conditions.

3. DIVERSITY OF *CALOCYBE INDICA*

There is huge diversity of milky mushroom prevalent in the rich fauna of Chhattisgarh, Madhya Pradesh, Tamil Nadu, Western Ghats and Rajasthan (Fig. 1). *Calocybe indica* and *Calocybe gambosa* are predominantly the two species that are documented and characterized. *Tricholoma giganteum* Heim, a new edible mushroom pure white in colour resembling the morphology of *Calocybe indica,* was reported growing widely in summer in Indo-Gangetic Plains of Howrah district, Hooghly in India (Chakravarty and Sarkar, 1982).

4. STRAIN IMPROVEMENT (BREEDING PROGRAMME)

The reservoir of edible mushrooms, like other micro-organisms of use for industrial purposes, is limited. It is generally recognised that in order to maintain and breed high-yielding strains, the techniques employed in mushroom breeding should now and then be modified and improved in accordance with new findings and progress in the scientific world as a whole, and in microbiology and genetics in particular.

4.1. By Selection

Tissue culture of selected sporophores has also been used to fix desired variants. In the short term, selection alone appears to have some role to play in mushroom strain improvement, but genetic improvement by selection becomes increasingly difficult. Therefore, strain improvement through hybridization has become a recourse and also necessity.

4.2. By Hybridization

In addition to the conventional method of matings between two genetically compatible strains through which dikaryon mycelia and fruiting bodies are formed, steps towards a broader spectrum of hybridisation can be achieved in strain improvement of edible mushrooms by the following ways.

4.3. Use of Auxotrophs

Auxotrophs can be obtained naturally or induced by mutagenesis. The contrasting auxotrophs can be paired and the products can be screened for hybridity on minimal medium. Certainly, the feasibility of auxotrophs to be used as a tool for hybridization depends on how easily auxotrophs can be obtained in the strains of the mushroom.

4.4. Use of Resistance Markers

Mutants resistant to antimetabolites have been suggested recently as alternatives to auxotrophs for use in mushroom breeding programmes. The treated spores or hyphal fragments, which can grow on a medium containing an inhibitory concentration of the anti-metabolite, would be considered to possess the marker. Complementary resistant strains would be grown together and, hybridity can be confirmed by transferring it to a medium containing the two appropriate antimetabolites.

4.5. Protoplast Fusion

One of the most effective barriers to sexual reproduction is the inability of hyphae from two selected strains to fuse. Several researchers have reported that protoplasts can be isolated from plant and microbial cells by enzymatic breaking of the cell wall, in the presence of an osmotic stabilizer. Such protoplasts can be effectively induced to fuse in the presence of polyethylene glycol (PEG). After a short period of time protoplasts can regenerate their cell walls and start to propagate as normal cells or hyphae. These cells are heterokaryotic if fusion occurred between cells from genetically different strains. This can serve to increase the frequency of intraspecies crosses in organisms in which natural matings rarely occur. The technique has even a much wider application, and can be used for interspecies and intergeneric crosses in some organisms, which normally cannot be crossed. Although such an approach has been carried out in several laboratories, until now, no clear and economically applicable results have been reported in edible mushrooms.

Although it seems likely that selection from existing commercial strains of edible mushrooms may continue to enjoy some improvement, by analogy with other crop species, hybridization seems to offer the best prospects for real progress, especially with regard to multiple gene transfers mediated by protoplast fusion, during the next few decades.

5. PRODUCTION TECHNOLOGY CROP MANAGEMENT AT DIFFERENT STAGES OF CULTIVATION OF MILKY MUSHROOM (*CALOCYBE INDICA*)

5.1. Spawn Production

The word spawn drawn from an old French verb espandre which means to expand and it has been defined in various ways by different scientists. According to Sinden (1972) vegetative growth of fungus is called the mycelium and material used for planting on beds is called as spawn. Thus, mycelium of mushroom and supporting medium which has nutrition to fungus during its growth called as spawn. Hence, it is equivalent to seed unlike other crops. It plays an important role in mushroom cultivation because failure or success of crop depends on quality of spawn. The spawn should be silky or strandy type not cottony type and color should be more or less white not brown. Fresh spawn give better yield than older spawn. Spawn older than three weeks after complete run will give lesser yield as compared to fresh spawn.

5.2. Isolation of Mushroom Mycelium from Fruit Body

For isolation of *C. indica* mycelium select a young healthy sporophore and wash thoroughly with tap water then cut into small pieces preferably from adjoining portion of stem and gills. Aseptically disinfect the pieces with 95% ethanol or sodium hypochlorite solution and transfer individually on agar slant or agar plate. Thereafter, incubate at 25 ± 1°C for 8-10 days and watch the plate slants time to time. Inoculated plate should be observed for white mycelial growth originating from the excised tissue within 2-3 days of inoculation.

5.3. Preparation of Mother Spawn

Process the wheat or sorghum grain as per procedure and fill in flask or glucose bottle up to ¾ capacity (Fig. 2). Subsequently plug with non absorbent cotton and sterilized at 121°C for 2 hrs then cool them. Aseptically inoculate these flask/bottle with pure culture of *C. indica* and incubate at 30°C for 2-3 weeks. During incubation period observed all inoculated flask/bottle for any type of contamination. After sufficient period all the grains should be covered by mushroom mycelium. This will serve as a mother spawn.

5.4. Preparation of Planting Spawn

Process the grains as prepared for mother spawn and fill in polypropylene bags, plug it and sterilized as per standard procedure (Fig. 3). After cooling aseptically inoculate the bags with mother spawn. Incubate the bag at 30°C for 15-20 days or till all the grains are completely colonized by mycelium. This will act as planting spawn for spawning or seeding the bed. Wheat and sorghum grains are generally used for raising spawn of *C. indica*. Different cereal grains have been used as substrate for spawn preparation to increase the yield of milky mushroom (Purkayastha and Nayak, 1981a). However, boiled wheat grains mixed with 2 percent $CaCO_3$ and $CaSO_4$ found as better substrate for earlier spawn preparation and better yield of *C. indica* (Prasuna, 2002). According to Shukla (2003) wheat grain as better substrate for spawn preparation of *C. indica*.

5.5. Substrate and Its Preparation

The mushroom can be grown on a wide range of substrates. Substrates exposed to rain or harvested prematurely (green colour) are prone to various weed moulds, which may result in crop failure. A lot of work have been carried out in India to evaluate the suitability of different kinds of substrate for growing of *C. indica*. These substrates are wheat straw, paddy straw maize straw, sorghum straw, cowpea straw bajra straw maize meal, rice meal, sorghum meal, wheat bran, coir pith, mango logs, euphorbia logs, sugarcane bagasse (Doshi *et al.*, 1987; Trivedi *et al.,* 1991; Doshi *et al.*, 1993; Sharma *et al.,* 1995; Krishnamurthy, 1995; Krishnamurthy and Muthusamy, 1997; Prasuna, 2002; Shukla, 2003). Many researchers have tried various substrate to their suitability for cultivation of *C. indica*. Among these, Purkayastha and Chandra (1985), Krishnamurthy (1995) suggested rice straw (1" length) for cultivation of *C. indica* whereas, Trivedi *et al.* (1991), Doshi *et al.* (1993), Prasuna (2002) and Shukla (2003) reported wheat straw as best substrate to obtain higher yield of *C. indica.*

However, cereal straw (paddy/wheat), which are easily available in abundance are favoured. Straw is chopped in small pieces (2-4 cm size) and soaked in fresh water for 8-16 hours. This period can be reduced when pasteurization is to be done by steam. Main purpose of soaking is to saturate the substrate with water. It is easier to soak if straw is first filled in gunny bag and dipped in water. The substrate harbours many active/inactive stages of micro-organism and insects which compete for growth and nutrition with mushroom mycelium. Therefore it is essential to inhibit or check the growth of undesirable organism. The growth of such types of organism can be reduced or eliminate by treatment of substrate. Methods generally used for substrate treatment in cultivation of milky mushroom include:

5.5.1. Pasteurization/sterilization

Pasteurization/sterilization can be achieved by any one of the following ways.

5.5.1.1. *Hot water treatment*

Water is boiled in wide mouth container and chopped wet straw filled in gunny bag is submersed in hot water for 40 minutes at 80-90°C to remove microorganisms. This is very popular method particularly with small growers.

5.5.1.2. *Steam pasteurization*

Wet straw is filled inside insulated room either in perforated shelves or in wooden trays. Steam is released from a boiler and temperature inside substrate is raised to 65°C and maintained for 5-6 hours. The air inside the room should be circulated to have uniform temperature in the substrate.

5.5.1.3. *Steam sterilization*

Substrate is filled in polypropylene bags (35 x 45 cm, holding 2-3 kg wet substrate) and sterilized at 15 psi for 1 hour. Once pasteurization/sterilization is over straw is shifted to spawning room for cooling and spawning.

5.5.1.4. *Chemical sterilization technique*

Chemical treatment of straw soaked in solution having 75 ppm bavistin and 500 ppm formalin is most commonly used. In South India many farmers are using this technique (Fig. 4). If the substrate is more wet coupled with high temperature then there are chances of *Coprinus* contamination.

5.6. Spawning and Spawn Running

Spawning methods are similar to that mentioned for oyster mushroom. However, layer spawning is most commonly used in milky mushroom. Higher spawn dose of 4-5% (wet wt. basis) is used. After spawning bags are shifted to spawn running room and kept in dark where temperature between 25-35°C with 80% RH is maintained. It takes about 20 days for substrate to get colonized and after that bags are ready for casing (Fig. 5).

5.7. Casing

Casing means covering the top surface of fully colonized bags, with pasteurized casing material. Pond soil/soil (75%) + sand (25%) or Coir pith + soil or FYM + soil can be used as casing material. Casing thickness is between 3-4 cm. Casing provides physical support, moisture and allows gases to escape from the substrate (Fig. 6). Casing material, pH adjusted to 7.8-7.9 with chalk powder, is sterilized in autoclave at 15 p.s.i/inch for one hour or chemically treated with formaldehyde solution (2%) about a week in advance of casing. Treated casing is covered with polythene sheet to facilitate the action of formaldehyde and also to avoid its escape in the atmosphere. Soil is turned at an interval of 2 days so that at the time of casing, it is free from formalin fumes. For casing, bags top is made uniform by ruffling top surface and spraying with carbendazim (0.1%) + formaldehyde (0.5%) solution. Casing material is sprayed with above chemicals to saturation level. Temperature 30- 35°C and RH 80-90% are maintained thereafter for entire cropping cycle. Various kinds of casing materials are used for fruit body induction and their suitability and availability. Generally loam soil + farm yard manure (1:1) is used as casing material for its cultivations. Sharma *et al.* (1997) recommended the use of 2 years old cow dung patties. Pandey and Tewari (1994) suggested use of garden soil, coir dust and FYM as casing material for cultivation of *C. indica*. Prasuna (2002) recommended clay loam garden soil with 12% calcium carbonate under A.P. conditions. Shukla (2003) studied different casing materials and obtained higher yield of milky mushroom from biogas slurry + soil (1:1) and cow dung manure + soil (1:1). Sharma *et al.* (1997) and Shukla (2003) have studied the effect of casing thickness on yield of *C. indica* and they obtained best results from 1-1.5 inch thick casing soil. After completing the process of casing, a light irrigation should be given on bed surface.

5.8. Cropping

After casing, the beds are to be incubated over racks in a partially sunken chamber lined with blue coloured high-density polythene sheet as roofing material. Optimum relative humidity of 80-95%, room temperature of 24-28°C and light intensity of about 1600 –3200 lux should be maintained in the cropping room. Proper ventilation for gaseous exchange is also essential in this chamber. The beds are regularly sprayed with water to maintain 50-60% moisture level on the casing surface. Pinheads appear in 8-10 days after casing and the first harvest can be made in 6-8 days after pinhead formation (Fig. 7). After obtaining the first harvest the casing medium is gently ruffled, slightly compacted back and sprayed regularly with water. Second and third harvest may be obtained within 45-50 days of bed preparation. Then the beds are removed and fresh beds may be kept for cropping. Mushrooms with 7-8 cm diameter are harvested by twisting, cleaned and packed in perforated polythene/polypropylene bags for marketing (Figs. 8 and 9).

6. CROP MANAGEMENT

In crop management three aspects are included *i.e.*, temperature, humidity and aeration.

6.1. Temperature

Mycelial growth of *C. indica* can take place in between 20-35°C but after casing, temperature of cropping room should be maintained at 25-35°C for whole cropping period.

6.2. Humidity

The humidity in the cropping room is maintained more there 80 percent during cropping period. Less humidity in cropping room increase the evaporation of moisture from fruit body which gives scaly surface of fruit bodies and killing of pin heads. The humidity is maintained by spraying of water on bed surface, wall and floors. If humidity in cropping room is very low then it can be maintained by spraying of water 4-5 times in a day.

6.3. Aeration

Ventilation is required and keep it at an optimum level that will maintain proper O_2-CO_2 concentration, desired humidity and adequate light is essential in cropping room.

6.4. Harvesting and Post-Harvest Technology

The primordial formation will be started after 10-12 days of casing on upper surface of beds and they will mature within 5-7 days. The mature sporophores should be harvested when they start releasing the spores. The fruit bodies harvested just by twisting so that stubs are not left on the straw and small fruit body will remain unaffected. After harvesting, remove adhering soil particle and straw debris, cut the lower portion of the stalk by a knife. The fresh mushrooms are packed in polythene bag for marketing. The fruit bodies can be stored in a refrigerator for 7-8 days in good condition. This mushroom has shelf life superior to other mushrooms. The mushroom can be dehydrated in the sun or in hot air oven at 45-50°C without change in colour. The dried product can be stored for 2-3 years in an air tight container. The dried produce can be rehydrated by dipping in lukewarm water within 15-20 minutes and give 80-90 percent original weight. Approximately 600-750 g fresh sporophores can be obtained per kg of dry substrate.

6.5. Diseases and Pests

During spawn-run and fructification several competitors moulds and fungi are encountered which may hamper mycelial growth and yield is adversely affected. The planting spawn also get contaminated (Fig. 10).

6.5.1. *Competitor fungi*

A number of fungi *i.e., Aspergillus niger, A. fumigatus, Aspergillus* spp., *Penicillium digitatum, Fusarium solani, Rhizopus stolonifer, Rhizoctonia solani, Trichoderma* spp. and *Coprinus* spp. have been observed during spawn run of *C. indica*

6.5.2. *Pathogens*

A bacterium sometimes infect the button stage of *C. indica* due to bacterial attack most of the buttons are killed and yellowing stipes can be noticed. To prevent this condition infected buttons may be removed from the beds or chlorinated water can be used for irrigation. Sufficient irrigation and ventilation may also be helpful to avoid or reduced the disease incidence.

7. USE OF SPENT STRAW OF *CALOCYBE INDICA*

The spent straw from *Calocybe indica* can be used.

7.1. In Biogas

Spent substrate can be used for biogas production. It has been experimentally proved that production of biogas increases by use of spent substrate 192.27 lit/kg dry matter in 40 days retention period.

7.2. As Biofertilizers

Spent substrate can also be used as manure in kitchen garden and in field. It is hoped that the avocation of mushroom farming will become a very important cottage industry activity in the integrated rural development programme, which will lead to the economic betterment of not only small farmers but also of landless labourers and other weak sections of communities. The advantages of mushroom cultivation can be summarized as:

1. Wastes such as cereal straws are largely burnt by the farmers, which causes air pollution. However, these raw materials can actually be

used for the cultivation of mushrooms. This kind of bioconversion exercise can greatly reduce environmental pollution.

2. Mushroom cultivation can be a labour intensive activity. Therefore, it will serve as means of generating employment, particularly for rural women and youths in order to raise their social status. It will also provide additional work for the farmers during winter months when the farming schedule is light.
3. It will provide the people with an additional vegetable of high quality, and enrich the diet with high quality proteins, minerals and vitamins which can be of direct benefit to the human health and fitness. The extractable bioactive compounds from medicinal mushrooms would enhance human s immune systems and improve their quality of life.
4. Mushroom cultivation is a cash crop. The harvested fruiting bodies can be sold in local markets for additional family income or exported for an important source of foreign exchange that will definitely improve the economic standards of the people.
5. Some warm mushrooms, *e.g., Volvariella volvacea* (straw mushrooms) and *Pleurotus sajor-caju* (oyster mushrooms) are relatively fast growing organisms and can be harvested in 3 to 4 weeks after spawning. It is a short return agricultural business and can be of immediate benefit to the community.

8. CONSTRAINTS IN PRODUCTION OF MILKY MUSHROOM

However, there are few constraints in production of tropical mushrooms and further expansion of mushroom industry in India.

(i) Unawareness about nutritional and medicinal values of mushrooms through mass media. Lack of awareness to incorporate in mid day meals, in regular diet of school children, in daliya, in Dal Bhat Centres, in Anganwadi Programme, to pregnant ladies, old age people, in Govt. canteens and hospital Wards to alleviate undernourishment and malnourishment.

(ii) Lack of good quality mushroom spawn laboratory, cold storage facility, and small scale processing units in a district place to facilitate mushroom production, processing and marketing.

(iii) Unawareness to blend mushroom powder in Atta, dalia, or dal, papad, badi, murku to improve the nutritional qualities of the food/food product.

(iv) Lack of mushroom processing units to enhance the shelf life, prolong the availability of mushroom products and promote indirect marketing of mushrooms.

(v) Unavailability of cheap and easy mushroom production technology to people who are mainly dependent on forest produce including mushrooms.

(vi) Absence of proper mushroom marketing channels to dispose off of the fresh, dehydrated and processed mushrooms involving Self Help Groups, NGO s and Private Organizations.

(vii) Lack of dissemination of generated mushroom technology/knowledge through recent means of communication like internet etc. to the entrepreneurs, tribal women, school dropouts, consumers and preferably middle class families.

(viii) Lack of efforts to trap solar energy for drying and processing of mushrooms. Improper development of cultivation methodology of mushroom in a identified areas looking for the availability of local resources.

(ix) Lack of proper designing and construction of mushroom huts/farms for small, medium and commercial scale cultivation of different mushrooms under different climates. Appropriate agencies are not involved in taking care of local ethos and culture and also considering maximum community involvement and community participation.

9. CONCLUSIONS

Mushroom cultivation represent one of the economically viable processes for the bioconversion of agricultural and agro-industrial wastes into protein rich food making it a potent weapon against malnutrition in developing countries like India, which consume lowest per capita protein in the world. Edible mushroom production represented an attractive method of improving the nutritional quality of lignocellulosic wastes. Mushrooms are cultivated in controlled biological environment and it has been extensively used as food since ancient time, due to nutritive and medicinal values. Although mushrooms contain low amount of calories and fat, they are rich in protein and fibre. Mushrooms are the intermittent food between vegetable and fish. Mushrooms derive all of their energy and growth materials from their growth medium, through biochemical decomposition processes. In view of the current energy crisis it has become most important to make a substantial breakthrough in the technology of food production to meet a serious food deficit situation. The cultivation of mushroom is a way to overcome the problem of pollution caused by the different crop residues, because the fungi have the ability to recycle these wastes and convert them into edible protein. So the cultivation of mushroom can also be helpful to overcome the shortage of food. Protein deficiency is not only a future problem but also an existing reality. Mushrooms provide a rich addition of nutrients to the diet in the form of protein, carbohydrate, valuable salts and vitamins.

10. REFERENCES

Alam, N., Amin, R., Khana, A., Ara, I., Shim, M.J., Lee, M.W. and Lee, T.S. 2008. Nutritional analysis of cultivated mushrooms in Bangladesh: *Pleurotus ostreatus, Pleurotus sajor-caju, Pleurotus florida* and *Calocybe indica*. *Mycobiol*., 36: 228–232.

Anandh, K. and Prakasam, V. 2002. *Tricholoma lobayense*, A new edible mushroom for commercial exploitation. *In:* Abstracts of paper presented in *3rd Indian Mushroom Conference Held during 6-7, March-2002 at Tamil Nadu Agricultural University, Coimbatore*, p. 65.

Breene, W. 1990. Nutritional and medicinal value of specialty mushroom. *J. Food Prot.,* 53: 883–894.

Chakraborty, U. and Sikdar, S.R. 2010. Intergenic protoplast fusion between *Calocybe indica* (milky mushroom) and *Pleurotus florida* aids in the qualitative and quantitative improvement of sporophore of milky mushroom. *World J. Microbial. Biotechnol.,* 26: 213–225.

Chakravarty, D.K. and Sarkar, B.B. 1982. *Tricholoma lobayense* – A new edible mushroom from India. *Curr. Sci.*, 53: 531–532.

Chiu, S.W. and Moore, D. 2001. Threats to biodiversity caused by traditional mushroom cultivation in China. *In:* Fungal conservation (Moore Nauta and Rotheroe (*eds.*)), The 21st Century Issue, Cambridge.

Doshi, A., Sharma, S.S. and Trivedi, A. 1993. A promising edible mushroom for the tropics: *Calocybe indica* P&C. *Mushroom Information*, 86(5): 14–22.

Doshi, A., Sidana, N. and Chakraborty, B.P. 1987. Cultivation of summer mushroom *Calocybe indica* P&C in Rajasthan. *Proceedings of the Twelfth International Congress on the Science and Cultivation of edible fungi, Braunschweig, Germany*. *Mushroom Science*, XII(II): 395–400.

Krishnamoorthy, A.S. 1995. Studies on the cultivation of milky mushroom, *Calocybe indica* P&C., Ph.D. Thesis, Tamil Nadu Agricultural University, Coimbatore, India, p. 124.

Krishnamoorthy, A.S. and Amutha, G. 2007. Potential of milky mushroom in the mushroom crop diversification in the tropical regions. *In: Mushroom Biology and Biotechnology,* Singh Rai, Yadav and Tewari, (*eds.*), MSI, Solan (H.P.), India, pp. 215–227.

Lakshmipathy, G., Jayakumar, A., Abhilash, M. and Raj, S.P. 2012. Optimization of growth parameters for increased yield of the edible mushroom *Calocybe indica*. *African Journal of Biotechnology,* 11(11): 7701–7710.

Li, Yu. 2012. Present development situation and tendency of edible mushroom industry in China. *Proceedings of the 18th Congress of the International Society for Mushroom Science* (Zhang Jinxia, Hexiang, W. and Mingjie, C. (*eds.*)), China Agriculture Press, pp. 1–9.

Mallavadhani, U.V., Sudhakar, A.V., Satyanarayana, K.V.S., Mahapatra, A., Li, W. and Van-Breemen, R.B. 2006. Chemical and analytical screening of some edible mushrooms. *Food Chem.,* 95: 58–64.

Martinez-carrera, D., Aguilar, A. and Martinez, Z. 2000. Commercial production and marketing of edible mushrooms cultivated on coffee pulp in Mexico. *In: Coffee Biotechnology and Quality* (Soccol Serra and Pandey *(eds.),* Kluwer Academic Publishers, Dordrecht.

Pandey, M. and Tewari, R.P. 1994. Evaluation of casing materials for *Calocybe indica* cultivation. *Paper Presented in National Symposium on Mushrooms held at NCMRT, Solan from 8-10th*, p. 74.

Pani, B.K. 2012. Sporophore production of milky mushroom (*Calocybe indica)* as influenced by depth and time of casing. *Int. J. Advanced Bot. Res.*, 2(1): 168–170.

Prasuna, K. 2002. Studies on nutritional and environmental factors affecting growth and yield of *Calocybe indica*. Ph.D. Thesis submitted to Osmania University, Hyderabad, p. 176.

Purkayastha, R.P. and Chandra, A. 1974. A new technique for *in vitro* production of *Calocybe indica,* an edible mushroom from India. *The Mushroom Journal*, 40: 1–2.

Purkayastha, R.P. and Chandra, A. 1974. New species of edible mushroom from India. *Trans. Br. Mycol. Soc.*, 62 (2): 415–418.

Purkayastha, R.P. and Chandra, A. 1985. *Manual of Indian Edible Mushrooms.* Today and Tomorrow's Printers and Publishers, New Delhi, p. 267.

Purkayastha, R.P. and Nayak, D. 1981a. Development of cultivation method and analysis of proteins of a promising edible mushroom *Calocybe indica. Mushroom Science*, XI (2): 697–713.

Purkayastha, R.P. and Nayak, D. 1981b. Analysis of protein patterns of an edible mushroom by gel-electrophoresis and its amino acid composition. *J. Food. Sci. Technol.*, 18: 89.

Sharma, S.S., Doshi, A. Trivedi, A. and Kothari, K.L. 1995. Studies on spawn research of *Calocybe indica* P&C. *Indian Bot. Reptr.,* 14(1+2): 97–100.

Sharma, S.S., Doshi, A., Trivedi, A. and Kothari, K.L. 1997. Evaluation of the suitability of different casing materials for *Calocybe indica* P&C. *Mushroom Res.*, 6(2): 81–82.

Shukla, C.S. 2003. Role of agronomical and biochemical parameters on growth and yield of *Calocybe indica* (P&C) Ph.D. Thesis submitted to IGAU, Raipur, p. 168.

Shukla, C.S. 2004. Role of agronomical and biochemical parameters on growth and yield of on *Calocybe indica* (P&C). Ph.D. Thesis submitted to Dept. of Plant Pathology, IGKV, Raipur, p. 148.

Sinden, J.W. 1972. Mushroom spawn and method of making the same. *U.S. Patent*, 1(869): 517.

Singh, M. 2011. Mushroom Production: An Agribusiness Activity. *In: Mushroom-Cultivation, Marketing and Consumption* (Singh Manjeet, Bhuvnesh, V., Shwet Kamal and Wakchaure, G.C. (*eds.*)), Directorate of Mushroom Research, ICAR, Solan (H.P.), pp. 1–10.

Thakur, M.P. 1998. Food and medicinal values of mushrooms. *In: Health Care and Development of Medicinal Plants* (Puri, S. and William, W.J. (*eds.*)), Baba Printers, Raipur, pp. 107–119.

Thakur, M.P. and Mohapatra, K.B. 2013. Presented a key note address on tropical mushrooms: Present status, constraints and success story in Technical Session IV on Indian Mushroom Conference-2013 organized by Mushroom Society of India in collaboration with Directorate of Mushroom Research, Solan and Punjab Agricultural University, Ludhiana from 16-17th April, 2013, pp. 42–43 (Abs.).

Thakur, M.P. and Singh, H.K. 2013. Mushrooms, their bioactive compounds and medicinal uses: A review. *Medicinal Plants*, 5(1): 1–20.

Trivedi, A., Sharma, S.S. and Doshi, A. 1991. Cultivation of *Calocybe indica* under semi-arid conditions. *Indian Mushrooms*, pp. 167–168.

Verma, R.N. 2013. Indian mushroom industry- past and present. Bulletin 8 of World Society for Mushroom Biology and Mushroom Products, p. 16.

Verma, R.N. Uphadhyay, R.C., Singh, S.K. and Rai, R.D. 2003. Genetic resources of commercial mushrooms, their conservation, characterization and improvement. *In: Current Vistas in Mushroom Biology and Production.* Mushroom Society of India, pp. 1–9.

Wakchaure, G.C. 2011. Production and marketing of mushrooms: Global and National Scenario. *In:* Mushroom-Cultivation, Marketing And Consumption (Singh Manjeet, Bhuvnesh, V., Shwet Kamal and Wakchaure, G.C. (*eds.*)), Directorate of Mushroom Research, ICAR, Solan (H.P.), pp. 15–22.

Section-3

Pre-Harvest and Post-Harvest Management of Mushroom Crop

3.1

Storage and Processing of Edible Mushrooms

BERNAŚ EMILIA[1*] AND JAWORSKA GRAŻYNA[2]

ABSTRACT

Fresh mushrooms are highly perishable and cannot be stored for long periods; moreover, during processing they readily change colour and texture. Among the most frequently used methods of preserving mushrooms are freezing, canning (either in brine or marinade), drying, frying and braising. In order to ensure that processed mushrooms are of acceptable quality, fresh fruiting body should be kept at low temperatures, preferably in a Modified Atmosphere (MAP). Among the various methods of processing, freezing causes the greatest difficulty. The most challenging technological problem in freezing mushrooms, particularly Agaricus bisporus, is the prevention of changes in colour (mainly enzymatic). In preliminary processing (washing, soaking, blanching or vacuum impregnation) the food industry generally uses metabisulphites; however, their residues are potentially harmful to human health, causing allergic-type reactions. Other substances used are table salt, citric acid, L-ascorbic acid, hydrogen peroxide, EDTA, sodium isoascorbate and cysteine hydrochloride.

Key words: Edible mushrooms, Storage, Preliminary treatment, Processing.

1. INTRODUCTION

Edible mushrooms are a popular food product in many countries because of their unique taste. They are also valued for their medicinal properties,

[1] Department of Fruits, Vegetables and Mushrooms Technology, Agricultural University of Krakow, 122 Balicka Street, 30-149 Krakow, Poland.

[2] Department of Plant Food Technology and Crop Quality, University of Rzeszow, 1 Ćwiklińskiej Street, 35-601 Rzeszow, Poland.

**Corresponding author:* E-mail: emilia.bernas@urk.edu.pl

which makes their use as a functional food, particularly significant in certain Asian countries, such as China and Japan. A number of studies have documented the role of mushrooms in the treatment of cancer and other diseases (Solomon *et al.*, 1999; Wasser and Weis, 1999; Manzi and Pizzoferrato, 2000; Barros *et al.*, 2008 a, b). The global consumption of mushrooms is increasing each year; in 1997 average global consumption was 1 kg per capita per year, but by 2012 this had risen to 3-4 kg.

Mushroom consumption is highest in China, at 10 kg per capita per year. In the USA, Canada, Japan and Australia mushrooms are mainly sourced internally, but also imported (Royse, 2014). The species most widely consumed in Europe and the USA are *Agaricus bisporus* (button) and *Pleurotus ostreatus* (Oyster) (Walde *et al.*, 2006; Aida *et al.*, 2009) while in Asian countries the main species are *Pleurotus ostreatus* (Oyster), *Lentinula edodes* (Shiitake), *Agaricus bisporus* (button), *Flammulina velutipes* and *Auricularia polytricha* (Wood ear, black fungus) (Wu *et al.*, 2013).

In order to keep pace with increasing demand, the global production of mushrooms has also been rising steadily, from over 5 mln tonnes in 2004 to 10 mln tonnes in 2013. For many years China has been the leading producer of mushrooms (71% of global production in 2013), followed by Italy (8%), the USA (4%), the Netherlands (3%) and Poland (2%). Trade in mushrooms is most common in the EU and the Russian Federation. The main exporters of mushrooms are Poland and the Netherlands, while the chief importers are the UK and Germany (FAOStat, 16/02/2015).

With the advance of globalization and increased competition, producers of fresh and processed mushrooms are having to devise new methods of cultivating and processing mushrooms which will enable them to sell their products at competitive prices, even in distant markets. A fundamental problem affecting the distribution of fresh mushrooms is their perishability; they retain their freshness for only a few days, even when chilled (Zivanivic *et al.*, 2000). Consequently, considerable efforts have been made to develop methods of extending their freshness while maintaining high quality.

2. STORAGE OF FRESH MUSHROOMS

Edible mushrooms are highly perishable; the rapid deterioration in quality is mainly caused by high moisture content and enzymatic activity, but also by microflora and biological processes (Burton and Noble, 1993; Burton *et al.*, 1995; Evered *et al.*, 1995; Jolived *et al.*, 1995; Plaza *et al.*, 1995; Czapski, 2000; Jiang, 2013). Fresh mushrooms undergo a number of adverse changes, which manifest themselves mainly in a deterioration of their sensory qualities, chiefly colour and texture, but also in a fall in moisture content through evaporation, cap opening, stipe elongation and a decrease in nutritional value, including health-promoting components (Lopez-Briones *et al.*, 1992; Burton and Noble, 1993; Minato *et al.*, 1999; Czapski, 2000; 2001).

2.1. Respiration

Edible mushrooms have a high respiratory coefficient, one consequence of which is rapid post-harvest quality deterioration, particularly when stored at ambient temperature (Burton, 1991; Kar *et al.*, 2004; Walde *et al.*, 2006; Kotwaliwale *et al.*, 2007; Sapata *et al.*, 2009; Argyropoulos *et al.*, 2011; Kumar *et al.*, 2013). This results in a decrease in nutritional value and a deterioration in organoleptic qualities, as well as the potential development of toxins and carcinogens (Burton, 1993). The growth of microorganisms and deterioration of quality depend, to a large extent, on harvest and storage conditions (Sapers *et al.*, 2001). Another potentially significant factor is the embellishment of mushrooms by removing any remaining substrate or peeling the caps (Nichols, 1985). Such procedures damage the tissue, which, together with high moisture content, enzymatic activity and microbiological contamination, can lead to extensive decomposition, rendering mushrooms unfit for consumption or processing (Beelman *et al.*, 1989; Brennan and Gormley, 1998).

Like fruits and vegetables, edible mushrooms have a high post-harvest respiration rate, which, through the mediation of O_2 intake, results in the decomposition of carbohydrates, fats and organic acids into simple compounds while giving off CO_2, moisture and energy (Foncesca *et al.*, 2002; Ares *et al.*, 2007). Respiration rate is usually expressed in terms of the rate of O_2 consumption or CO_2 production and is described as Respiratory Quotient (RQ) (metabolic activity). RQ is significantly affected by the temperature and length of storage. An RQ value of about 1.0 signifies aerobic respiration, while a value of over 1.0 indicates anaerobic metabolism. Villaescausa and Gil (2003) found that after 12 hours of storage, the respiratory quotient of *Pleurotus* mushrooms was 0.50, 0.99 and 1.23 μmol CO_2/kg/s at 0, 4 and 7°C, respectively. These initial high respiration rates were associated with harvest stress caused by the cutting process, which in this experiment was carried out in the laboratory just before storage. After a further 2 days of storage, the RQ decreased by about 50%, remaining stable until the 7th day, with a further decrease between the 7th and 10th days. At the end of the storage period, the respiration rates were very similar for all the assayed temperatures: 0.37 μmol CO_2/kg/s. The RQ of edible mushrooms is higher than that of fruits and vegetables (Foncesca *et al.*, 2002). The data in the literature indicate a disparity in respiration rates between different species of mushrooms; for example, the RQ for shiitake mushrooms was 10 times higher than that for *A. bisporus* during storage at 10°C (Ares *et al.*, 2006; Parantelli *et al.*, 2007).

2.2. Colour and Texture Changes

The deterioration of colour and texture of fresh mushrooms which occurs during storage is a major factor in the overall deterioration of mushroom quality (Lopez-Briones *et al.*, 1992; 1993; Braaksma *et al.*, 1994; Villaescusa and Gil, 2003; Ares *et al.*, 2006; Parentelli *et al.*, 2007). Browning results from the transmutation of phenolic compounds to melanin mediated by Polyphenol Oxidase Enzymes (PPO) (Jolivet *et al.*, 1998; Kotwaliwale *et al.*, 2007; Kumar *et al.*, 2013). According to Weijn *et al.* (2013), there are four pathways to melanin synthesis in mushrooms, with the transformation of chorismate to γ-L-glutaminyl-4-hydroxybenzene (GHB) -melanins being the most important factor in browning. Initially, enzymes including γ-glutamyltransferase are involved in the transformation of chorismate to γ-L-glutaminyl-4-hydroxybenzene (GHB), which is the main phenolic compound in *A. bisporus*. GHB is then oxidized into the corresponding diphenol (γ-L-glutaminyl-3,4-dihydroxybenzene, GDHB) and subsequently into o-quinone (γ-L-glutaminyl-3,4-benzoquinone, GBQ), which polymerizes into GHB-melanins.

Enzymatic darkening in mushrooms is mainly the result of tissue damage caused by rough handling, as well as ageing and bacterial contamination. Polyphenol Oxidase Enzymes (PPO) are metalloenzymes, which include tyrosinases, catechol oxidases and laccases. They are most active at pH levels of 5.0-7.0, which, in view of their alkaline nature, also affects mushrooms. In *Agaricus bisporus* laccase occurs at very low levels, while tyrosinase is dominant (Ratcliffe *et al.*, 1994; Kolcuoglu *et al.*, 2007). Mushroom tyrosinase plays an important role in the production and stability of spores, is involved in anti-viral defence mechanisms and imparts colour to fruiting bodies (Halaouli *et al.*, 2006). In plants, polyphenol oxidase is located in the chloroplast thylakoids, while in mushrooms it is found in the cytoplasm. In a study on *Agaricus bisporus*, Espin and Wichers (1999) showed that mushroom tyrosinase is synthesized mainly in its latent form (67 kDa), which is gradually converted into the active form (43 kDa) by endogenous protease. According to Leeuwen and Wichers (1999), the latent form may comprise as much as 95% of total polyphenol oxidase activity in mushrooms. Latent PPO can be activated by different treatments or agents, which include detergents (sodium dodecyl sulphate) (Czapski, 1994 b; 1998), fatty acids (Sugumaran and Nellaiappan, 1991), trypsin (Sanchez-Ferrer *et al.*, 1989), and by divalent cations such as Ca^{2+}, Mn^{2+} and Mg^{2+} (Jimenez and Garcia-Carmona, 1993). Flurkey and Ingebrigtsen (1989) showed that the activity of the latent from of tyrosinase is more closely linked to the activity of dopa oxidase than catechol oxidase or tyrosine hydroxylase. They also found that a decrease in dopa oxidase activity may be linked to an increase in the activity of the latent form.

The general view at present is that the rapid post-harvest browning of mushrooms is caused by the mechanisms described above, leading to adverse effects on their sensory and nutritional qualities, fitness for consumption and market value. Melanin formation can be minimized mainly by inhibiting PPO, by reducing o-quinones to diphenols or Cu^{2+} to Cu^{+}, by interacting with the formation of o-quinone product and by decreasing the uptake of O_2 for the reaction (Kermasha *et al.*, 1993; Czapski, 1994 a). AA-2-phosphates are effective against tissue browning in apples and potatoes but not in mushrooms, probably due to their high acidic phosphatase activity (Sapers and Miller, 1992). Czapski (1999) found that salicylhydroxamic acid (SHAM) is a strong inhibitor of catechol oxidase in mushrooms. According to Sapers *et al.* (1994), consecutive rinsing in solutions of sodium erythorbate, cysteine and EDTA provides good protection against browning in minimally processed *Agaricus bisporus*. β cytokinin can also be an inhibitor of PPO by chemical modification of the phenolic substrates of PPO (Hicks *et al.*, 1996). Sapers *et al.* (1989) showed that cinnamic acid and benzoic acid are very good inhibitors of PPO. L-ascorbic acid and sulphur-containing compounds such as L-cysteine and sodium metabisulphite are also widely used as inhibitors of browning in mushrooms (Sapers *et al.*, 1989; Sapers, 1993; Vamos-Vigyazo, 1995). Other potential inhibitors of enzymatic browning may be honey and propolis, which are used with apple slices and grape juice (Oszmiański and Lee, 1990). Billaud *et al.* (2003) found that products resulting from the Maillard reaction can be strong inhibitors of PPO, their inhibitory capacity increasing with the time and temperature of thermal treatment.

According to Palacios *et al.* (2011), enzymatic reactions leading to browning involve not only the enzymes themselves but also their substrates, such as amino acids or phenolic compounds. Using a liquid chromatography coupled to a DAD detector, it can be observed that the dominant phenolic compounds in mushrooms are phenolic acids, which, depending on the species, include caffeic acid, chlorogenic acid, gallic acid, p-coumaric acid, ferulic acid, gentisic acid, homogentisic acid and protocatechuic acid, while other polyphenols include pyrogallol, catechin and γ-L-glutaminyl-4-hydroxybenzene (GHB). It should, however, be noted that there are only a few studies in this field. Authors conducting research into PPO activity in mushrooms consider the following to be the main substrates of this enzyme: γ-L-glutaminyl-4-hydroxybenzene (GHB), L-phenylalanine, L-tyrosine, L-DOPA, 4-methylcatechol, chlorogenic acid, catechin, coumaric acid and phlorizin (Espin *et al.*, 1998; Rodriguez-Lopez *et al.*, 1999; Chang, 2007; Weijn *et al.*, 2013).

Apart from polyphenol oxidase, mushrooms also contain peroxidase and catalase (Kadiri and Fasidi, 1990; Choi and Sapers, 1994; Jiménez and Garcia-Carmona, 1997; Espin and Wichers, 1999; Bernaœ and Jaworska, 2015), whose activity, according to Kadiri and Fasidi (1990), increases with the age of the mushroom. Lipases, inverting enzymes and proteolytic enzymes can also occur (Ratcliffe *et al.*, 1994). According to Ratcliffe *et al.* (1994), the activity of this group of enzymes varies according to mushroom species: for example, polyphenol oxidase activity is higher in *Agaricus bisporus* than in shiitake or oyster mushrooms. Moreover, Rajarathnam *et al.* (2003) maintain that while tyrosinase is characteristic in *Agaricus bisporus,* it does not occur in *Pleurotus florida*.

Changes in texture occurring during post-harvest storage are connected with losses in cell membrane barrier function, which has been described as a loss of softness and firmness (Beelman *et al.*, 1987). Zivanowic *et al.* (2000) hold that textural changes are linked to the decomposition of proteins and polysaccharides, damage to cell vacuoles, shrinking of hyphae and an increase in intercellular spaces. During storage, changes occur in the cohesiveness/stiffness of mushrooms (Zivanowic *et al.*, 2000; Parentelli *et al.*, 2007), which are caused by an increase in chitin in mushroom tissue and the formation of covalent bonds with β -glucans, resulting in increased stiffness of the tissue (Zivanowic *et al.*, 2000).

2.3. Microbiological Changes

Fresh edible mushrooms are prone to high levels of microbiological contamination; microorganism populations found on mushroom surfaces range from 1.9×10^6 to 1.5×10^7/g fresh weight (Dores *et al.*, 1986). *Pseudomonas* occur in the highest numbers (54% of total bacterial count), followed by *Flavobacterium* (10%). According to Chikthimmah and Beelman (2006), the following may also be present on mushrooms: *Campylobacter jejuni, Listeria monocytogenes, Salmonella* spp. and *Escherichia coli.* Moreover, *Clostridium botulinum* may proliferate in anaerobic conditions, causing food poisoning (Austin *et al.,* 1998; Farber *et al.*, 2003). Halamia *et al.* (1999) isolated 25 strains of *Lactobacillus* spp. and *Pediococcus* spp., while Chikthimmah and Beelman (2006) detected the presence of yeasts and moulds. Brennan *et al.* (2000) and Guillamon *et al.* (2010) state that microorganisms, notably *Pseudomonas tolaasii, Pseudomonas fluorescens,* moulds and *Candida sake* yeasts, are responsible for mushroom tissue decomposition. On the other hand, *Pseudomonas tolaasii* bacteria are responsible for changes in mushroom colour known as bacterial blotch (Soler-Rivas *et al.*, 2009). Parenteli *et al.* (2007) observed that in shiitake mushrooms packed in a modified atmosphere (5% O_2 and 2.5% CO_2) and stored for 20 days at 5°C, the growth of aerobic mesophilic bacteria was slowed for the first five days, and increased only slightly thereafter. The above authors also found that the increase in lactic bacteria counts was also small, and levels did not exceed 10^4 cfu/g.

2.4. Methods Used to Extend the Shelf Life of Fresh Mushrooms

2.4.1. *Controlled and modified atmosphere*

Storage conditions are an important factor in maintaining mushroom quality. The best results are achieved by storing fresh mushrooms in a cold room at 0-2°C and relative humidity of 90% (Lopez-Briones *et al.*, 1992; Minato *et al.*, 1999). Mushrooms can also be packed in a Modified Atmosphere (MAP) (Amodio *et al.*, 2003; Choi and Kim, 2003; Kim *et al.*, 2006). This technology allows fresh or minimally processed foods to retain their freshness for longer, provided that the operation uses MAP gases of the appropriate composition and packaging film with the correct permeability of oxygen, carbon dioxide and water vapour (Barron *et al.*, 2002).

A modified atmosphere has a composition different from atmospheric air (Philips, 1996; Farber *et al.*, 2003). There are two methods of storing foods in a modified atmosphere: Controlled Atmosphere Storage (CAS), and Modified Atmosphere Packaging (MAP). In the former case, foods are stored in chilled conditions at atmospheric pressure and at a constant temperature, and in an atmosphere whose composition is constantly controlled. In the latter, foods are packed in gas-permeable polymeric films which allow respiration of the product, thus providing a modified atmosphere within the packaging (Foncesca *et al.*, 2002; Farber *et al.*, 2003; Ares *et al.*, 2007). A modified atmosphere within the packaging may be achieved actively or passively. In the former, the desired equilibrium atmosphere is generated and the packaged product is immediately exposed to it, while in the latter, the equilibrium atmosphere is achieved by respiration during storage (Farber *et al.*, 2003).

A modified atmosphere with increased levels of CO_2 and decreased levels of O_2 potentially lowers the respiration rate of the product and the production of ethylene, and retards organoleptic changes, decomposition processes and physiological changes (Saltveit, 1997; Gorris and Tauscher, 1999; Farber *et al.*, 2003), as well as slowing the growth of microorganisms (Farber *et al.*, 2003). Lower levels of oxygen lower the respiration rate of fresh produce and inhibit the activity of oxidases such as polyphenol oxidase, ascorbic acid oxidase and glycolic acid oxidase (Kader, 1986). The composition of the modified atmosphere depends on the product, the packaging material and the storage temperature. Oxygen concentration in MAP systems is maintained within the range 1-10% (Table 1), and sometimes below 1%. An oxygen level of below 1% may result in a deterioration of product texture, the generation of unpleasant odours (Lee *et al.*, 1995; Austin *et al.*, 1998) and an increase in the anaerobic bacteria, *Clostridium botulinum* (Austin *et al.*, 1998; Farber *et al.*, 2003). Instead of reducing oxygen levels, products are sometimes packed in a 100% oxygen atmosphere, referred to as 'oxygen shock' treatment, which is intended to inhibit bacterial growth, prevent enzymatic discolouration and inhibit the growth of anaerobic microorgnisms (Day, 1996; Kader and Ben-Yehoshua, 2000; Jiang *et al.*, 2013).

Carbon dioxide levels in a modified atmosphere and their effect on packed food depend on the stage of development of the raw material, the concentration and the length of exposure (Foncesca *et al.*, 2002). Foncesca *et al.* (2002) stated that carbon dioxide lowers respiration rates of mushrooms, which is explained by two hypotheses: simple feedback inhibition and elevated CO_2 concentrations, which might affect the Krebs cycle intermediates and enzymes (Ares *et al.*, 2007). Carbon dioxide is used in modified atmospheres mainly for its ability to inhibit microorganism growth (Philips, 1996). According to Faber (1991), carbon dioxide causes changes in the functioning of microorganism cell membranes, impeding the absorption/assimilation of basic nutritional compounds. It inactivates certain enzymes, thus retarding enzymatic reactions. Moreover, CO_2 changes the pH of microorganism cell membranes as it permeates them and affects the physiochemical properties of proteins. The proliferation of the aerobic bacteria, *Pseudomonas*, is inhibited by carbon monoxide concentrations of 10 to 20%; however, high concentrations of CO_2 may stimulate the growth of lactic fermentation bacteria (Faber, 1991; Lee *et al.*, 1995; Faber *et al.*, 2003). Nitrogen may be used to replace the oxygen in modified atmosphere packages (Parry, 1993).

Table 1. The modified atmosphere composition (MAP) for mushroom storage

	Atmosphere composition [%]		
	O_2	CO_2	
Earth's atmosphere	>0.01	21	
Agaricus bisporus	5	5	Cliffe-Byrnes and Beirne (2007)
	100	0	Jiang *et al.* (2013)
	10	5	Koushki *et al.* (2011)
	5–10	2.5–5	Lopez-Briones *et al.* (1992)
	3.6	11.5	
	6	2–6	Roy *et al.* (1995)
	5	3	Tao *et al.* (2006)
Agrocybe chaxingu	5	10	Li *et al.* (2008)
Lentinus edodes	5	2.5	Parentelli *et al.* (2007)
	2	2	Kim *et al.* (1989)
	1	40	Minamide *et al.* (1980)
	5	2.5	Parentelli *et al.* (2007)
Pleurotus florida	10	5	Jafri *et al.* (2013)

The use of MAP with specific products should take into account the type of packaging, its shape and particularly the permeability of the film with regard to carbon dioxide, oxygen and water vapour (Faber *et al.*, 2003). The permeability of specific packaging films is designated at 2°C for carbon dioxide and O_2, and 38°C for water vapour (Greengrass, 1995; Schlimme and Rooney, 1997; Faber *et al.*, 2003). Zanderighi (2001) noted that the permeability of films depends on the type of polymer (number of layers), and the size and shape of the perforations. Apart from permeability, the selection of packaging film is also determined by its durability, air tightness, transparency, resistance to mechanical damage and possibilities for labelling and product presentation (Zagory, 1995). Previously, every effort was made to avoid the product being affected by the packaging material; however, this view has been somewhat modified in the case of so-called 'intelligent packaging', which is designed to have a positive effect on the product (Philips, 1996; Antmann *et al.*, 2008). Antmann *et al.* (2008) showed that in the case of shiitake mushrooms, the selection of packaging should be based on the intended storage period; when stored for up to six days at 5°C, better results were obtained using macroperforated polyethylene bags with 9.0 x 10^3perforations/m^2 (A) than those with 17.0 x 10^3 perforations/m^2 (B). When stored for 12 days, on the other hand, only packaging B could be used since the use of packaging A resulted in an unpleasant and unacceptable odour. According to Parentelli *et al.* (2007), the type of packaging film used for shiitake mushrooms packed in a passively modified atmosphere and stored for 20 days at 5°C had little effect on the growth of yeasts and moulds. However, the type of film did make a difference when the mushrooms were packed in an actively modified atmosphere: samples stored in polypropylene bags always showed low counts of yeast and mould. Similar patterns of fungal growth were observed when air was present (packages with passive modified atmosphere or perforated bags).

The following may be used to pack mushrooms in a modified atmosphere:

1. *Agaricus bisporus*
   a. Perforated PVC film of 12 µm thickness (1 pore of 1 mm of diameter for every 2.5 cm^2 of film surface) (Simon *et al.*, 2010),
   b. Non-perforated PVC film of 12 µm thickness with an O_2 permeability of 25000 ml^{-2} day^{-1} atm^{-1} at 25°C and water vapour transmission rate of 200 g m^{-2} day^{-1} at 25°C (60),
   c. Biaxially Oriented Polypropylene (BOPP) bags (35 µm thickness; O_2 permeance 38x10^{-14} mol s^{-1} m^{-2} Pa^{-1} at 23°C (Simon *et al.*, 2010),
   d. Micro-perforated OPP film (PA-190, 35 µm thick)(Jiang *et al.*, 2011).
2. *Agrocybe chaxingu* (Li *et al.*, 2008)
   a. Polystyrene packaging trays silicon with silicon gum film windows (O_2 permeability of 4.08 x 10^{-9}/mol/s/m^2/Pa and CO_2 permeability of 12.24 x 10^{-9}/mol/s/m^2/Pa at 20°C and 90% RH).
3. *Hypsizygus marmoreus* (Xing *et al.*, 2008)
   a. Polyvinyl chloride film (thickness 11 µm), polyethylene film (thickness 6 and 11 µm), polyoletin film (thickness 10 µm) and biaxially oriented polypropylene film (thickness 20 µm).
4. *Pleurotus florida* (Jafri *et al.*, 2013)
   a. Poly bags made from polyamide 20 µm – polyethylene 70 µm with ethylene vinyl alcohol (EVA) sealant layer. Oxygen and carbon dioxide permeability: 5.8×$10^{"9}$ and 1.7×10^{-8} mL/m^2 /s /Pa (both at 23°C and 0% RH), respectively, and water vapour transmission rate for the poly pouches: 3.2×10^{-5} g/m^2/s (at 23°C and 85% RH).

5. *Pleurotus ostreatus* (Villaescusa and Gil, 2003)
   a. PVC film (0_2 permeance 0.28 10^{-10} mol/ s/m^2/ Pa),
   b. Low-density polyethylene film (0_2 permeance 0.48 10^{-10} mol/ s/m^2/ Pa),
   c. Microperforated polypropylene film (33 holes of 2 mm/dm^2 and 35 µm thickness, (0_2 permeance 9.34 or 3.36 10^{-10} mol/s/m^2/ Pa).
6. *Lentinus edodes*
   a. Polyethylene film ((60 µm thickness, Lee *et al.* (1991) or 30 µm thickness, 350 x 400 mm, Minato *et al.* (1999)),
   b. Polyethylene film (0.04 mm thickness, gas transmission the PE gas transmission rates were: 1078 × 10^{-18} mol/s/m^2/ Pa for O_2, 4134 × 10^{-18} mol/s/m^2/ Pa for CO_2 (both at 20°C and 100% RH), and 2.8 × 10^{-5} to 6.5 × 10^{-5} g/m/ s for H_2O (at 37°C and 90% RH) (Jaing *et al.*, 2011),
   c. Polypropylene film (Ares *et al.*, 2007; Parentelli *et al.*, 2007).

Despite numerous studies, there is still no consensus as to the optimal atmospheric composition for storing the most popular mushrooms, especially *A. bisporus* (Ares *et al.*, 2007, Table 1). Lopez and Briones (1992) maintain that during mushroom storage the concentration of oxygen is less important than that of carbon dioxide, and that the former should be 5 to 10 % and the latter 2.5 to 5%. They argue that a carbon dioxide concentration of 5% delays the ripening of mushrooms, browning and textural changes, while concentrations above 5% can lead to yellowing of the caps. Ye *et al.* (2012) showed that shiitake mushrooms packed in a modified atmosphere containing 2% O_2 and 10% CO_2 and stored for 17 days at 4ºC maintained better quality than those packed in an atmosphere of 2% O_2 and 7% CO_2 or 2% O_2 and 13% CO_2; browning was delayed, the texture was firmer and the content of vitamin C and soluble sugars was higher.

A different method of packing food products is vacuum packing. The reduced level of oxygen in the package results in a lowering of the respiration rate and polyphenol oxidases activity (Kader, 1986). The low accessibility of oxygen also reduces the likelihood of mushroom browning (Ares *et al.*, 2007). However, Lee *et al.* (1995) and Austin *et al.* (1998) are of the opinion that oxygen concentrations of below 1% create anaerobic conditions which can cause damage to the cells and natural metabolism of mushrooms, leading to changes in flavour and unpleasant odours. Low levels of oxygen can also give rise to the toxic pathogenic bacteria, *Clostridium botulinum* (Austin *et al.*, 1998; Faber *et al.*, 2003).

2.4.2. *Treatment with various substances before storage*

Another method of mushroom storage is to combine MAP with the use of added substances which maintain the quality characteristics of fresh mushrooms, mainly colour and firmness, as well as lowering microorganism counts (Aguirre *et al.*, 2008). Koushki *et al.* (2011) showed that spraying the surface of *A. bisporus* with 0.3% $CaCl_2$ solution for 5 minutes prior to packing them in a modified atmosphere (5% CO_2, 10% O_2) enabled mushrooms to maintain good quality for 11 days of storage at 4ºC. Similar results were obtained for *A. bisporus* stored for 6 days at 12ºC by spraying whole or sliced mushrooms with 2% chitosan solution or 2% $CaCl_2$ solution prior to MAP (Kim *et al.* , 2006). *A. bisporus* can also be rinsed in 0.1% citric acid solution for 5 minutes prior to MAP and 17-day storage at 5ºC (Simon *et al.*, 2010). This method reduces microorganism counts by 2.5 log units on day 0 and minimizes tissue browning. Cliffe-Brynes and Beirne (2008) found that rinsing *A. bisporus* for 60 s in 3% H_2O_2 (hydrogen peroxide) solution prior to spraying with 4% sodium D-isoascorbate monohydrate or 1% H_2O_2 had a beneficial effect on the sensory quality of mushrooms (acceptability and colour) during 7-day storage at 4ºC. Moreover, H_2O_2 solution was more effective in reducing *Pseudomonas* bacteria counts than ClO_2 (chlorine oxide) solution. According to Jiang (2013), dipping *A. bisporus* in 2% alginate for 2 minutes prior to storage in an atmosphere of pure oxygen enabled mushrooms to maintain good sensory quality, including firmness, delayed changes in the levels of soluble solids, total sugars and ascorbic acid, and also inhibited polyphenol oxidase and peroxidase activity. Jiang (2012) showed that dipping *A. bisporus* for 10 minutes in 5 mM natamycin solution and then applying pure oxygen prior to storage in jars at 4ºC enabled mushrooms to retain their firmness, lowered the respiration rate, delayed browning and cap opening, and reduced yeast and mould counts during the 16 days storage period. Treating *A. bisporus* with 100 µM methyl jasmonate has a beneficial effect on mushroom quality during 7-day storage at 7ºC (Meng *et al.*, 2012). Mushrooms undergoing this treatment had higher levels of protein, total sugars, phenolics and flavonoids; the activity of polyphenol oxidase was lower, while that of antioxidant enzymes (catalase and superoxide dismutase) was higher, resulting in delayed darkening, a lower respiratory coefficient and less cell leakage. Other treatments used to extend the shelf life of fresh *A. bisporus* mushrooms are: soaking in sodium metabisulphite for 10 minutes (1 g/l) (Brennan *et al.*, 1998); rinsing in an organic acid solution, such as malic acid (20 g/l) (Singla *et al.*, 2012) or citric acid (10 g/l) (Simon *et al.*, 2010) for 5 minutes; and dipping in 2,2'-(hydroxynitrosohydrazino)-bisethanamine (DETANO) (1 mM) solution for 10 minutes (Jiang *et al.*, 2011). The shelf-life *A. bisporus* can also be extended by fumigation, with CO_2 or 1-metylocyklopropenu (1-MCP), prior to packing in PVC film. Jamjumroon *et al.* (2013) reported that applying 250 ppb 1-MCP for 6 hours at 25°C substantially reduced browning and extended shelf life during storage at 15ºC.

The shelf life of fresh shiitake mushrooms can be extended by coating with chitosan-glucose complex (CGC) prior to MAP (Jiang *et al.*, 2012). After undergoing this pretreatment, mushrooms stored for 16 days at 4°C showed improved sensory quality, firmness and microbiological stability, as well as greater retention of ascorbic acid and soluble solids. Similar results can be obtained with the same storage conditions by coating with chitosan + thyme oil (1.5%) for 30 s (Jiang *et al.*, 2011).

Jafri *et al.* (2013) noted that good results may be obtained by dipping *Pleurotus florida* for 2 minutes in a solution containing chemicals: sorbitol (0.05%), calcium chloride (1%) and citric acid (3%). After pretreatment, mushrooms were MAP packed (10% O_2 and 5% CO_2 or 5% O_2 and 10% CO_2) prior to storage for 25 days at 4°C. These authors showed that chemical treatment of mushrooms followed by modified atmosphere packaging using 10% O_2 and 5% CO_2 provided better retention of quality characteristics and received higher sensory ratings compared to other samples.

2.4.3. *Addition of the moisture absorbent in the package*

Moisture levels in packaged fresh mushrooms can increase during storage as a result of intensive respiration and the use of packaging materials with low water vapour permeability, which can lead to an increase in microorganism growth. On the other hand, low moisture inside the packaging can cause an excessive decrease in mushroom mass, making the product uneconomical. The level of water vapour within the packaging must be controlled if the quality of fresh mushrooms is to be maintained. Among the compounds used to achieve this are sorbitol, silica gel, xylitol, $CaCl_2$, KCl, CaO and bentonite (Roy *et al.*, 1995; Villaescusa and Gil, 2003; Mahajan *et al.*, 2008), either singly or in combination. Mahajan *et al.* (2008) found that for *A. bisporus* stored at 4°C the best combination was bentonite, sorbitol and $CaCl_2$ in the following proportions 0.55, 0.25 and 0.20 g/g desiccant. On the other hand, Villaescusa and Gil (2003) found that in the case of *P. ostreatus* stored at 0, 4 and 7°C, sorbitol promoted tissue leakage, while the use of silica gel did not improve mushroom quality despite controlling moisture inside the packaging.

2.4.4. *Gamma and electron beam irradiation*

Food processing employing different forms of irradiation is well established as a physical, non-thermal mode of food preservation (cold-pasteurization) that processes foods at or nearly at ambient temperature. Among the sources of radiation are gamma rays (^{60}Co), electron-beam (e-beam) and UV irradiation. Irradiation of mushrooms extends shelf life by delaying tissue browning, cap opening, stipe elongation and microorganism growth (Lescano, 1994; Xiong *et al.*, 2009). Research has mainly been carried out on the most widely consumed species, *Agaricus bisporus*, *Pleurotus ostreatus* and *Lentinus edodes* (Jasinghe and Perera, 2006; Ko *et al.*, 2008; Jiang *et al.*, 2010; Fernandes *et al.*, 2012). Other mushrooms species which are irradiated are *Agaricus campestris, Cantharellus tubaeformis, Hypsizygus marmoreus, Pleurotus cystidus, Pleurotus nebrodensis, Lactarius deliciosus,* and *Volvariella volvacea* (Xiong *et al.*, 2009). Duan *et al.* (2010) showed that 2 kGy electron beam irradiation of *A. bisporus* limited adverse textural changes and reduced levels of malondialdehyde (MDA) during 16-day storage at 4°C. Fernandes *et al.* (2012) noted that low-dose gamma irradiation (up to 1.0 kGy) enabled good quality to be maintained in *Lactarius deliciosus* stored for 8 days at 4°C; nutritional quality and antioxidant activity were less affected by irradiation than by storage. Jiang *et al.* (2010) found that ionizing irradiation of fresh shiitake mushrooms improved shelf life during 21-day storage at 4°C. The mushrooms were MAP packed in biodegradable BOPP film; the optimal irradiation dose was 1.0 kGy. The shelf life of fresh mushrooms can also be extended by UV-C treatment. Jiang *et al.* (2012) found that applying UV-C radiation prior to MAP packing can delay softening and enhance antioxidant capacity in shiitake mushrooms during 15-day storage at 1°C. Mushrooms undergoing such treatment had higher levels of total flavonoids and ascorbic acid.

2.4.5. *Other methods*

Less widely used methods of extending the shelf life of fresh mushrooms include pulsed light treatment (PL) (Oms-Oliu *et al.*, 2010) and hyperspectral imaging (HSI) (Gowen *et al.* 2008).

The former is an emerging non-thermal technology for the rapid inactivation of pathogenic and spoilage microorganisms in foods. The significant microbial reductions in very short treatment times, the limited energy cost, the lack of residual compounds, and its great flexibility are some of the major benefits of that technique. PL treatment involves the use of intense pulses of short duration and a broad spectrum to ensure microbial inactivation on the surface of either foods or packaging materials (Elmnasser *et al.*, 2007). Oms-Oliu *et al.* (2010) found that exposing fresh sliced *A. bisporus* to pulsed light by flashing at 4.8, 12 and 28 J/cm^2 reduced native microflora counts from 0.6 to 2.2 log after 15 days of refrigerated storage. Pulsed light treatments allowed extension of the microbiological shelf life of fresh-cut mushrooms by 2–3 days in comparison to untreated samples, while providing a high quality product. Application of pulsed light at doses of 4.8 J/cm^2 could extend the shelf life of fresh-cut mushrooms without dramatically affecting texture and antioxidant properties.

3. MAIN TRENDS IN MUSHROOM PROCESSING

Due to high moisture content and enzymatic activity, edible mushrooms are highly perishable, making them unsuitable for lengthy storage in their fresh state. When supply exceeds demand or when forest mushrooms come into season, mushrooms should be processed without delay using methods which are both simple and modern. The food processing industry provides a wide range of preserved mushroom products, applying methods of preservation and pretreatment which are appropriate to the species. The most common methods of preserving edible mushrooms are drying, freezing, canning, salting and fermenting. Other processed products include mushroom flour, pastes, concentrates and extracts.

3.1. Pretreatment Before Preservation

Ensuring appropriate quality in preserved mushrooms is made possible by eliminating or inhibiting undesirable changes in colour and texture which can occur when preparing mushrooms for preservation, during preservation itself or during the storage of finished products. Mass loss resulting from pretreatment and preservation should also be minimised. The pretreatments most frequently applied are washing (Źródłowski, 1995; Sapers *et al.*, 1999; 2001), blanching (Murr and Morris, 1975; Czapski, 1994 a,b; Coskuner and Ozdemir, 1997; Vivar-Quintana *et al.*, 1999; Coskuner and Ozdemir, 2000; Jaworska *et al.*, 2008; Jaworska and Bernaś, 2009 a,b;), soaking or vacuum soaking (Czapski, 1994 b; Kukura *et al.*, 1998; Vivar-Quintana *et al.*, 1999; Jaworska *et al.*, 2003, 2008; Jaworska and Bernaś, 2009 a,b) and high isostatic pressure (Matser *et al.*, 2000). The type of pretreatment and the parameters of application should be selected according to the species of mushroom, method of preservation and the storage time of the finished product (Jaworska *et al.*, 2008; Jaworska and Bernaś, 2009 a,b; Jaworska *et al.*, 2014).

3.1.1. *Washing*

Washing is a fundamental operation in the pretreatment of mushrooms. Washing mushrooms promotes hygiene by removing mineral contamination and reducing microorganism counts, but also damages the cell membranes separating polyphenol oxidase from its substrates, which results in rapid darkening of the tissue. This is particularly conspicuous in a light-coloured mushroom like *A. bisporus* (Ponting, 1960; Burton and Noble, 1993). In order to avoid browning, various compounds are added to the water used to wash mushrooms. These are generally sulphur compounds, especially sodium metabusulphite, which is regarded as the most effective substance for maintaining light colour in mushrooms but whose residues may cause allergic reactions in consumers (Linn and Gong, 1999). Sodium metabisulphite owes its effectiveness primarily to its ability to inhibit polyphenol oxidase activity and reactions with products remaining after enzymatic reactions, including o-quinones, which are converted to diphenols by sodium metabisulphite (Taylor and Higley, 1986). According to Czapski (1994a), and Czapski and Szudyga (2000) the whiteness of frozen *A. bisporus* depends on the concentration, sorption and sorption dynamics of sodium metabisulphite in the solution used to wash the mushrooms. The optimal concentration of sodium metabisulphite in solutions used prior to freezing *A. bisporus* is 4000 mg/dm^3 (Czapski and B¹kowski, 1995). Other compounds used include hydrogen peroxide, sodium erythorbate, cysteine hydrochloride, disodium edetate (Na_2EDTA) and sodium hydroxide (Park *et al.*, 1991; Sapers *et al.*, 1999; Czapski, 2002). Burton and Noble (1993) found that washing *A. bisporus* mushrooms in H_2O_2 solution and then in a solution containing sodium erythorbate, cysteine hydrochloride and Na_2EDTA had a positive effect on their colour and lowered the level of polyphenol oxidase. Similar results were obtained by McEvily *et al.* (1992) and Sapers *et al.* (2001) using a solution of H_2O_2 (3-5%), sodium erythorbate (2.25% and 4.50%) and NaCl (0.1%). Sapers *et al.* (1999) showed that *A. bisporus* washed in a 5% solution of H_2O_2 followed by spray application of 4% sodium erythorbate solution contained lower levels of soluble phenols than mushrooms washed in water, which was unfavorable. A 0.01% solution of sodium hypochlorite can also be used in pretreatments to induce darkening of mushrooms by oxidising L-3,4-dihydroxyphenylalanine (L-DOPA) to brown-hued quinones (Choi and Sapers, 1994).

3.1.2. *Blanching*

Apart from washing, another procedure frequently applied prior to thermal preservation is blanching. According to Fuster *et al.* (1982), Gormley and Walsh (1982), and Murr and Morris (1975) blanching is essential prior to freezing *A. bisporus* in order to prevent browning during frozen storage and defrosting. Moreover, Vivar-Quintana *et al.* (1999) stated that blanching prior to canning reduces mass loss due to sterilisation by half. Blanching is normally carried out in water or aqueous solutions containing compounds which prevent changes in mushroom colour. In the case of mushrooms with dark caps, blanching in water is sufficient pretreatment; however, light-coloured species, such as *A. bisporus*, are usually blanched in aqueous solutions with added compounds, most commonly 0.2% sodium metabisulphite or potassium metabisulphite solutions (Czapski, 1994 b; Czapski and Szudyga, 2000; Czapski, 2003; Walde *et al.*, 2006), or 0.05-0.5% citric acid solutions (Okereke and Beelman, 1990; Rodrigo *et al.*, 1999; Martin-Belloso and Llanos-Barriobero, 2001; Jaworska *et al.*, 2008; Jaworska and Bernaś, 2009 a,b). In situations where mushrooms radically change colour, such as boiling, braising or frying, blanching may be entirely omitted or limited to water as the medium (Jaworska *et al.*, 2003). Other compounds used in blanching are L-ascorbic acid, hydrogen peroxide, EDTA, sodium erythorbate, cysteine hydrochloride, methyl jasmonate, ethanol, sodium chloride, and calcium chloride (Beelman *et al.*, 1973; Kukura *et al.*, 1998; Coskuner and Ozdemir, 2000; Czapski, 2001; 2002; Walde *et al.*, 2006).

Blanching is usually carried out for 1-3 minutes at 85-95ºC; however, the time can vary from several seconds to 15 minutes (Czapski, 1994 a,b; Coskuner and Ozdemir, 1997; Vivar-Quintana *et al.*, 1999; Zivanovic and Buescher, 2004; Jaworska *et al.*, 2008; Jaworska and Bernaś, 2009 a,b), depending primarily on the specific purpose. One important purpose of blanching is to remove air from the raw material, decreasing its volume, inactivating enzymes and destroying vegetative microflora. As a result of blanching, mushrooms are more elastic, less brittle and, according to Zivanovic and Buescher (2004) firmer. Lespinard *et al.* (2009) showed that blanching leads to a 64-80% softening of mushrooms, and that an increase in blanching temperature from 50°C to 90°C improves retention of light colours, indicating complete inactivation of polyphenol oxidase. Vivar-Quintana *et al.* (1999), and Zivanovic and Buescher (2004) demonstrated that for canning, mushrooms can be blanched for 8 or 10 minutes at 95-98°C. Coskuner and Ozdemir (1997) and (2000) consider that prior to canning, *A. bisporus* can be blanched for 15 minutes at 95-100°C in one of the following solutions: sodium bisulphate - $NaHSO_4$ (0.05%); citric acid (0.05%) and NaCl (0.1%); or EDTA (0.05, 0.1, 0.2, 0.4 or 0.8%). Glucono-δ-lactone can be used instead of organic acids in blanching solutions, which enables good quality to be maintained in canned mushrooms without the product having an unusual or acidic after taste (Rodrigo *et al.*, 1999). Frozen *A. bisporus* can be blanched for 3 minutes at 96-98°C (Jaworska *et al.*, 2008), according to Steinbuch (1986), Lee and Lee (1988) and Czapski (1995) blanching for longer than 3 minutes results in mushroom texture becoming distinctly harder and more rubbery, especially after lengthy frozen storage. According to Jaworska *et al.* (2008) frozen *A. bisporus* previously blanched in water should be stored for a maximum of 4 months to avoid the deterioration of sensory quality and tissue darkening. For 8 months' storage, mushrooms should be blanched in a citric acid (0.5%) and L-ascorbic acid (0.1%) solution, or a lactic acid (1.0%) and L-ascorbic acid (0.1%) solution. In the case of *P. ostreatus* and *B. edulis*, on the other hand, blanching in water is sufficient pretreatment for frozen mushrooms stored for up to 12 months at –25°C (Jaworska and Bernaś, 2009 a,b). Czapski (1995) suggested washing *A. bisporus* in a sodium metabisulphite solution then blanching in water as a pretreatment to retain their light colour and prevent hardening.

Blanching is not recommended if mushrooms are to be dried. Martinez-Soto *et al.* (2001) and Jaworska *et al.* (2014) showed that blanching *A. bisporus* and *B. edulis* in water leads to a considerable deterioration in the sensory quality of dried mushrooms.

Despite having a beneficial effect on colour, blanching can result in significant mass loss in mushrooms, including loss of water soluble constituents (Jaworska *et al.*, 2008; Jaworska and Bernaś, 2009 a,b). At the same time, Vivar-Quintana *et al.* (1999) found that blanching time was not a factor in mass loss in canned mushrooms, including loss of total polyphenols. Jaworska *et al.* (2008; 2009 a,b) found that blanching *A. bisporus, P. ostreatus* and *B. edulis* in water, or in aqueous solutions containing citric acid (0.5%) and L-ascorbic acid (0.1%), or lactic acid (1.0%) and L-ascorbic acid (0.1%) brought about a 10-22% mass loss, including losses of dry matter (3-8%), ash (8-35%), vitamin B_1 (14-68%) and vitamin B_2 (15-60%). Gormley and Walsh (1982) showed that blanching *A. bisporus* in a citric acid (1.5%) solution reduced mass loss caused by blanching from 11.9 to 10.9%, and that raising the pH of the blanching solution from 2.0 to 8.0 reduced such losses from 12.0 to 6.6%. Coskuner and Ozdemir (2000) found that blanching *A. bisporus* in a citric acid solution minimised the losses of iron, copper, manganese and zinc, while blanching in EDTA resulted in decreased iron and copper. Sobkowska and Woźniak (1974) reported that blanching *Tricholoma equestre* in a citric acid and table salt solution led to decreases in dry matter (20%), total sugars (5%) and protein (12%).

Instead of traditional blanching in water, adverse quality changes in mushrooms can be prevented by steam or microwave blanching, or the application of high isostatic pressure. Bernaś *et al.* (2015) showed that in the case of *A. bisporus*, microwave blanching for 5 minutes (1000 W) prior to freezing resulted in a frozen product which, after 8 months of frozen storage, had higher levels (9-49%) of dry matter and, in most cases, ash and vitamins B_1 and B_2 compared with mushrooms blanched in water prior to freezing. Rodriguez-Lopez *et al.* (1999) reported that microwaving *A. bisporus* at a frequency of 2.45 GHz for 20 seconds completely inactivated polyphenol oxidase, whereas achieving the same effect would require 6 minutes of water blanching. High isostatic pressure can inactivate micro-organisms and enzymes without significantly affecting flavour or vitamins. Matser *et al.* (2000) showed that pressure of at least 959 MPa is required to inactivate polyphenol oxidase; however, this had an adverse effect on mushroom colour. Retaining mushroom colour to the same degree as when using traditional water blanching required prior vacuum soaking in tap water.

An alternative to traditional blanching in water is a combination of microwave blanching and blanching in hot water (Devece *et al.*, 1999; Bernaœ and Jaworska, 2015). Devece *et al.* (1999) argue that, in contrast to water blanching, the combined method results in complete inactivation of polyphenol oxidase in a short time, as well as reducing mass loss, shrinkage and tissue browning. However, Bernaœ *et al.* (2015) found that blanching in water for 1 minute at 96-98°C combined with microwaving for 3 minutes at 1000 W prior to freezing and 8-month frozen storage resulted in only a marginal improvement in mushroom colour compared with traditional blanching in water. Moreover, the combined method results in considerable loss of mass and phenolic compounds in mushrooms.

3.1.3. *Soaking*

Vacuum soaking or saturation of mushrooms can be applied prior to blanching and preservation in order to reduce mass loss and increase yield (Beelman *et al.*, 1973; Czapski, 1994 b; Martinez-Soto *et al.*, 2001; Jaworska *et al.*, 2008; Jaworska and Bernaś, 2009 a,b). Food vacuum impregnation makes it possible to introduce dissolved or suspended substances directly into the product porous structure in a controlled way, allowing fast product compositional and structural changes. Gormley and Walsh (1982) reported that mass loss in *A. bisporus* due to blanching was 17% lower when the following pretreatment (known as the 3S process) was applied prior to 5-minute blanching in citric acid (1.5%) solution instead of simply washing mushrooms: soaking in water for 20 minutes, followed by 72 h storage at 2-4°C and further soaking in water for 2 hours. Similar results were obtained by Lin *et al.* (2001) using the same pretreatment, with the additional observation that the 3S process also had beneficial effects on mushroom colour. Prior to canning, *A. bisporus* can also be soaked for 15 minutes in 0.05% sulphur dioxide solution followed by steam blanching for 5 minutes at 98°C (Martin-Belloso and Llanos-Barriobero, 2001). Using this procedure, mushrooms retain 89% of protein, 65% of carbohydrates, 65% of potassium, 51% of thiamine, 54% of riboflavin and 59% of vitamin C found in the raw material. Jaworska *et al.* (2008) and (2009 a,b) showed that, 1-hour soaking in a citric acid (0.5%) and L-ascorbic acid (0.1%) solution, or a lactic acid (1%) and L-ascorbic acid (0.1%) solution prior to blanching in the same solutions resulted in greater losses of vitamin B_1 in *P. ostreatus* and *B. edulis*, vitamin B_2 in *A. bisporus*, and ash in *A. bisporus* and *B. edulis* compared with blanching alone. Moreover, Bernaś *et al.* (2007), Jaworska and Bernaś (2010) and Jaworska *et al.* (2010) all concluded that the above pretreatment had a limited effect on the texture of frozen and canned *A. bisporus* and *B. edulis*, with only canned *B. edulis* showing any difference between soaked and blanched mushrooms and those which were only blanched. An alternative to soaking in water or in sulphur dioxide solution is soaking in curd or fermented whey. Walde *et al.* (2006) found that applying this pretreatment prior to microwave drying shortens the drying time of *A. bisporus* and *P. ostreatus* compared with blanching in 2% saline solution for 3-4 minutes.

3.2. Preservation

3.2.1. *Drying*

Dehydration is one of the important methods of preserving mushrooms and dehydrated mushrooms are valuable ingredients in a variety of sauces and soups. As mushrooms are very sensitive to temperature, selecting the appropriate drying method is essential to the success of the process. Correctly dried mushrooms should have a pleasant aroma, a crisp consistency and a moisture content not exceeding 10% (Jaworska *et al.*, 2014). Mushrooms may be dried entire (usually with the stipes shortened) or in slices. The most commonly used method is hot air drying: mushrooms should initially be dried at 40°C for 3 h, then at 60°C. In the case of *Boletus edulis*, the total drying time should not exceed 15 h. Fluidized bed drying can also be part of the blast drying process. Walde *et al.* (2006) reported that in the case of *A. bisporus* and *P. ostreatus*, this method provided a better quality product than microwave drying.

Other methods of drying mushrooms include freeze-drying, freeze-drying combined with microwave drying, freeze-drying combined with vacuum drying, vacuum drying, vacuum drying combined with microwave drying, and vacuum drying or freeze-drying in an adsorbent fluidized bed (Le Loch-Bonazzi *et al.*, 1991; Giri and Prasad, 2007; Jaworska *et al.*, 2014; Pei *et al.*, 2014).

Freeze-drying produces a high quality product, but is an expensive process (Martinez–Soto *et al.*, 2001). According to Fang *et al.* (1976), this method restricts polyphenol oxidase activity, thus retaining a light colour in the dried product. It also enables the mushroom to retain its natural shape and size. In the case of *A.bisporus*, freeze-drying has a significant effect on 8-carbon aromatic compounds, mainly1-octen-3-ol, 3-octanol and 3-octanone, drastically reducing their levels by 72, 76 and 78% respectively compared with fresh mushrooms (Le Loch-Bonazzi *et al.*, 1991).

Vacuum drying is another alternative method and is especially suitable for products that are prone to heat damage; however, in vacuum processes requiring heat, the transfer of heat energy to the workload becomes difficult as convection is ineffective at low pressure. Most conventional vacuum dryers rely on conduction heat transfer from hot plates, which is slow, difficult to control and requires a large surface area; therefore, conventional vacuum drying has high operating and installation costs.

The desire to prevent significant quality loss and to achieve fast and effective dehydration has resulted in increasing use of microwave heating for food drying. Giri and Prasad (2007) showed that microwave-vacuum drying resulted in a 70–90% decrease in the drying time and the dried products had better rehydration characteristics compared to convective air drying. Button mushroom slices could be microwave-vacuum dried to 6% moisture content (w.b.) in 25–30 min, avoiding any burning of the product and with good rehydration characteristics.

Comparing hot air drying, vacuum drying, freeze-drying in an adsorbent fluidized bed and vacuum freeze-drying of *A.bisporus*, Le Loch-Bonazzi and Wolf (1991) concluded that the freeze dried product had the best quality in terms of rehydration capacity, colour and flavour. Moreover, the authors cited also reported that the greatest losses of aromatic compounds (1-octen-

3-ol) occurred during pretreatment and freezing, and that the most important parameter for ensuring good quality dried products is pressure. In contrast, Pei *et al.* (2014) compared freeze-drying (FD), freeze-drying combined with hot air drying (FD+AD), freeze-drying combined with vacuum drying (FD+VD) and freeze-drying combined with microwave vacuum drying (FD+MVD), finding that the best quality dried *A. bisporus* product was obtained using the FD+MVD method. Furthermore, losses of reducing sugar, total sugar, protein and vitamin C were lower in FD+VD and FD+MVD products than in FD+AD products.

A long drying time and overheating of the surface may lead to problems of colour darkening, loss of flavour and a decrease in rehydration capacity. In addition, levels of free amino acids, including glutamic acid and aspartic acid, decrease during drying (Lobaszewski and Paczyñska, 1990). Jaworska *et al.* (2014) found that dried *Boletus edulis* contained less total flavonoids (4-7%), vitamin C (2–36%), beta-carotene (26–32%) and total tocopherols (72–81%) than fresh mushrooms, but exhibited higher antioxidant activity (1-33%). The increased antioxidant activity was probably caused by the production of melanins, which can result from the Maillard reaction (amino acids with sugars) (Manzocco *et al.*, 2001; Morales and Jimenez-Perez, 2004). Losses of water-soluble compounds following rehydration are lower in freeze-dried than in convection-dried mushrooms. On the other hand, Jaworska *et al.* (2014) found that whereas water content and water activity were lower in freeze-dried than in air-dried *Boletus edulis*, rehydration capacity showed the opposite tendency.

An alternative drying method is osmo-convective drying. Osmotic dehydration is an inexpensive method of partially removing moisture from food products; however, secondary drying is required since the level of moisture remaining in the product is too high. Osmo-convective drying of mushrooms may be carried out for 5 h in the following aqueous solutions: sugar-salt, salt or sugar-salt plus $MgHCO_3$ (Shukla and Singh, 2007). This procedure removes approximately 60% of water from the tissue and is followed by secondary drying at 60ºC to a moisture content of 6-8%. According to the authors cited, quality parameters like texture, moisture content, colour, hydration as well as rehydration ratios and microbiological loads were evaluated and found acceptance to a satisfactory level.

3.2.2. *Freezing*

Freezing is one of the best methods of preserving food. During freezing most of the liquid water changes into ice, which greatly reduces microbial and enzymatic activity. Oxidation and respiration are also weakened effectively by low temperatures. The biggest challenge in freezing mushrooms is preventing changes in colour and texture, as well as mass loss after defrosting; therefore, appropriate pretreatments are applied, which should be selected according to the mushroom species, as well as the conditions and period of frozen storage. Pretreatment often includes blanching in water, or in aqueous solutions containing sodium metabisulphite or organic acids (Vivar-Quintana *et al.*, 1999; Czapski and Szudyga, 2000; Jaworska *et al.*, 2008; Jaworska and Bernaś, 2009 a,b). Jaworska *et al.* (2008) and (2009 a,b) showed that soaking prior to blanching improves the colour of frozen mushrooms; however, it also increases leakage of cell fluid after defrosting. An alternative to blanching in an aqueous environment is microwave blanching (5 min., 300 kJ) or a combination of blanching in water (1 min., 96-98°C) and microwave blanching (3 min., 180 kJ) (Bernaś and Jaworska, 2015). The authors cited showed that microwave blanching of *A. bisporus* results in distinctly lower leakage of cell fluid and vitamin B_2 after defrosting than blanching for 3 minutes in water or in a sodium metabisulphite (0.2%) and citric acid (0.5%) solution.

An important consideration when freezing mushrooms is the method of freezing. The most commonly used method is blast freezing at –35°C (Czapski and Szudyga, 2000; Jaworska *et al.*, 2008; Jaworska and Bernaś, 2009 a,b). Good results have been obtained from cryogenic freezing, which can be used to freeze various foods, including mushrooms (Worka *et al.*, 1997; Kidmose and Kaack, 1999; Agnelli and Mascheroni, 2002). According to Kondratowicz and Kowałko (2000), cryogenic freezing of mushrooms results in a higher quality product than air-blast freezing, particularly with regard to consistency as well as intensity of taste and aroma, and enables frozen mushrooms to be stored for up to 12 months. Angeli and Mascheroni (2002), on the other hand, reported that despite retaining light colour in frozen mushrooms, cryomechanical freezing resulted in 33% greater mass loss than conventional air-blast freezing. On the basis of as yet unpublished data, the authors of this subsection have shown that the freezing method used (air-blast or cryogenic in liquid nitrogen) affects levels of B vitamins in mushrooms to varying degrees depending on the individual vitamin and species of mushroom. Levels of vitamin B_1 in *Pleurotus ostreatus, Agaricus bisporus, Almirallia mellea* and *Lactarius deliciosus* were higher using the air-blast method, while in *Boletus edulis, Xerocomus badius* and *Cantharellus cibarius* higher levels were found after cryogenic freezing. In the case of vitamin B_2, however, the only difference was found in *P. ostreatus*, in which levels were higher after air-blast freezing.

12-month frozen storage of *A. bisporus, P. ostreatus* and *B. edulis* causes a deterioration in mushroom quality, mainly colour, taste, aroma and texture (Jaworska *et al.*, 2008; Jaworska and Bernaś, 2009 a,b), as well as reducing levels of vitamin C and, in the case of *P. ostreatus*, vitamins B_1 and B_2. The maximum period of frozen storage at –25°C which ensures acceptable quality is 8 months for *A. bisporus* and 12 months for *P. ostreatus* and *B. edulis*, on condition that prior to freezing mushrooms are blanched in sodium metabisulphite solution (*A. bisporus*) or water (*P. ostreatus* and *B. edulis*).

3.2.3. *Canning*

Canning is another widely used method of preserving mushrooms. The raw material for canning may be fresh or desalted mushrooms; fermented mushrooms are less frequently used. Canned mushrooms can be pickled or sterilised (ENV/JM/MON, 2007). The former are generally preserved in glass jars in a sweet-salt-acidic brine solution containing acetic acid (1.5 to 4%), table salt and sugar. Sterilised products, on the other hand, are packed in glass jars or metal cans in salt or salt-acidic brine solutions containing approximately 2% sodium chloride and occasionally 0.05% citric acid, L-ascorbic acid, sodium metabisulphite or potassium metabisulphite (Coskuner and Ozdemir, 1997; 2000; Lin *et al.*, 2001; Jaworska *et al.*, 2003; 2011). Such products have a shelf life of up to 24 months, with relatively low storage costs.

Achieving good quality in canned mushrooms depends to a large extent on the selection of appropriate pretreatments (ENV/JM/MONO, 2007), which have been described in an earlier section. Blanching is essential since it enables oxygen to be removed from mushroom tissue. Moreover, Okereke and Beelman (1990) showed that blanching *A. bisporus* for 7 minutes in a citric acid (0.05M) solution or brine solution containing table salt and calcium disodium ethylenediaminetetraacetate ($CaNa_2EDTA$) has a beneficial effect on the colour and texture of sterilised mushrooms as well as improving microbiological stability. Following thermal treatment, mushrooms are cooled and placed in containers to about ¾ capacity; the containers are then filled with brine, usually hot (90°C). Pickled mushrooms can be pasteurised at 85°C (Zivanovic *et al.*, 2003) and mushrooms in brine are sterilised at 118-121°C for 10-20 minutes, depending on the type and size of the container, size of mushroom pieces, composition of the brine and type of equipment used (Okereke and Beelman, 1990; Rodrigo *et al.*, 1999; Vivar-Quintana *et al.*, 1999; Zivanovic *et al.*, 2003; Jaworska *et al.*, 2011). Coskuner and Ozdemir (1997) suggested that *A. bisporus* should be sterilised in 170 cm^3 glass containers for 25 minutes at a slightly lower temperature (112°C). After cooling, the product should be stored in a cool, dark place.

The process of canning mushrooms results in a decrease in dry weight, partly caused by the addition of brine, with 12-36% reductions observed in total sugars, total acids and total nitrogen (Jaworska *et al.*, 2003). Jaworska *et al.* (2007) also reported a decrease in vitamin content; compared with frozen mushrooms, vitamin B_1 and B_2 content was 67-90% and 17-33% lower respectively. The sterilisation process also affects the protein content of mushrooms, including amino acid composition. Larousse (1986), Martin-Belloso and Llanos-Barriobero (2000) recorded a 10.4 -11.0% decrease in protein content in canned *A. bisporus*, while Nagy (1989) found an average decrease of 13.9% in canned forest mushrooms such as *Cantharellus cibarius*, *Boletus edulis* and mixed mushrooms. The greatest losses of amino acids were found to occur in alanine, asparagine, glutamine, methionine, cysteine, proline and valine (Jaworska *et al.*, 2012). A different view was expressed by Vetter (2003) who maintains that sterilisation does not have a significant effect on protein and fat content in mushrooms but does cause decreases in dry weight, ash and certain mineral components such as potassium, phosphorus and magnesium, as well as increasing levels of sodium, calcium and iron.

3.2.4. *Fermentation*

One of the best, though little used, methods of preserving mushrooms is fermentation. The lactic acid bacteria involved in the fermentation process give fermented mushrooms their characteristic and pleasant aroma as well as having health benefits. Lactic acid of up to 1.5% is the preserving agent. The species most suitable for fermenting are *Lactarius deliciosus*, *Pleurotus ostreatus*, *Armillaria mellea*, *Agaricus bisporus* and *Auricularia auricula*. Joshi *et al.* (1996) and Khaskheli *et al.* (2015) showed that prior to fermenting *A. bisporus* and *Auricularia auricula* should be blanched at 96-98°C for 4-5 minutes in a solution of NaCl (2%) and potassium metabisulphite (KMS) (0.1%). The brine used for fermenting can contain NaCl (2%) and lactose (1%) (Joshi *et al.*, 1996); mustard oil, vinegar and salt; or vinegar, soft water and salt (Khaskheli *et al.*, 2015). Red chilli powder, cumin seed powder, carom seed (ajwain), nigella seed (kalonji), fennel seed powder (saunf) may be used for additional flavour (Khaskheli *et al.*, 2015). Fermented mushrooms may be produced through spontaneous fermentation, but in order to achieve pH<4.0, starter cultures inoculated with bacteria such as *Lactobacillus plantarum, Lactobacillus delbrueckii* and *Streptococcus lactis* should be used. Mushrooms can be fermented at 26°C for 3-4 weeks (Khaskheli *et al.*, 2015). Kreb and Lelley (1991), on the other hand, maintain that good results may be obtained by fermenting *P. ostreatus* with fresh shredded cabbage for 10 days at 21°C. The resulting product has a mild, pleasant flavour similar to that of fermented cabbage. Moreover, the product may be stored for 6 months without pasteurising the mushrooms. Fermented mushrooms may be used as finished product or in the production of pickled mushrooms

3.2.5. *Salting*

One of the least popular methods of preserving mushrooms is salting (Yang and Maguer, 1992) this is due to the adverse effect on mushroom quality resulting from the substanial loss of soluble constituents, such as those determining flavour (free amino acids and nucleotides), and vitamins during desalination. Compared with frozen and canned mushrooms, salted mushrooms have an inferior flavour, contain more dry matter, ash and total carbohydrates, and less protein and fat (Liu *et al.*, 2014). The preserving agent is salt, which comprises 15-25% of the finished product (Yang and Maguer, 1992; Liu *et al.*, 2014). According to Liu *et al.* (2014), salting takes about 20 days. Salted mushrooms may be stored for 6 months at 20-25°C.

3.2.6. *Other methods*

Further methods of preserving mushrooms include cooking, frying, braising and smoking. According to Barros *et al.* (2007) mushrooms may be cooked in extra virgin olive oil with salt and onions added. Cooked mushrooms have lower nutrient concentrations and lower antioxidant activity than dried or frozen mushrooms; they also contain higher levels of monounsaturated fatty acids and lower levels of polyunsaturated fatty acids and sugars. Pan frying *Cantharellus cibarius* and *Cantharellus tubaeformis* mushrooms without oil for 5 minutes had only a slight effect on vitamin D content, with losses not exceeding 10% compared with fresh mushrooms (Mattila *et al.*, 1999). According to Jaworska *et al.* (2015) *Agaricus bisporus* and *Pleurotus ostreatus* mushrooms can be braised in a covered saucepan with a small amount (10%) of 100% rapeseed oil and NaCl (0.5%) for 10 min. The authors cited found that *A. bisporus* prepared for consumption was a valuable source of vitamins B_1, B_2, B_3 and B_6, while *P. ostreatus* contained only vitamins B_2 and B_3. Braised mushrooms were particularly rich in tocopherols; other antioxidant compounds, such as total polyphenols and total flavonoids (expressed as catechin equivalents), vitamin C and carotenoids were also found to be present. *A. bisporus* prepared for consumption showed higher antioxidant activity than *P. ostreatus*.

Smoking is one of the oldest methods of food preservation, but at present food is smoked to enhance its sensory quality rather than for its preservative effect. In general, smoking infuses high protein food with aromatic components, which lend flavour and colour to the food as well as having a bacteriostatic and antioxidant effect. According to Eissa *et al.* (2009), *Pleurotus ostreatus* can be smoked for 1.0-1.5 h at 65°C and prior to smoking should be dipped in sodium metabisulfite, EDTA and H_2O_2 solution, blanched in water or steam, salted and dried. The authors cited showed that the most effective pretreatments ensuring the quality and safety of smoked mushrooms were those using H_2O_2 and steam blanching. Thermal and chemical treatments followed by smoking reduce enzyme activity and are suitable for preserving mushrooms.

4. CONCLUSIONS

Edible mushrooms are a challenging raw material for the food processing industry. In view of their high moisture content and enzymatic activity, fresh mushrooms have a very limited shelf life. They are best kept refrigerated at 0-2°C with a relative air humidity of 90%. The shelf life of fresh mushrooms may be extended by exposure to various compounds, including chitosan, $CaCl_2$, malic acid, citric acid, H_2O_2, sodium D-isoascorbate monohydrate, alginate, natamycin, methyl jasmonate and metabisulphite, followed by packing in a modified atmosphere. Other possible treatments include exposure to gamma and electron beam irradiation, pulsed light treatment and hyperspectral imaging.

When supply exceeds demand or when forest mushrooms come into season, mushrooms should be preserved as soon as possible. The food processing industry provides a wide range of preserved mushroom products, applying methods of preservation and pretreatment which are appropriate to the species. The most common methods of preserving edible mushrooms are drying, freezing, canning, salting and fermenting.

5. REFERENCES

Agnelli, M.E. and Mascheroni, R.H. 2002. Quality evaluation of foodstuffs frozen in a cryomechanical freezer. *J. Food Eng.*, 52: 257–263.

Aida, F.M.N.A., Shuhaimi, M., Yazid, M. and Maaruf, A.G. 2009. Mushrooms as a potential source of prebiotics, a review. *Trends Food Sci. Technol.*, 20: 567–575.

Amodio, M.L., Colelli, G., De Cillis, F.M., Lovino, R. and Massignan, L. 2003. Controlled-atmosphere storage of fresh-cut, Cardoncello mushrooms (*Pleurotus eryngii*). *Acta Hort.*, 599: 731–735.

Antman, G., Area, G., Lema, P. and Lareo, C. 2008. Influence of modified atmosphere packaging on sensory quality of shiitake mushrooms. *Posth. Biol. Techn.*, 49: 164–170.

Ares, G., Lareo, C. and Lema, P. 2007. Modified atmosphere packaging for postharvest storage of mushrooms. A review. Global Science Books, Fresh Produce, pp. 32–40.

Argyropoulos, D., Heindl, A. and Muller, J. 2011. Assessment of convection, hot-air combined with microwave-vacuum and freeze-drying methods for mushrooms with regard to product quality. *Int. J. Food. Sci. Tech.*, 46: 333–342.

Austin, J.W., Dodds, K.L., Blanchfield, B. and Faber, J.M. 1998. Growth and toxin production *Clostridium butulinum* on inoculated fresh cut packaged vegetables. *J. Food Protect.,* 61: 324–328.

Barron, C., Varoquaux, P., Guilbert, S., Gontard, N. and Gouble, B. 2002. Modified atmosphere packaging of cultivated mushroom (*Agaricus bisporus* L.) with hydrophilic films. *J. Food Sci.,* 67: 251–255.

Barros, L., Baptista, P., Correia, D.M., Morais, J.S. and Ferreira, I.C.F.R. 2007. Effects of conservation treatment and cooking on the chemical composition and antioxidant activity of portuguese wild edible mushrooms. *J. Agric. Food Chem.,* 55: 4781–4788.

Barros, L., Cruz, T., Baptista, P., Estevinho, L.M. and Ferreira, I.C.F.R. 2008a. Wild and commercial mushrooms as source of nutrients and nutraceuticals. *Food Chem. Toxic.*, 46: 2742–2747.

Barros, L., Venturini, B.A., Baptista, P., Estevinho, L.M. and Ferreira, I.C.F.R. 2008b. Chemical composition and biological properties of Portuguese wild mushrooms: A comprehensive study. *J. Agric. Food Chem.,* 56: 3856–3862.

Beelman, R.B., Guthrie, B.D. and Royse, D.J. 1989. Influence of bacterial populations on postharvest deterioration of fresh mushroom. *Mushroom Sci.*, 12: 655–665.

Beelman, R.B., Kuhn, G.D. and McArdle, F.J. 1973. Influence of post-harvest storage and soaking on the yield and quality of canned mushrooms. *J. Food Sci.*, 38: 951–953.

Beelman, R.B., Okereke, A. and Guthrie, B. 1987. Evaluation of textural changes related to postharvest quality and shelf life of fresh mushrooms. *In:* Wuest, P.J. Royse. D.J., and Beelman, R.B. (*eds.*), Cultivating Edible Fungi, Elsevier, Amsterdam, pp. 251–258.

Bernaś, E. and Jaworska, G. 2015. Effect of microwave blanching on the quality of frozen *Agaricus bisporus*. *Food Sci.Techn. Int.*, 21: 245–255.

Bernaś, E., Jaworska, G., Maciejaszek, I. and Biernacka, A. 2007. The influence of preliminary treatment, freezing and frozen storage on *Agaricus bisporus* texture. *Food Sci. Techn. Qual.*, 5: 165–172 [in Polish].

Billaud, C., Roux, E., Brun-Mérimee, S., Maraschin, C. and Nicolas, J. 2003. Inhibitory effect of unheated and heated d-glucose, d-fructose and l-cysteine solutions and Maillard reaction product model systems on polyphenoloxidase from apple. I. Enzymatic browning and enzyme activity inhibition using spectrophotometric and polarographic methods. *Food Chem.*, 81: 35–50.

Braaksma, A., Schaap, D.J., de Vrije, T., Jongen, W.M.F. and Woltering, E.J. 1994. Ageing of mushroom (*Agaricus bisporus*) under post-harvest conditions. *Posth. Biol. Techn.*, 4: 99–110.

Brennan, M., Le Port, G. and Gormley, R. 2000. Post-harvest treatment with citric acid or hydrogen peroxide to extend the shelf life of fresh sliced mushrooms. *LWT.*, 33: 285–289.

Brennan, M.H. and Gormley, T.R. 1998. Extending the shelf life of fresh sliced mushrooms. Research Report n 2. Dunsinea, Castleknock, Dublin: The National Food Centre.

Burton, K.S. 1991. Modified atmosphere packaging of mushrooms— Review and recent developments. *In:* Maher, M.J. (*ed.*), Science and cultivation of edible fungi. Rotterdam: Balkema, pp. 683–688.

Burton, K.S. and Noble, R. 1993. The influence of flush number, bruising and storage temperature on mushroom quality. *Posth. Biol. Technol.*, 3: 39–47.

Chang, T.S. 2007. Two potent suicide substrates of mushroom tyrosinase: 7,8,4'-trihydroxyisoflavone and 5,7,8,4'-tetrahydroxyisoflavone. *J. Agric. Food Chem.,* 55: 2010–2015.

Chikthimmah, N. and Beelman, R.B. 2006. Microbial ecology of mushroom casing soils and preharvest strategies to enhance safety and quality of fresh mushrooms. Doctor of Philosophy (Food Science). The Pennsylvania State University, p. 161.

Choi, M.H. and Kim, G.H. 2003. Quality changes in *Pleurotus ostreatus* during modified atmosphere storage as affected by temperatures and packaging material. *Acta Hort.,* 628: 357–362.

Choi, S.W. and Sapers, G.M. 1994. Effects of washing on polyphenols and polyphenol oxidase in commercial mushrooms (*Agaricus bisporus*). *J. Agric. Food Chem.*, 42: 2286–2290.

Cliffe-Byrnes, V. and O' Beirne, D. 2007. Effects of gas atmosphere and temperature on the respiration rates of whole and sliced mushrooms (*Agaricus bisporus*)-implications for film permeability in modified atmosphere packages. *J. Food Sci.,* 72: E197–204.

Coskuner, Y. and Ozdemir, Y. 1997. Effects of canning processes on the elements content of cultivated mushrooms (*Agaricus bisporus*). *Food Chem.*, 60: 559–562.

Coskuner, Y. and Ozdemir, Y. 2000. Acid and EDTA blanching effects on the essential element content of mushrooms (*Agaricus bisporus*). *J. Sci. Food Agric.*, 80: 2074–2076.

Czapski, J. 1994a. Dynamics of sorption of sodium metabisulfite solution in aspect of sulfur dioxide residue in mushrooms. *Veg. Crops Res. Bull.*, 41: 169–176 [in Polish].

Czapski, J. 1999. Some properties of active and latent catechol oxidase of mushroom. *Acta Agrobot.,* 52: 139–147.

Czapski, J. 2000. Stored mushrooms quality as affected by specificity of film packaging. *Veg. Crops Res. Bull.,* 53: 111–124.

Czapski, J. 2001. The effect of methyl jasmonate and ethyl alcohol vapours on storage of mushrooms. *Veg. Crops Res. Bull.,* 54: 219–222.

Czapski, J. 2002. Mushrooms quality as affected by washing and storage conditions. *Veg. Crops Res. Bull.,* 56: 121–128.

Czapski, J. 2003. Evaluation of chemical composition of commercially canned mushrooms processed from fresh and desalted mushrooms and derived from different geographic regions. *Veg. Crops Res. Bull.,* 58: 135–141.

Czapski, J. and B[1]kowski, J. 1995. Investigation of quality of frozen mushrooms and methods reducing their darkening, and residues of sulfur dioxide. Part 1. The influence of different concentrations of sodium metabisulfite and time of storage on whiteness and sulfur dioxide residues in frozen mushrooms. *Veg. Crops Res. Bull.*, 43: 103–108 [in Polish].

Czapski, J. and Szudyga, K. 2000. Frozen mushrooms quality as affected by strain, flush, treatment before freezing, and time of storage. *J. Food Sci.,* 65: 722–725.

Czapski, J. 1994b. The effect of some technological treatments on the yield and quality of blanched mushroom stored in brine. *Veg. Crops Res. Bull.*, 42: 101–119 [in Polish].

Day, B.P.F. 1996. High oxygen modified atmosphere packaging for fresh prepared produce. *Posth. News Inf.*, 7: 32–34.

Devece, C., Rodriguez-Lopez, J.N., Fenoll, L.G., Tudela, J., Catala, J.M., De los Reyes, E. and Garcia-Canovas, F. 1999. Enzyme inactivation analysis for industrial blanching applications: comparison of microwave, conventional and combination heat treatments on mushroom polyphenoloxidase activity. *J. Agric. Food Chem.*, 47: 4506–4511.

Duan, Z., Xing, Z., Shao, Y. and Zhao, X. 2010. Effect of electron-beam irradiation on postharvest quality and selected enzyme activities of the white button mushroom, *Agaricus bisporus*. *J. Agric. Food Chem.*, 58: 9617–9621.

Eissa, H.A., Fouad, G.M. and Shouk, A.E.A. 2009. Effect of some thermal and chemical pre-treatments on smoked oyster mushroom quality. *Int. J. Food Sci. Techn.*, 44: 251–261.

Elmnasser, N., Guillou, S., Leroi, F., Orange, N., Bakhrouf, A. and Federighi, M. 2007. Pulsed-light system as a novel food decontamination technology: A review. *Canadian J. Microb.*, 53: 813–821.

ENV/JM/MONO 2007. Consensus document on compositional considerations for new varieties of the cultivated mushroom *Agaricus bisporus*: Key food and feed nutrients, anti – Nutrients and toxicants. OECD Environment, Healt and Safety Publications. *Series on the Safety of Novel Foods and Feeds,* 15: 1–42.

Espin, J.C. and Wichers, H.J. 1999. Kinetics of activation of latent mushroom (*Agaricus bisporus*) tyrosinase by benzyl alcohol. *J Agric. Food Chem.*, 47: 3503–3508.

Espin, J.C., Jolivet, S. and Wichers, H.J. 1998. Inhibition of mushroom polyphenol oxidase by agaritine. *J. Agric. Food Chem.,* 46: 2976–2980.

Fang, T.T., Ou, S.M. and Lin, C. 1976. Effects of chemical treatments before and after blanching and freezing methods on the quality of frozen mushrooms. *Mush. Sci.*, 9: 319–331.

FAOStat 2015. Mushrooms and truffles. Rome: Food and Agriculture Organization of the United Nations. *http: / / faostat3.fao.org/*(16/02/ 2015).

Farber, J.M. 1991. Microbiological aspects of modified-atmosphere packaging technology – A review. *J. Food Prot.*, 54: 58–70.

Farber, J.N., Harris, L.J., Parish, M.E., Beuchat, L.R., Suslow, T.V. and Gorney, J.R., Garrett, E.H. and Busta, F.F. 2003. Microbiological safety of controlled and modified atmosphere packaging of fresh and fresh-cut produce. *Compreh. Rev. Food Sci. Food Safety*, 2: 142–160.

Fernandes, A., Antonio, A.L., Oliveira, M.B.P.P., Martins, A. and Ferreira, I.C.F.R. 2012. Effect of gamma and electron beam irradiation on the physico-chemical and nutritional properties of mushrooms: A review. *Food Chem.,* 135: 641–650.

Flurkey, W.H. and Ingebrigtsen, J. 1989. Polyphenol oxidase activity and enzymatic browning in mushrooms. *In*: Quality Factors of Fruit and Vegetables, American Chemical Society, Washington, DC.

Fuster, C., Prestamo, G. and Espinosa, J. 1982. Frozen mushrooms. I. Choice of optimum treatment to prevent browning. *Alimentaria,* 131: 33–35.

Giri, S.K. and Prasad, S. 2007. Drying kinetics and rehydration characteristics of microwave-vacuum and convective hot-air dried mushrooms. *J. Food Eng.,* 78: 512–521.

Gormley, T.R. and Walsh, P.E. 1982. Reducing shrinkage in canned and frozen mushrooms. *Irish J. Food Sci. Techn.,* 6: 165–175.

Gorris, L. and Tauscher, B. 1999. Quality and safety aspects of novel minimal processing technology. *In:* Oliveira, FAR and Oliviera, J.C. (*eds.*), *Processing of Foods Quality Optimisation and Process Assrsment. CRC Press, Boca Raton, USA,* pp. 325–339.

Gowen, A.A., O'Donnell, C.P., Taghizadeh, M., Gaston, E., O'Gorman, A., Cullen, P.J., Frias, J.M., Esquerre, C. and Downey, G. 2008. Hyperspectral imaging for the investigation of quality deterioration in sliced mushrooms (*Agaricus bisporus*) during storage. *Sens. Instr. Food Qual.*, 2: 133–143.

Greengrass, J. 1995. Films para envasado en atmosfera modificada. *In*: Parry, R.T. (*ed.*), Envasado de los Alimentos en Atmósfera Modificada, A. Madrid Vicente Ediciones, Madrid, España, pp. 79–118.

Guillamon, E., Garcia-Lafuente, A., Lozano, M., D'Arrigo, M., Rostagno, M.A., Villares, A. and Martinez, J.A. 2010. Edible mushrooms: Role in the prevention of cardiovascular diseases. *Fitoteraphy,* 81: 715–723.

Halamia, P.M., Chandrashekara, A. and Josepha, R. 1999. Characterization of bacteriocinogenic strains of lactic acid bacteria in fowl and fish intestines and mushroom. *Food Biotech.*, 13: 121–136.

Halaouli, S., Asther, M., Sigoillot, J.C., Hamdi, M. and Lomascolo, A. 2006. Fungal tyrosinases: New prospects in molecular characteristics, bioengineering and biotechnological applications. *J. Appl. Microbiol.,* 100: 219–232.

Hicks, K.B., Haines, R.M., Tong, C.B.S., Sapers, G.M., El-Atawy, Y., Irwin, P.L. and Seib, P.A. 1996. Inhibition of enzymatic browning in fresh fruit and vegetable juices by soluble and insoluble forms of β-cyclodextrin alone or in combination with phosphates. *J. Agric. Food Chem.*, 44: 2591–2594.

Jafri, M., Jha, A., Bunkar, D.S. and Ram, R.C. 2013. Quality retention of oyster mushrooms (*Pleurotus florida*) by a combination of chemical treatments and modified atmosphere packaging. *Posth. Biol. Techn.*, 76: 112–118.

Jamjumroon, S., Wongs-Aree, C., McGlasson, W.B., Srilaong, V., Chermklin, P. and Kanlayanarat, S. 2013. Alleviation of cap browning of 1-MCP/High CO_2- treated straw mushroom buttons under MAP. *Int. Food Res. J.*, 20: 581–585.

Jasinghe, V.J. and Perera, C.O. 2006. Ultraviolet irradiation: The generator of Vitamin D2 in edible mushrooms. *Food Chem.*, 95: 638–643.

Jaworska, G. and Bernaœ, E. 2009 a. The effect of preliminary processing and period of storage on the quality of frozen *Boletus edulis* (Bull.: Fr.) mushrooms. *Food Chem.,* 113: 936–943.

Jaworska, G. and Bernaś, E. 2009b. Qualitative changes in *Pleurotus ostreatus* (Jacq.: Fr.) Kumm. mushrooms resulting from different methods of preliminary processing and periods of frozen storage. *J. Sci. Food Agric.*, 89: 1066–1075.

Jaworska, G. and Bernaś, E. 2010. Effects of pre-treatment, freezing and frozen storage on the texture of *Boletus edulis* (Bull: Fr.) mushrooms. *Int. J. Refr.*, 33: 877–885.

Jaworska, G., Bernaś, E. and Biernacka, A. 2012. Effect of pretreatments and storage on the amino acid content of canned mushrooms. *J. Food Proc. Preserv.*, 36: 242–251.

Jaworska, G., Bernaś, E. and Mickowska, B. 2011. Effect of production process on the amino acid content of frozen and canned *Pleurotus ostreatus* mushrooms. *Food Chem.*, 125: 936–943

Jaworska, G., Bernaś, E., Biernacka, A. and Maciejaszek, I. 2010. Comparison of the texture of fresh and preserved *Agaricus bisporus* and *Boletus edulis* mushrooms. *Int. J. Food Sci. Techn.*, 45: 1659–1665.

Jaworska, G., Bernaś, E., Cichoń, Z. and Possinger, P. 2008. Estabilishing the optimal period of storage for frozen *Agaricus bisporus*, depending on the preliminary processing applied. *Int. J. Refr.*, 31: 1042–1050.

Jaworska, G., Biernacka, A., Wybraniec, S. and Bernaś, E. 2007. Comparison of the content of vitamin B_1 and B_2 in frozen and canned *Pleurotus ostreatus, Boletus edulis* and *Agaricus bisporus*. *Food. Sci. Techn. Qual.*, 6: 177–185. [in Polish].

Jaworska, G., Gębczyński, P. and Gołyszny, A. 2003. The use of mushrooms (*Agaricus bisporus*) in the production of frozen and canned stuffings. *Food. Sci. Techn. Qual.*, 36: 63–71 [in Polish].

Jaworska, G., Pogoń, K., Bernaś, E. and Duda-Chodak, A. 2015. Nutraceuticals and antioxidant activity of prepared for consumption commercial mushrooms *Agaricus bisporus* and *Pleurotus ostreatus*. *J. Food Qual.*, 38: 111–122.

Jaworska, G., Pogoń, K., Bernaś, E. and Skrzypczak, A. 2014. Effect of different drying methods and 24-month storage on water activity, rehydration capacity and antioxidants in *Boletus edulis* mushrooms. *Drying Techn.*, 32: 291–300.

Jiang, T. 2012. Effect of natamycin in combination with pure oxygen treatment on postharvest quality and selected enzyme activities of button mushroom (*Agaricus bisporus*). *J. Agric. Food Chem.*, 60: 2562–2568.

Jiang, T. 2013. Effect of alginate coating on physicochemical and sensory qualities of button mushrooms (*Agaricus bisporus*) under a high oxygen modified atmosphere. *Posth. Biol. Techn.*, 76: 91–97.

Jiang, T., Feng, L. and Li, J. 2012. Changes in microbial and postharvest quality of shiitake mushroom (*Lentinus edodes*) treated with chitosan–glucose complex coating under cold storage. *Food Chem.*, 131: 780–786.

Jiang, T., Jahangir, M.M., Jiang, Z., Lu, X. and Ying, T. 2010. Influence of UV-C treatment on antioxidant capacity, antioxidant enzyme activity and texture of postharvest shiitake (*Lentinus edodes*) mushrooms during storage. *Posth. Biol. Techn.*, 56: 209–215.

Jiang, T., Zheng, X., Li, J., Jing, G., Cai, L. and Ying, T. 2011. Integrated application of nitric oxide and modified atmosphere packaging to improve quality retention of button mushroom (*Agaricus bisporus*). *Food Chem.*, 126: 1693–1699.

Jimenez, M. and Garcia-Carmona, F. 1997. 4-substituted resorcinols (sulfite alternatives) as slow-binding inhibitors of tyrosinase catecholase activity. *J. Agric. Food Chem.*, 45: 2061–2065.

Jolived, S., Voiland, A., Pellon, G. and Arpin, N. 1995. Main factors involved in the browning of *Agaricus bisporus*. *Mushr. Sci.*, 14: 695–702.

Joshi, V.K., Kaur Mohinder and Thakur, N.S. 1996. Lactic acid fermentation of mushroom (*Agaricus bisporus*) for preservation and preparation of sauce. *Acta Aliment.*, 25: 1–11.

Kader, A.A. 1986. Biochemical and physiological basis for effects of controlled and modified atmospheres on fruits and vegetables. *Food Techn.*, 40: 99–104.

Kader, A.A. and Ben-Yehoshua, S. 2000. Effects of superatmosferic oxygen levels on postharvest physiology and quality os shitake mushrooms *Lentinus edodes* during storge. *Korean J. Food Sci. Techn.*, 21: 461–467.

Kadiri, M. and Fasidi, I.O. 1990. Studies on enzyme activities of *Pleurotus tuber-regium* (Fries) Singer and *Tricholoma lobayensis* Heim at various fruiting body stages. *Die Nahrung*, 34: 695–699.

Kar, A., Chandra, P., Parsad, R. and Dash, S.K. 2004. Microwave drying characteristics of button mushroom (*Agaricus bisporus*). *J. Food Sci. Tech.*, 41: 636–641.

Kermasha, S., Goetgheber, M., Monfette, M. and Rovel, B. 1993. Studies on inhibition of mushrooms polyphenol oxidase using chlorogenic acid as substrate. *J. Agric. Food. Chem.,* 41: 526–531.

Khaskheli, S.G., Zheng, W., Sheikh, S.A., Khaskheli, A.A., Liu, Y., Wang, Y.F. and Huang, W. 2015. Effect of processing techniques on the quality and acceptability of *Auricularia auricula* mushroom pickle. *J. Food Nutr. Res.*, 3: 46–51.

Kidmose, U. and Kaack, K. 1999. Changes in texture and nutritional quality of green asparagus spears (*Asparagus officinalis* L.) during microwave blanching and cryogenic freezing. *Acta Agric. Scand., Section B - Plant Soil Sci.*, 49: 110–116.

Kim, D.M., Baek, H.Y., Yoon, H.H. and Kim, K.H. 1989. Effect of CO_2 concentration in CA conditions on the quality of shiitake mushroom (*Lentinus edodes*) during storage. *Korean J. Food Sci. Techn.*, 21: 461–467.

Kim, J.W., Yoo, I.D. and Kim, W.G. 2006. Free radical-scavenging δ-lactones from *Boletus calopus*. *Planta Med.*, 72: 1431–1432.

Ko, J.A., Lee, B.H., Lee, J.S. and Park, H.J. 2008. Effect of UV-B exposure on the concentration of vitamin D_2 in sliced shiitake mushroom (*Lentinus edodes*) and white button mushroom (*Agaricus bisporus*). *J. Agric. Food Chem.*, 56: 3671–3674.

Kolcuoglu, Y., Colak, A., Sesli, E., Yildirim, M. and Saglam, N. 2007. Comparative characterization of monophenolase and diphenolase activities from a wild edible mushroom (*Macrolepiota mastoidea*). *Food Chem.*, 101: 778–785.

Kondratowicz, J. and Kowałko, P. 2000. Freezing mushrooms in liquid nitrogen. *Refrigeration*, 35: 48–50 [in Polish].

Kotwaliwale, N., Bakane, P. and Verma, A. 2007.Changes in textural and optical properties of oyster mushroom during hot air drying. *J. Food Eng.*, 78: 1207–1211.

Koushki, M., Abras, S.K., Mohammadi, M., Hadian, Z., Poorfallah, N.B., Sharayei, P. and Mortazavian, A.M. 2011. Physicochemical properties of mushrooms as affected by modified atmosphere packaging and $CaCl_2$ dipping. *African J. Agric. Res.*, 6: 5414–5421.

Kreb, M. and Lelley, J. 1991. Preservation of oyster mushrooms by lactic acid fermentation. *In: Science and Cultivation of Edible Fungi', M.J. Maher Conference. Dublin, Republic of Ireland.* 1-6 Sept. Rotterdam, Netherlands.

Kukura, J.L., Beelman, R.B., Peiffer, M. and Walsh, R. 1998. Calcium chloride added to irrigation water of mushrooms (*Agaricus bisporus*) reduces postharvest browning. *J. Food Sci.*, 63: 454–457.

Kumar, A., Singh, M. and Singh, G. 2013. Effect of different pretreatments on the quality of mushrooms during solar drying. *J. Food Sci. Tech.*, 50: 165–170.

Larousse, J. 1986. Probleme nach der Ernie und Konservierungstechniken beim Champignon. Verkürze Übersetzung aus Bulletin de la FNSACC (Verband der französischen Champignonanbauer), pp. 25–35.

Le Loch-Bonazzi, C. and Wolff, E. 1991. Characterization of the flavour properties of the cultivated mushroom (*Agaricus bisporus*) and the influence of drying process. *LWT,* 24: 386–390.

Lee, L., Arul, J., Lencki, R. and Castaigne, F. 1995. A review on modified atmosphere packaging and preservation of fresh fruits and vegetables; Physiological basis and practical aspects – Part 1. *Packaging Techn. Sci.,* 8: 315–331.

Lee, S.E., Kim, D.M. and Kim, K.H. 1991. Changes in quality of shiitake mushroom (*Lentinus edodes*) during modified atmosphere (MA) storage. *J. Korean Soc. Food Nutr.,* 20: 133–138.

Lee, Y.C. and Lee, K.H. 1988. Effects of blanching, chemical dipping, freezing methods and storage period on quality of frozen mushrooms. *Korean J. Food Sci. Techn.*, 20: 536–540.

Leeuwen, J. and Wichers, H.J. 1999. Tyrosinase activity and isoform composition in separate tissues during development of *Agaricus bisporus* fruit bodies. *Mycol. Res.*, 103: 413–418.

Lescano, G. 1994. Extension of mushroom (*Agaricus bisporus*) shelf life by gamma irradiation. *Posth. Biol. Techn.*, 4: 255–60.

Lespinard, A.R., Goni, S.M., Salgado, P.R. and Mascheroni, R.H. 2009. Experimental determination and modelling of size variation, heat transfer and quality indexes during mushroom blanching. *J. Food Eng.*, 92: 8–17.

Li, T., Zhang, M. and Wang, S. 2008. Effects of temperature on *Agrocybe chaxingu* quality stored in modified atmosphere packages with silicon gum film windows. *LWT,* 41: 965–973.

Lin, H., Chen, H. and Lin, F. 2001. Influence of pretreatments and blanching treatments on the yield and color of canned mushrooms. *J. Food Proc. Preserv.*, 25: 381–388.

Linn, W.S. and Gong, H. Jr. 1999. The 21st century environment and air quality influences on asthma. *Curr. Opin. Pulm. Med.*, 5: 21–26.

Liu, Y., Huang, F., Yang, H., Ibrahim, S.A., Wang, Y.F. and Huang, W. 2014. Effects of preservation methods on amino acids and 5'-nucleotides of *Agaricus bisporus* mushrooms. *Food Chem.*, 149: 221–225.

Lopez-Briones, G., Varoquaux, P., Chambroy, Y., Bouquant, J., Bureau, G. and Pascat, B. 1992. Storage of common mushroom under controlled atmospheres. *Int. J. Food Sci. Techn.*, 27: 493–505.

Łobaszewski, E. and Paczyńska, B. 1990. The effect of selected technological processes on mushrooms aroma-traste profile. *Food Industry*, 11: 280–281 [in Polish].

Mahajan, P.V., Oliveira, F.A. and Macedo, I. 2008. Effect of temperature and humidity on the transpiration rate of the whole mushrooms. *J. Food Eng.*, 84: 281–288.

Manzi, P. and Pizzoferrato, L. 2000. Beta-glucans in edible mushrooms. *Food Chem.,* 68: 315–318.

Manzocco, L., Calligaris, S., Mastrocola, D., Nicoli, M.C. and Lerici, C.R. 2001. Review of nonenzymatic browning and antioxidant capacity in processed foods. *Trends Food Sci. Techn.*, 11: 340–346.

Martin-Belloso, O. and Llanos-Barriobero, E. 2001. Proximate composition, minerals and vitamins in selected canned vegetables. *Europ. Food Res. Techn.*, 212: 182–187.

Martinez-Soto, G.R., Ocanna-Camacho, R. and Paredes-Lopez, O. 2001. Effect of pretreatment and drying on the quality of oyster mushrooms (*Pleurotus ostreatus*). *Drying Techn.*, 19: 661–672.

Matser, A.M., Knot, E.R., Teunissen, P.G.M. and Bartels, P.V. 2000. Effects of high isostatic pressure on mushrooms. *J. Food Eng.*, 45: 11–16.

Mattila, P., Ronkainen, R., Lehikoinen, K. and Piironen, V. 1999. Effect of household cooking on the vitamin d content in fish, eggs and wild mushrooms. *J. Food Comp. Anal.*, 12: 153–160.

McEvily, J.A., Iyengar, R. and Otwell, W.S. 1992. Inhibition of enzymatic browning in foods and beverages. *Critic. Rev. Food Sci. Nutr.,* 32: 253–273.

Meng, D., Song, T., Shen, L., Zhang, X. and Sheng, J. 2012. Postharvest application of methyl jasmonate for improving quality retention of *Agaricus bisporus* fruit bodies. *J. Agric. Food Chem.*, 60: 6056–6062.

Minamide, T., Nishikawa, T. and Ogata, K. 1980. Influence of CO_2 and O_2 on the keeping freshness of shiitake (*Lentinus edodes* (Berk.) Sing.) after harvest. *Nippon Shokuhin Kogyo Gakkaishi*, 27: 505–510.

Minato, K., Mizuno, M., Terai, H. and Tsuchida, H. 1999. Autolysis of lentinan an antitumor polysaccharide, during storage of *Lentinus edodes*, shiitake mushroom. *J. Agric. Food Chem.,* 47: 1530–1532.

Morales, F.J. and Jimenez-Perez, S. 2004. Peroxyl radical scavenging activity of melanoidins in aqueous systems. *Europ. Food Res. Techn.*, 218: 515–520.

Murr, D.P. and Morris, L.L. 1975. Effect of storage temperature on postharvest changes in mushrooms. *J. American Soc. Hort. Sci.*, 100: 16–19.

Nagy, M. 1989. Eiweißvelust von Pilzen in Konserven, die aus eingesalzenen und anschließend gewässerten Pilzen hergestellt Worden sind. *Industrielle- Obst - und – Gemuseverwertung*, 74: 291–293.

Nichols, R. 1985. Post-harvest physiology and storage. *In*: Flegg, P.B., Spencer, D.M. and Wood, DA (*eds.*), The Biology and Technology of the Cultivated Mushroom. New York: John Wiley & Sons, pp. 195–210.

Okereke, A. and Beelman, R.B. 1990. Acid-blanching and EDTA effects on yield, quality and microbiological stability of canned mushrooms. *J. Food Sci.*, 55: 1327–1330.

Oms-Oliu, G., Aguiló-Aguayo, I., Martín-Belloso, O. and Soliva-Fortuny, R. 2010. Effects of pulsed light treatments on quality and antioxidant properties of fresh-cut mushrooms (*Agaricus bisporus*). *Posth. Biol. Techn.*, 56: 216–222.

Oszmiański, J. and Lee, C.Y. 1990. Inhibition of polyphenol oxidase activity and browning by honey. *J. Agric. Food Sci.,* 38: 1892–1895.

Palacios, I., Lozano, M., Moro, C., D'Arrigo, M., Rostagno, M.A., Martínez, J.A., García-Lafuente, A., Guillamón, E. and Villares, A. 2011. Antioxidant properties of phenolic compounds occurring in edible mushrooms. *Food Chem.*, 128: 674–678.

Parentelli, C., Ares, G., Corona, M., Lareo, C., Ga´mbaro, A. and Soubes, M. 2007. Sensory and microbiological quality of shiitake mushrooms in modified-atmosphere packages. *J. Sci. Food Agric.*, 87: 1645–1652.

Park, D.L., Sam, M., Rua, J.R. and Robert, F.A. 1991. Direct application of a new hypochlorite sanitizer for reducing bacterial contamination on foods. *J. Food Prot.*, 54: 960–965.

Parry, R.T. 1993. Introduction. *In*: Parry, R.T. (*ed.*), *Principles and Applications of Modified Atmosphere Packaging Foods, Blackie Academic and Professional*, New York, pp. 1–18.

Pei, F., Yang, W.J., Shi, Y., Sun, Y., Mariga, A.M., Zhao, L.Y., Fang, Y., Ma, N., An, X.X. and Hu, O-H. 2014. Comparison of freeze-drying with three different combinations of drying methods and their influence on colour, texture, microstructure and nutrient retention of button mushroom (*Agaricus bisporus*) slices. *Food Biop. Techn.*, 7: 702–710.

Philips, C.A. 1996. Rewiev Modified atmosphere packaging and its effects on microbiological quality and safety of produce. *Int. J. Food Sci. Techn.*, 31: 463–479.

Ponting, J.D. 1960. The control of enzymatic browning of fruits. *In*: Food enzymes. Schultz, H.W. (*ed.*), AVI Publ. Westport, Conn, p. 105.

Rajarathnam, S., Shashirekha, M.N. and Rashmi, S. 2003. Biochemical changes associated with mushroom browning in *Agaricus bisporus* (Lange) Imbach and *Pleurotus ostreatus* (Block & Tsao): Commercial implications. *J Sci. Food Agric.*, 83: 1531–1537.

Ratcliffe, B., Flurkey, W.H., Kuglin, J. and Dawley, R. 1994. Tyrosinase, laccase and peroxidase in mushrooms (*Agaricus*, Crimini, Oyster, and Shiitake). *J. Food Sci.*, 59: 824–827.

Rodrigo, M., Calvo, C., Sanchez, T., Rodrigo, C. and Martínez, A., 1999. Quality of canned mushrooms acidified with glucono-δ-lactone, *Intern. J. Food Sci. Technol.*, 34: 161–166.

Rodriguez-Lopez, J.N., Fenoll, l.G., Tudela, J., Devece, C., Sanchez-Hernandez, D., De los Reyes, E. and Garcia-Canovas, F. 1999. Thermal inactivation of mushroom polyphenoloxidase employing 2450 MHz microwave radiation. *J. Agric. Food Chem.*, 47: 3028–3035.

Roy, S., Anantheswaran, R.C. and Beelman, R.B. 1995. Fresh mushroom quality as affected by modified atmosphere packaging. *J. Food Sci.*, 60: 334–340.

Royse, D.J. 2014. A global perspective on the high five: *Agaricus, Pleurotus, Lentinula, Auricularia* and *Flammulina*. *In:* Singh Manjit (*ed.*), *Proceedings of the 8th International Conference on Mushroom Biology and Mushroom Products, New Delhi, India, pp.* 1–6.

Saltveit, M.E. 1997. A summary of CA and MA requirements and recommendations for harvested vegetables. *In*: *Seventh International Controlled Atmosphere Research Conference. Vegetables and Ornamentals, Univ. Calif., Davis, Postharvest Horticulture Series,* 18: 98–117.

Sanchez-Ferrer, A., Bru, R. and Garcia-Carmona, F. 1989. Novel procedure for extraction of latent grape polyphenol oxidase using temperature – Induced phase separation in Trition X-114. *Plant Physiol.*, 91: 1481–1487.

Sapata, M.L., Ramos, A.C., Ferreira, A., Andrada, L. and Candeias, M. 2009. Changes of quality of *Pleurotus* ssp. carpophores in modified atmosphere packaging. *Acta Sci. Pol., Technol. Aliment.*, 8: 17–22.

Sapers, G.M. 1993. Browning of foods. Control by sulfites, antioxidants and other means. *Food Technol.*, 47: 75–84.

Sapers, G.M. and Miller, R.L. 1992. Enzymatic browning control in potato with ascorbic acid-2-phosphates. *J. Food Sci.*, 57: 1132–1135.

Sapers, G.M. Hicks K.B., Phillips J.G., Garzarella L., Pondish D.L. and Matulaitis R.M. 1989. Control of enzymatic browning in apple with ascorbic acid derivatives, polyphenol oxidase inhibitors, and complexing agents, J. Food Sci. 54: 997–1002.

Sapers, G.M., Miller, R.L., Choi, S.W. and Cooke, P.H. 1999. Structure and composition of mushrooms as affected by hydrogen peroxide wash. *J. Food Sci.*, 64: 889–892.

Sapers, G.M., Miller, R.L., Miller, F.C., Cooke, P.H. and Choi, S.W. 1994. Enzymatic browning control in minimally processed mushrooms. *Food Sci.,* 59: 1042–1047.

Sapers, G.M., Miller, R.L., Pilizota, V. and Kamp, F. 2001. Shelf-life extension of fresh mushrooms (*Agaricus bisporus*) by application of hydrogen peroxide and browning inhibitors. *J. Food Sci.,* 66: 362–366.

Schlimme, D.V. and Rooney, M.L. 1997. Envasado de frutas hortalizas mínimamente procesadas. *In*: Wiley, R.C. (*ed.*), Frutas Hortalizas Mínimamente Procesadas y Refrigeradas, Editorial Acribia S.A., Zaragoza, pp. 131–178.

Shukla, B.D. and Singh, S.P. 2007. Osmo-convective drying of cauliflower, mushroom and greenpea. *J. Food Eng.*, 80: 741–747.

Simon, A., González-Fandos, E. and Vázquez, M. 2010. Effect of washing with citric acid and packaging in modified atmosphere on the sensory and microbiological quality of sliced mushrooms (*Agaricus bisporus* L.). *Food Cont.*, 21: 851–856.

Singla, R., Ganguli, A. and Ghosh, M. 2012. Physicochemical and nutritional characteristics of organic acid-treated button mushrooms (*Agaricus bisporus*). *Food Biopr. Techn.*, 5: 808–815.

Soler-Rivas, C., Ramirez-Anguiano, A.C., Reglero, G. and Santoyo, S. 2009. Effect of cooking, *in vitro* digestion and $CaCO_2$ cells absorption on the radical scavenging activities of edible mushrooms. *Int. J. Food Sci. Techn.*, 44: 2189–2197.

Solomon, P., Wasser, S.P. and Weis, A. 1999. Therapeutic effects of substances occurring in higher Basidiomycetes mushrooms: A modern perspective. *Critical Rev. Immunol.*, 19: 65–96.

Sugumaran, M. and Nellaiappan, K. 1991. Lysolecithin– A potent activator of prophenoloxidase from the hemolymph of the lobster, *Homarus americanas*. *Bioch. Bioph. Res. Com.*, 176: 1371–1376.

Tao, F., Zhang, M., Hangqing, Y. and Jincai, S. 2006. Effects of different storage conditions on chemical and physical properties of white mushrooms after vacuum cooling. *J. Food Eng.,* 77: 545–549.

Taylor, S.L., Higley, N.A. and Bush, R.K. 1986. Sulfites in food. Uses, analytical methods, residues, fate, exposure assessment, metabolism, toxicity and hypersensitivity. *Adv. Food Res.*, 30: 1–76.

Vamos-Vigyazo, L. 1981. Polyphenol oxidase and peroxidase in fruits and vegetables. *Crit. Rev. Food Sci.,* 15: 49–127.

Vamos-Vigyazo, L. 1995. Prevention of enzymatic browning in fruits and vegetables. A Review of principles and practice. Enzymatic browning and its prevention. American Chemical Society, Washington, pp. 49–62.

Vetter, J. 2003. Chemical composition of fresh and conserved mushroom. *Europ. Food Res. Techn.*, 217: 10–12.

Villaescausa, R. and Gil, M.I. 2003. Quality improvment of *Ploreatus* mushrooms by modified atmosphere packaging and moisture. *Posth. Biol. Techn.*, 28: 169–179.

Vivar-Quintana, A.M., González-San José, M.L. and Collado-Fernández, M. 1999. Influence of canning process on colour, weight and grade of mushrooms. *Food Chem.*, 66: 87–92.

Walde, S.G., Velu, T., Jyothirmayi and Math, R.G.. 2006. Effects of pretreatments and drying methods on dehydration of mushroom. *J. Food Eng.*, 74: 108–115.

Wasser, S.P. and Weis, A.L. 1999. Medicinal properties of substances occurring in Higher Basidiomycetes Mushrooms: Current perspectives (Review). *Int. J. Med. Mush.*, 1: 31–62.

Weijn, A., Van den Berg-Somhorst, D.B., Slootweg, J.C., Vincken, J.P., Gruppen, H., Wichers, H.J. and Mes, J.J. 2013. Main phenolic compounds of the melanin biosynthesis pathway in bruising-tolerant and bruising-sensitive button mushroom (*Agaricus bisporus*) strains. *J. Agric. Food Chem.*, 6: 8224–8231.

Worka, M.T., Bushwaya, A.A., Worka, RW. and Raoul, C. 1997. Pelletiera the effect of cryogenic freezing on the quality of soft and hard shell lobsters (*Homarus americanus*). *J. Aquatic Food Prod. Technol.*, 6: 53–64.

Wu, S.R., Zhao, C.Y., Hou, B., Tai, L.M. and Gui, M.Y. 2013. Analysis on chinese edible fungus production area layout of nearly five years. *Edible Fungi China*, 1: 51–53.

Xing, Z., Wang, Y., Feng, Z. and Tan, Q. 2008. Effect of different packaging films on postharvest quality and selected enzyme activities of *Hypsizygus marmoreus* mushrooms. *J. Agric. Food Chem.*, 56: 11838–11844.

Xiong, X.L., Xing, Z.T., Feng, Z., Tan, Q. and Bian, Y.B. 2009. Effect of ^{60}Co g-irradiation on postharvest quality and selected enzyme activities of Pleurotus nebrodensis, LWT, 42: 157–161.

Yang, D.C. and Maguer, M. 1992. Mass transfer kinetics of osmotic dehydration of mushrooms. *J. Food Proc. Preserv.*, 16: 215–231.

Ye, J.J., Li, J.R., Han X.X., Zhang, L., Jiang, T.J. and Xia, M. 2012. Effects of active modified atmosphere packaging on postharvest quality of shiitake mushrooms (*Lentinula edodes*) stored at cold storage. *J. Integ. Agric.*, 11: 474–482.

Zagory, D. 1995. Principles and practice of modified atmosphere packaging horticultural commodities. *In:* Farber, J.M. and Dodds, K.L. (*eds.*), Principles of Modified Atmosphere and Sous-vide Product Packaging, Technomic Publishing Co Inc., Lancaster, pp. 175–204.

Zanderighi, L. 2001. How to design perforated polymeric films for modified atmosphere packs (MAP). *Packaging Techn. Sci.*, 14: 253–266.

Zivanovic, S. and Buescher, B. 2004. Changes in mushroom texture and cell wall composition affected by thermal processing. *J. Food Sci.,* 69: SNQ44–SNQ49.

Zivanovic, S., Buescher, R. and Kim, S.K. 2003. Mushroom texture, cell wall composition, color and ultrastructure as affected by pH and temperature. *J. Food Sci.,* 68: 1860–1865.

Zivanovic, S., Buescher, R.W. and Kim, K.S. 2000. Textural changes in mushrooms (*Agaricus bisporus*) associated with tissue ultrastructure and composition. *J Food Sci.*, 65(8): 1404–1408.

Źródłowski, Z. 1995. The influence of washing and peeling of mushrooms *Agaricus bisporus* on the level of heavy metal contaminations. *Polish J. Food and Nutr. Sci.*, 4(45): 26–33.

3.2

Nematode Pests of Mushrooms and Their Management

H.K. Bajaj[1]*, R. S. Kanwar[2] and Surjeet Singh[3]

ABSTRACT

Mushroom cultivation is challenged by a number of biotic factors such as insects, mites, bacteria, fungi and nematodes. Among these, the nematode pests of mushroom are the most dreaded ones. Several species of mycophagous, bacteriophagus, predatory nematodes are found in compost. Once they get entry into the mushroom house, it becomes almost impossible to get rid of these worms. Agaricus bisporus (button or temperate mushroom) is more susceptible to these pests. Maintenance of hygienic conditions is undoubtedly the best way of protecting the mushroom crop from nematode infestation, but is better said than done, in developing countries. Nematode control is difficult in this crop as chemicals can not be applied due to residual problem. This chapter deals with nematode pests of button mushroom, on which most of the investigations have been concentrated. Preventive measures are suggested for their management.

Key words: Mushroom, Nematodes, Management, Preventive measures.

1. INTRODUCTION

Mushrooms contain numerous species that are edible, of medicinal value or even poisonous. *Agaricus bisporus* (Button or temperate mushroom), *Auricularia auricula* (Black ear mushroom), *Calocybe indica* (White milky mushroom- 'dudh chatta, safed dudhiya mushroom), *Ganoderma lucidum* (Reishi mushroom), *Lentinus edodes* (Shiitake or Japanese mushroom),

[1] CCS Haryana Agricultural University, Hisar- 125 004, Haryana.
**Corresponding author:* E-mail: kanwarrambir@gmail.com

Pleurotus species (*P. florida*, *P. flabellatus*, *P. sajor-caju*, *P. cornucopiae*, *P. ostreatus*, *P. eryngii*- Oyster mushroom- 'dhingri') and *Volvariella volvacea* (Paddy straw mushroom, Chinese or subtropical mushroom) are different species presently being cultivated in India. However, *A. bisporus* (Lange) Singer contributes more than 90% of total mushroom production in India and most of the work on nematodes problems has been conducted on this edible fungus only.

Mushroom cultivation is challenged by a number of biotic factors such as insects, mites, bacteria, fungi and nematodes. Among these, the nematode pests of mushroom are the most dreaded ones. Once they get entry into the mushroom house, it becomes almost impossible to get rid of these worms. Their life cycle is very short (8-10 days) due to which multiplication is very fast. Mushroom nematodes are fungivorous and feed on a variety of fungi, including plant pathogenic ones. Like plant parasitic nematodes, they are stylet-bearing and have the same feeding mechanism. They devitalize the growing mushroom hyphae by sucking cell sap and thus adversely affect its yield. Their presence in mushroom beds is associated with severe losses, often leading to crop failure. Bukowski (1964) recorded 60% of total crop failure in mushroom due to nematode infestation. In India, 84% of nematode infested beds showed poor initial flushes and no button production at later stages (Khanna, 1994). Authors have recorded a total crop loss in several mushroom houses in Haryana, due to *Aphelenchoides swarupi* and *A. asterocaudatus* (Fig. 4). Unlike the management practices commonly used for plant parasitic nematodes like crop rotation, summer ploughings, use of chemicals *etc.*, are not applicable here due to different nature of the crop.

2. NEMATODES

Mushroom beds harbour a variety of microflora and fauna. Nematodes belonging to different trophic groups, commonly encountered in mushroom houses include - fungal feeder, saprophagous or bacteriophages, predators and insect parasites.

2.1. Fungivorous Nematodes

Fungal feeding nematodes belonging to genera *Aphelenchoides* Fischer, *Aphelenchus* Bastian (Order Aphelenchida) and *Ditylenchus* Filipjev (Tylenchida) are common in mushroom beds (Fig. 1 and Table 1). Occasionally, species of *Paraphelenchus* (Micol.) (Aphelenchida), *Hexatylus* Goodey, *Deladenus* Thorne and *Pseudhalenchus* Tarjan (Tylenchida) are also encountered. Common characteristics of fungal feeding nematodes of mushrooms are:

1. *Wide host range*: They feed upon a variety of fungi. However, rate of multiplication varies with fungal species, some of which may even be non hosts.
2. *Short life cycle*: Fungal feeding nematodes of mushroom can reproduce and complete life cycle over a wide range of temperature. At optimum temperature of about 25°C, the life cycle is completed in 7-10 days.
3. *High fecundity*: The population of these nematodes increases exponentially in the mushroom beds due to high fecundity. One female lays up to three eggs per hour.

Table 1: Myceliophagous nematodes associated with button mushroom in India.

Nematode	***State***	***References***
Aphelenchoides agarici	Haryana and H.P.	Seth and Sharma (1986), Bajaj and Walia (1999)
Aphelenchoides asterocaudatus	Haryana	Bajaj and Walia (1999)
Aphelenchoides bicaudatus	Haryana	Anon (2003)
Aphelenchoides brushimucronatus, Bajaj and Walia	Haryana	Bajaj and Walia (1999)
Aphelenchoides composticola	Haryana, H.P. and Punjab	Bajaj and Walia (1999), Khanna and Sharma (1988)
Aphelenchoides delhiensis	Haryana	Bajaj and Walia (1999)
Aphelenchoides indicus, Chawla *et al.*	Haryana	Bajaj and Walia (1999)
Aphelenchoides minoris and	Haryana and H.P. Ebsary	Bajaj and Walia (1999), Seth Sharma (1986)
Aphelenchoides myceliophagus	Haryana and H.P.	Anon (2003), Seth and Sharma (1986)
Aphelenchoides neocomposticola	Haryana and H.P.	Seth and Sharma (1986)
Aphelenchoides sacchari	Delhi, Haryana, H.P., Punjab, J&K and U.P.	Sharma *et al.* (1981), Thapa *et al.* (1981)
Aphelenchoides spinohamatus, Bajaj and Walia	Haryana	Bajaj and Walia (1999)
Aphelenchoides swarupi	Haryana and H.P.	Anon (2003), Seth and Sharma (1986)
Aphelenchus avenae	Haryana	Anon (2003)
Aphelenchus radicicolus	Haryana	Anon (2003)
Hexatylus viviparous	Haryana	Anon (2003)
Deladenus sp.	Haryana	
Ditylenchus myceliophagus	Delhi, Haryana, H.P., Punjab, J&K and U.P.	Sharma *et al.* (1981), Thapa *et al.* (1981) and Anon. (2003)

2.1.1. *Aphelenchoides species*

Though several species of *Aphelenchoides* have been encountered from mushroom beds, but detailed studies have been conducted on *A. agarici*

Seth and Sharma, *A. asterocaudatus* Das, *A. composticola* Franklin, *A. neocomposticola* Seth and Sharma, *A. sacchari* Hooper and *A. swarupi* Seth and Sharma. These nematodes have a wide host range and feed upon a variety of fungi (Khanna and Sharma, 1989). Males are common in these species except for *A. asterocaudatus*) that reproduces uniparently. They are active and voracious feeders. They feed at different sites at intervals of 1 to 30 seconds on healthy, growing hyphae. Nematode may prick at the same cell during feeding with 1-3 stylet thrusts to accelerate the movement of protoplasm towards the nematode body. Distinct shrinkage of hyphal cell at feeding site and emptying of hyphae invariably occurs as a result of nematode feeding (Fig. 2). Bajaj *et al.* (2013) calculated the suction force of feeding of *A. swarupi* on *Fusarium oxysporum*.

A. composticola- Life cycle of *A. composticola* from egg to adult on button mushroom is completed in 22, 10 and 8 days at 13, 18 and 23°C, respectively (Cayrol, 1967). The nematode feeds upon a variety of fungi and can also survive freezing and slow desiccation (Hesling, 1977). Swarming and aggregation are common phenomena observed in this species. In India, this species has been recorded from Himachal Pradesh (H.P.), Punjab and Haryana from mushroom beds. Initial level of 10 nemas can deplete mycelium completely in 40 days and multiplying over 1000 times.

A. agarici- Described by Seth and Sharma in 1986 this species is prevalent in Himachal Pradesh. It completes life cycle in 8 days at 28 ± 1°C and multiplication rate is over 800 in 25 days at 28°C.

A. saprophilus and *A. swarupi*- These species also reproduce sexually. *A swarupi* can survive slow drying and starvation for at least two years in a culture flask where they acquire tight spiral, coiled to spring like postures and revive on getting moisture (Fig. 2).

A. myceliophagus - This amphimictic species is common in mushroom houses of H.P. One hundred nemas of this species resulted in populations of 179, 8017 and 10450 after 15, 30 and 40 days upon *A. bisporus in vitro* (Khanna and Sharma, 1988).

A. neocomposticola- This nematode could multiply at 12-25°C on mycelium of button mushroom and multiplication rate was 12.33, 677.3 and 942 after 15, 30 and 40 days, respectively (Khanna and Sharma, 1988).

A. sacchari- This species is prevalent in mushroom houses of Delhi, Haryana, H.P., Jammu and Kashmir, Punjab and Uttar Pradesh (Sharma *et al.*, 1981; Thapa *et al.*, 1981).

A asterocaudatus- Males are absent in this species which is common in Haryana in rhizosphere and mushroom houses. It can also survive slow drying and starvation in culture flasks for more than two years.

2.1.2. *Aphelenchus species*

Aphelenchus avenae Bastian and *A. radicicolus* (Cobb) Steiner are found frequently in large number in mushroom beds during mid crop season, but their populations decline drastically by the end of the crop. Like *Aphelenchoides*, *Aphelenchus* species are also active feeders of mushroom mycelia. Penetration of walls of fungal hyphae is usually accomplished in 6-7 seconds; the median bulb then pulsates usually for 8 seconds. Repeated feeding seems to be responsible for the collapse of fungal hyphae (Juhl, 1966; Fisher and Evans, 1967).

Aphelenchus avenae- Most of the populations of *A. avenae* reproduce by parthenogenesis. A female lays about 225 eggs at a rate of one egg in 3 hours. Generation time varies from 6-12 days depending upon temperature (Hechler, 1962; Jairajpuri, 1968). When inoculated on *A. bisporus* mycelia, growing on PDA, in 9 cm diameter petri dishes, a 1413 fold increase in population was recorded after 30 days of inoculation of five nematodes (Anon, 2003). This species enter cryptobiosis under adverse environmental conditions of anoxybioisis and slow drying and can be revived after 123 months (Cooper and Van Gundy, 1970).

A. radicicolus- This species is very similar in morphology to *A. avenae.* However, males are common in this species and reproduction is by amphimixis.

2.1.3. *Ditylenchus species*

D. myceliophagus Goodey is the only species of the genus investigated in details in context with button mushroom in several countries (Hesling, 1972). It is considered to be a major pest along with *A. composticola*. In India, this species has been recorded from the cropping beds of button mushroom in H.P., Punjab, Haryana, Delhi, Jammu and Kashmir, and U.P. It caused mycelial destruction of 15, 40 and 70% at 10, 100 and 10,000 nemas/ milk bottle *in vitro* with multiplication rates of 12253, 632 and 169 after 60 days (Sharma *et al.*, 1985). *D. myceliophagus* thrives and breeds optimally at temperature between 18 and 20°C but fails to multiply at 5°C. The life cycle is completed in 40 days at 13, 26 days at 18 and 11 days at 23°C (Cayrol, 1962; Evans and Fisher, 1969). The nematode can remain viable in cryptobiotic stage for at least 3 and half years, when dried slowly (Cayrol, 1970). When *D. myceliophagus* over colonizes its mycelial substrate, it emerges in swarm from it, moving in water films and aggregating as they dry out. In mushroom houses these aggregates form long stalactites of up to 450000 individuals hanging beneath boxes of compost, stalactites break off their weight or as they dry out, fall on to boxes beneath. This species can multiply on *Alternaria alternata, Agaricus bisporus, Fusarium solani, Mucor* species, *Pleurotus sajor-caju, Rhizoctonia solani, R. bataticola,*

Sclerotium sclerotiarum, and *Trichoderma harzianum,* but fails to reproduce on *Penicillium pinophilum, S. rolfsii* and *Aspergillus niger* (Das and Bajaj, 2007).

Feeding process (Fig. 2) consists of exploration, penetration of stylet, injection of dorsal gland secretions, ingestion of hyphal contents and stylet withdrawal phases. Nematode prefers new mycelium and generally feeds near the septum. Feeding in a particular cell continues for two to five and half an hour, and no shrinkage of the host cell is observed just after feeding (Das and Bajaj, 2009).

2.2. Other Nematodes

Females of *Hexatylus viviparous* Goodey are seldom encountered from the mushroom beds. Relationship of this nematode with mushroom cultivation has yet not been worked out.

Deladenus species have alternation of insectivorous and myceliophagous generations. Adult males and females of myceliophagous generation and gravid inseminated infective females of this genus are occasionally encountered in mushroom beds. Effect of myceliophagous generation on mushroom production or insect parasitic generation for the management of insects are the areas yet to be explored.

2.3. Saprophagous/ Bacteriophagous Nematodes

Mushroom beds constitute a common niche for mushroom, bacteria, other saprophages and nematodes. Effects of bacteriophagous nematodes vary from beneficial, neutral or antagonistic, depending upon quality of pasteurization of compost, a type of bacteria present in the compost or casing material, and time of introduction of nematodes and bacteria (Grewal, 1991).

Bacteriophagous nematodes belonging to orders Rhabditida, Cephalobida and Diplogasterida are frequently encountered in mushroom houses. They reproduce at a very fast rate. They are unable to establish themselves in well pasteurized compost because of (i) Quick growing mycelia absorb moisture from compost that consequently restricts nematode activity, (ii) *A. bisporus* mycelia create nematostatic environment by producing volatile substances with antibiotic activity, (iii) Temperature of 25°C required for spawn run and heat generated by actively growing mycelium may prove lethal to several species of saprobes. Absence or inadequate supply of bacteria hampers the reproduction and survival of saprophagous nematodes. On the other hand, inadequately pasteurized compost supports the multiplication of bacteria and indirectly saprophagous nematodes that by themselves are incapable of degrading organic matter or feed upon mushroom mycelium.

Their presence in great number in such inadequately pasteurized compost affects the spawn run that ultimately results in reduced mushroom yield. The effect of nematodes may not be due to feeding on mycelium but due to (i) Production and accumulation of nematode by products which ultimately inhibit mycelium / sporophore growth, (ii) Since many bacteria remain viable in nematode digestive system, their spread in mushroom beds is facilitated by nematodes.

Grewal and Richardson (1991) reported adverse effect on the quality and quantity of mushroom when *Caenorhabditis elegans* (Osche) were introduced along with casing material at the rate of 267 or more nema per g of casing material. Nematode rapidly colonized the casing layer, produced large populations on top of the casing material. Consequently, significant reduction in mushroom yield, disturbed flushing pattern and distorted buttons occurred. Gills of distorted mushrooms turned violet and brown and supported actively reproducing population. Swarming nematodes on sporophores were also recorded. Several other workers (Ingratta and Olthof, 1978; Hesling, 1979; Mc Leod and Nair, 1981; Sanderson *et al.*, 1981; Kaufman *et al.*, 1984; Chandran, 2000) have reported similar or even greater number of rhabditid nematodes for compost and casing material. However, in India, inert material like burnt rice husk or soil used as casing material is usually treated with formaldehyde solution prior to covering the mycelial run and, therefore, are unlikely to harbour such a large number of nematodes. Recently, *Trilabiatus* spp. have been reported to affect the growth of button mushroom (Zhi *et al.*, 2011). There are few reports wherein saprophagous nematodes have been found beneficial in increasing mushroom production (Cayrol *et al.*, 1981; Ingratta and Olthof, 1978; Grewal and Richardson, 1991)

2.4. Predatory Nematodes

Predacious nematodes belonging to Aphelenchida and Diplogasterida are common in mushroom beds in India. Amongst them species of *Seinura* Fuchs (Aphelenchida) are the most frequent and abundant.

Vats *et al.* (2001, 2004) observed frequent occurrence of two species of this genus, *S. paratenuicaudata* Geraert and *S. oxura* Goodey in high numbers in mushroom beds near the end of crop season in Haryana. The life cycle and feeding of *S. paratenuicaudata* was studied using *Aphelenchus avenae* as prey (Vats *et al.*, 2004). This species reproduced uniparentally. Eggs were laid only after feeding and life cycle was completed in 4-5 days. For feeding, the predator inserts its stylet deep into the prey body (Fig. 4) and this is followed by injection of the contents of the dorsal esophageal glands, which paralyze the prey. This is followed by ingestion of body contents (Bajaj *et al.*, 2014). Predator could feed and reproduce on several funigivorus nematodes like *Aphelenchoides bicaudatus* (Imamura) Filipjev and Sch. Stek., *A. composticola, A. avenae, A. radicicolus* and *D. myceliophagus*, common in mushroom houses.

Diplogasterids are also common in compost and mushroom beds. These nematodes have varied feeding habits ranging from saprophagy to predatory. *Fictor composticola* Khan *et al.*, was recovered from the mushroom compost. This nematode is a voracious feeder (Fig. 5) of myceliophagous nematodes (*Aphelenchus avenae*, *A.radicicolus*, *Aphelenchoides swarupi* and *A. asterocaudatus*), completes life cycle in 3-4 days and has high fecundity. It also survives and multiplies on bacteria and, seems to have a great potential for the biological control of mushroom nematodes (Keshari and Kanwar, 2015; Bajaj and Kanwar, 2015).

3. YIELD LOSSES AND PATHOGENICITY

Pathogenicity trials involving fungal feeders on mushroom have been conducted (i) *In vitro* (Mycelium grown in Petri plate/ milk bottle), (ii) Grown in polythene bags, and (iii) Mushroom beds. *In vitro* testing is very useful to test the suitability of a mushroom species/ strain to a particular nematode. Thapa *et al*. (1983) and Khanna and Sharma (1989) tested reproduction of some species of nematodes on different edible fungi. Khanna and Sharma (1988) compared the mycelial destruction of *A. bisporus* by *A. agarici*, *A. myceliophagus*, *A. neocomposticola* and *A. composticola in vitro* and found that all of them multiplied and caused complete destruction after 40 days of inoculation except for *A. neocomposticola* where only 80% of mycelium was damaged. *A. myceliophagus* and *A. composticola* were more pathogenic than *A. agarici* and *A. neocomposticola* after 30 days. In laboratory studies, five to 100 *Aphelenchus avenae* destroyed *A. bisporus* mycelia completely in 30 days *in vitro* (Vats *et al.*, 2002b). However, *in vitro* results may not give clear picture about comparative pathogenicity in mushroom house due to different environmental conditions.

Studies conducted in mushroom grown in polythene bags have shown that the extent of damage and yield losses depend upon time of entry, initial population and species of nematode. In general, nematode entering mushroom beds at casing are less harmful than those introduced at the time of spawning (Arrold and Blake, 1968; Anon, 2003) and this can be attributed to healthy growth of mycelium prior to attack by nematodes at the time of casing. Losses estimated for different nematodes in button mushroom are depicted in Table 2. Vats *et al.* (2002a) showed that fungal feeding nematodes are less pathogenic when introduced in large sized mushroom beds than in polythene bags due to different environmental factors prevailing there. Populations of *Ditylenchus myceliophagus* occurring in India are less pathogenic than occurring in Europe, where they caused cent percent crop loss if present at the time of spawning (Arrold and Blake, 1968; Khanna, 1993; Vats *et al.*, 2002a).

Table 2: Yield losses in *Agaricus bisporus* due to myceliophagous nematodes.

Nematode	*Inoculum level/ 100 g compost*	*Nematode entry at*	*% Yield loss*	*References*
Aphelenchoides saccharі	10	Spawning	98.2	Sharma *et al.* (1984)
	100–10000	Spawning	100	Sharma *et al.* (1984)
	100	Casing	92	Sharma *et al.* (1984)
	1000	Casing	87	Sharma *et al.* (1984)
	10000	Casing	100	Sharma *et al.* (1984)
A. composticola	1	Spawning	26.0	Arrold and Blake (1968)
	10	Spawning	30.0 35.86* 27.16**	Arrold and Blake (1968), Anon (2003) and Vats *et al.* (2002a)
	500	Spawning	95	Khanna (1991)
	500	Casing	64	Khanna (1991)
Ditylenchus myceliophagus	20, 100, 300	Spawning	100	Arrold and Blake (1968)
	10	Spawning	33.4* 18.9**	Anon (2003) and Vats *et al.* (2002a)
	100	Spawning	89 23.8*	Khanna (1993) and Anon (2003)
	500	Spawning	49.3*	Anon. (2003)
	1000	Spawning	62.2*	Anon (2003)
	10	Casing	30.7*	Anon (2003)
	20	Casing	50.0	Arrold and Blake (1968)
	100	Casing	68.0 40 22.0*	Arrold and Blake (1968), Khanna (1993) and Anon (2003)
	300	Casing	75.0	Arrold and Blake (1968)
	500	Casing	47.2*	Anon (2003)
	1000	Casing	64.7*	Anon (2003)
Aphelenchus avenae	10	Spawning	55.5* 23.3**	Anon (2003) and Vats *et al.* (2002a)
	100	Spawning	45.3*	Anon (2003)
	500	Spawning	57.7*	Anon (2003)
	1000	Spawning	61.4*	Anon (2003)
	10	Casing	25.8*	Anon (2003)
	100	Casing	32.9*	Anon (2003)
	500	Casing	53.7*	Anon (2003)
	1000	Casing	58.0*	Anon (2003)

*- 3 Kg bag, ** Mushroom bed with 70 Kg compost.

Peak populations of myceliophagous nematodes in mushroom beds also vary with nematodes species. *Aphelenchoides composticola* and *D. myceliophagus* are recorded in the greatest numbers 7 and 13 weeks after spawning, respectively (Arrold and Blake, 1968). *A. avenae* population declines very sharply after mid crop season so much so that very few are encountered at the end of crop season (Vats *et al.*, 2002b).

4. NEMATODE CONTAMINATION

Nematodes infecting mushroom crop find their way any time from initiation of composting till last harvest. But most of them enter either with ingredients of compost or casing material. At composting out of various ingredients used (wheat/ rice straw, inorganic/ organic fertilizer (Farm Yard Manure, chicken/ horse manure) *etc.*, animal refuses are the major source of contamination (Khanna *et al.*, 2006). Majority of seasonal growers prepare their compost in open by long method. Flies and other insects, as carriers, also introduce nematodes to compost. Since many aphelenchids are known to undergo anhydrobiosis and can be carried away by the wind, their entry through air is another mean of contamination. As the majority of mushroom feeding nematodes survive and multiply on a wide variety of fungi, they build up their populations in compost on such food source. Used racks and trays are other sources of contamination.

Other sources of contamination from spawning to casing, and casing to harvesting are insect carriers, mites, water or air.

Out of several components of casing material, spent compost, if used, is the major source of nematode contamination. Farm Yard Manure, garden soil or other fungi supporting material also introduce nematodes in the mushroom beds.

5. SYMPTOMS

Nematodes produce non-specific symptoms of poor mycelial growth and sporophore production. Symptom expression depends upon the time of entry of nematode in mushroom beds. Very poor to no mycelial network formation or patchy mycelial growth occurs in case nematodes are present in the compost at the time of spawning (Figs. 3 and 4). Such mushroom beds yield little or no buttons. In case nematode-contaminated casing material is used, first few flushes are almost normal, but little yield is obtained in subsequent flushes that become irregular. Other symptoms reported in the literature are: sinking of compost surface, soggy compost with foul smell, disturbed flush pattern, including delayed and reduced button production, and mis-shapening, notching and discoloration of fruiting bodies. Involvement of bacteria with decaying mycelium destroyed by nematodes has been suspected in production of such symptoms.

6. MANAGEMENT

6.1. Chemical Control

In addition to usual care in using chemical pesticides, mushroom require special precautions because cropping continues for 6-8 weeks, and mushroom is harvested and consumed soon after button appear. Therefore, chemicals used should also fulfill the following criteria.

- They should be non toxic and non persistent.
- They should reach to the target nematode.
- Once spawned, compost should not be disturbed. Therefore, any treatment should, preferable, be given either prior to spawning or at the time of casing.
- Chemicals that are safe in compost may deteriorate the quality of buttons. Special care has to be taken in case chemical is applied at the time of casing.
- Since a good quality compost contains 70% moisture and only slight deviation is permissible, nematicidal drenches must not add more than 2% moisture.

Unfortunately, most of the nematicides tested do not conform to these conditions. Wyatt (1977, '78a, '78b) studied the effect of pesticides and found that when applied in compost or casing; most of them adversely affect the mushroom production. Further, chemicals like aldicarb have shown a residual effect in sporophores whereas dazomet is highly toxic to mushroom mycelium (Cayrol and Ritter, 1971). Nevertheless, Thionazin has been advocated for managing *D. myceliophagus* and *A. composticola* and is non-fungitoxic up to 80 ppm (Hesling, 1966, '72). It improves mushroom yield by reducing the population of fungivorous nematodes. However, doubts have been raised about its toxic hazards to human beings (Hesling and Kempton, 1969; Cayrol and Ritter, 1971; Fiori *et al.*, 1987). McLeod and Khair (1978) found that incorporation of Fenamiphos @ 20 mg/kg in *Aphelenchoides composticola* infested compost is a good practical preventive measure.

6.2. Biological Control

6.2.1. *Microorganisms*

Control of nematodes of mushroom through microorganisms and nematode predators has been tried. Nematode trapping fungi (*Arthrobotrys oligospora*, *A. superba*, *A. irregularis* and *A. robusta*) have been tested against *A. composticola* and/or *D. myceliophagus in vitro* or *in vivo* (Kermarrec, 1973; Khanna and Sharma, 1990). These fungi do not compete with *A. bisporus* in compost and increase mushroom yield by checking the nematode population. A strain of *A. robusta* has been developed by Cayrol *et al.* (1978) and is marketed as Royal 300 for the management of *D. myceliophagus*. This strain is a good colonizer of compost and non antagonist to *A. bisporus*. It is introduced into compost in rye grain @ 1% with spawn. Another fungus *Candelariella musiformis* has been found effective against nematodes *in vitro* (Khanna and Sharma, 1990). Cayrol (1974) reported that bacterium, *Bacillus thuringiensis* suppress *D. myceliophagus* population and increase mushroom yield.

6.2.2. *Predatory nematodes*

Seinura and *Fictor* species are common predators of fungivorous nematodes commonly occurring in mushroom houses of northern India. Both the nematodes are voracious feeders, have a wide host range, high fecundity and potential to survive under adverse environmental conditions. There are strong indications that they are regulating the nematode pest populations as natural enemies. Detailed systematic studies are needed to exploit these predators for managing nematodes pests of mushroom.

6.2.3. *Botanicals*

Addition of selected plant parts or their products in compost improves the physicochemical properties of compost, increases the population of nematode trapping- antibiotic producing fungi, suppresses competitive pathogenic moulds and increases the yield. The nematodes are also killed due to plant extract toxicity, plant decomposition or activity of nematode trapping fungi or fungal antibiosis (Grewal, 1989a). A number of plant species have been screened for their nematicidal properties against mushroom nematodes as extract, dry powder form apart from botanical pesticide or cake (Sohi *et al.*, 1987; Khanna *et al.*, 1988, '96; Grewal, 1989a, 89b) and quite a few of them have been found effective. Addition of neem leaf powder, oil cakes of neem (*Azadirachta indica*), karanj (*Pongamia pinnata*), coconut (*Cocos nucifera*), castor (*Ricinus communis*) and groundnut (*Arachis hypogaea*) in nematode infested compost proved effective in managing *A. composticola* (Rao and Pandey, 1991; Khanna and Sharma, 1994). Vats *et al.* (2003b) evaluated Achook (0.15% Azadirachtin 60% w/w; emulsifier 5% + treated neem oil 35%), Neem Seed Kernel Water Extract (NSKWE) and Neem Seed Kernel Powder (NSKP) for their efficacy against *A. composticola* in button mushroom grown in polythene bags. Observations revealed that 64.4% losses caused by 100 *Aphelenchoides composticola* per Kg compost can be compensated by the addition of 2% NSKWE or 800 ppm achook. Application of 2% NSKP or 800 ppm achook at spawning or casing or 2% NSKE also resulted in more sporophore yield as compared to untreated, inoculated check. Nematode populations were significantly reduced in treatments receiving botanicals. Integration of 1% NSKWE + 400 ppm Achook at the time of spawning gave the best results in improving mushroom yield and in reducing *Aphelenchoides composticola, Aphelenchus avenae* and *Ditylenchus myceliophagus* populations drastically in 70 Kg mushroom beds (Vats *et al.*, 2003a). After detailed experiments done for three years, the addition of 4% NSKWE in compost at the rate of 7.5 l/q has been recommended in Haryana (Anon, 2011).

6.2.4. *Resistance*

Till date, no resistance in *A. bisporus* has been found against *Aphelenchoides* species or *D. myceliophagus*.

7. PREVENTION

Maintenance of hygienic conditions is undoubtedly the best way of protecting the mushroom crop from nematode infestation, but is better said than done, in developing countries. Some important steps to prevent the entry of nematodes are:

1. Cropping should be done in bricked, ventilated, 14-16 mesh- wire shuttered rooms to prevent the entry of flies.
2. Composting platform should preferably be raised, cemented and disinfected to avoid direct contact between compost and nematode-contaminated soil.
3. All raw materials used for composting and casing should be free of nematodes.
4. Persons engaged in composting should enter the cropping area only after cleaning their hands and feet properly.
5. Compost carrying baskets, equipments and tools should be thoroughly cleaned before and after use. Casing soil should be disinfected with some disinfectant like formalin before use.
6. Spent compost, the major source of contamination, should never be used for casing.
7. Fresh plastic bags should be preferred over wooden trays/ shelves which may harbour nematodes at corners. Wooden trays, if to be used, should be thoroughly washed and disinfected.

8. NEMATODE PROBLEMS OF OTHER MUSHROOMS

Meager information is available with respect to nematode problems of other mushrooms. *Pleurotus sajor-caju* is resistant to several species of *Aphelenchoides* like *A. composticola*, *A. agarici*, *A. asterocaudatus*, *etc.*, *Aphelenchus avenae*, *A. radicicolus* (Bajaj *et al.*, 2007) but is susceptible to *Ditylenchus myceliophagus* (Das and Bajaj, 2007). One species of *Paraphelenchus* found associated with milky mushroom, *Calocybe indica,* caused complete destruction of its mycelium growing on Potato Dextrose Agar in flasks in 30 days at an inoculum level as low as 10 nemas per flask (Fig. 6; Vats *et al.*, 2006). *A. composticola, Trilabiatus* spp. *Acrobeloides* sp. *Mesorhabditis* sp. damaged the hyphae of wood ear mushroom (*Auricularia auricular*) causing its browning and withering (Zhi *et al.*, 2011; Lin, *et al.*, 2011; Fei *et al.*, 2012).

9. FUTURISTIC APPROACHES

Maintenance of hygienic conditions during mushroom cultivation is undoubtedly the best way of minimizing the losses caused by nematodes, but is not feasible for seasonal growers considering their economic conditions. Government and Non-Government organizations or cooperatives may help in setting up pasteurization facilities for composting.

Resistance to a number of *Aphelenchoides* and *Aphelenchus* species has been identified in *Pleurotus sajor-caju* and *Agaricus edulis*. The basic mechanism involved in resistance needs to be worked out and efforts be initiated in incorporating resistance from closely related species in *A. bisporus* using modern biotechnical tools.

Chemical pesticides are not likely to be used in protecting mushrooms from fungal feeding nematodes due to various reasons. Therefore, safe botanical pesticides, plant extracts or products should be tried to find the alternatives. Neem seed kernel water extract and achook have already shown promising results in managing some species of nematodes. Identification of effective plant products and development of application technology that can take care of other insect pest and pathogens of mushroom in addition to nematodes will be most welcomed by mushroom growers.

Mushroom compost is a complex ecosystem. Presence of naturally occurring parasites and predators should be exploited. *In vitro* studies have clearly demonstrated the potential of *Seinura* and *Fictor* species in managing nematode problems in mushrooms. Systematic research needs to be taken to understand the finer details of the biodiversity, biology, survival and application strategies. Biochemical analysis of saliva of *Seinura* species that paralyses the prey instantly may go in long way in developing effective nematicides managing nematodes pests of mushroom and field crops.

10. CONCLUSIONS

Highest populations of nematodes in mushroom beds vary with nematodes species. Nematodes infecting mushroom crop find their way any time from initiation of composting till last harvest. But most of them enter either with ingredients of compost or casing material. Symptoms shown on mushroom beds depend upon the time of entry of nematode in mushroom beds. Effect of pesticides has found that when applied in compost or casing; most of them adversely affect the mushroom production. Further, chemicals like aldicarb have shown a residual effect in sporophores whereas dazomet is highly toxic to mushroom mycelium. Exploit different predators for management of nematodes pests of mushroom. Apart from chemical and biological control use of botanical pesticides or phytotherapeutic substances is another option for their management. Addition of these products in compost improves the physicochemical properties of compost, increases the population of nematode trapping-antibiotic producing fungi, suppress competitive pathogenic moulds and increases the yield.

11. REFERENCES

Anon 2003. Nematode problems of mushroom in Haryana and their management. Final Tech. Report, Department of Nematology, CCS HAU, Hisar, India.

Anon 2011. Package of practices for horticultural crops in Haryana. Directorate of Extension Education, CCS Haryana Agricultural University, Hisar.

Arrold, N.P. and Blake, C.D. 1968. Some effects of nematodes *Ditylenchus myceliophagus* and *A. composticola* on the yield of the cultivated mushroom. *Ann. Appl. Biol.,* 61: 161–166.

Bajaj, H.K. and Kanwar, R.S. 2015. Biology and predatory attributes of a diplogasterid nematode, *Fictor composticola,* Khan *et al. Helminthologia,* 52.

Bajaj, H.K. and Walia, K.K. 1999. Three new species of *Aphelenchoides* Fisher, 1894. (Aphelenchina: Nematoda) from India along with a compendium of Indian species. *Indian J. Nematol.,* 29: 154–170.

Bajaj, H.K., Kanwar, R.S and Vats, R. 2014. *http://www.youtube.com/watch?v=4Jar6dgw9w4*

Bajaj, H.K., Kanwar, R.S. and Garg, M.K. 2013. Video assessment of power of median oesophageal bulb of *Aphelenchoides swarupi. Indian J. Nematol.,* 43: 231–233.

Bajaj, H.K., Singh, S. and Nandal, S.N. 2007. Reaction of oyster mushroom, *Pleurotus sajor-caju* Singer, to species of *Aphelenchus* and *Aphelenchoides. Paper Presented at National Symposium on Nematology in 21st Century; Emerging Paradigms. 22-23 November, 2007.* Assam Agricultural University, Jorhat, Assam.

Cayrol, J., Frankowski, J.P. and Quiles 1981. Recherches sur les possibilities d'utilization des Nematodes libres pour inouler les terres de gobetage su moyen de bacteries inductrices de la fructification. *Bulletin de la Federation nationale des Syndicetes Agricoles de Cultivateurs de Champignons NS,* 11: 301–312.

Cayrol, J.C. 1962. Etude preliminaires sur le cycle evolutif de *Ditylenchus myceliophagus* Goodey. *Nematologica,* 10: 361–368.

Cayrol, J.C. 1967. Etude du cycle evolutif d' *Aphelenchoides composticola. Nematologica,* 13: 23–32.

Cayrol, J.C. 1970. Phenomene de survie du nematode mycophage *Ditylenchus myceliophagus*: Anabiose. *Bull. Fed. Natn. Synds Agric. Cult. Champ. No.,* 189: 2329–2333.

Cayrol, J.C. 1974. Action of *Bacillus thuringiensis* toxins on the mycophagus nematode *D. myceliophagus. Paper Presented at XII Simposis Internationale de Nematologia. Sociedad Europea de Nematologos. 1-7 September*, Granada, Spain.

Cayrol, J.C. and Ritter, M. 1971. Nematodes and nematicides in the culture of mushrooms. *Paper Presented at 8th International Congress on Mushroom Science, London.*

Cayrol, J.C., Frankowski, J.P., Laniece, A., D'Hardemare, G. and Talon, J.P. 1978. Contre les nematodes en champignoniere. Maise au point d'une methode de lutte biologique a l'aide d'un Hyphomycete predateur: Arthobotrys robusta souche 'Antipolis' (Royal 300). *Pepinieristes, Horticulteurs, Maraichers- Revue Horticole,* 184: 23–30.

Chandran, R. 2000. Studies on saprophagous nematodes associated with white button mushroom. MSc Thesis, Dept. of Entomology and Apiculture, COH, UHF, Nauni, Solan.

Cooper, Jr A.F. and Van Gundy, S.D. 1970. Metabolism of glycogen and neutral lipids by *Aphelenchus avenae* and *Caenorhabditis* sp. in aerobic, microaerobic and anaerobic environments. *J. Nematol.,* 2: 305–315.

Das, D. and Bajaj, H.K. 2007. Effect of host and temperature on multiplication of *Ditylenchus myceliophagus. Ann. Pl. Protec. Sci.,* 15: 278–279.

Das, D. and Bajaj, H.K. 2009. Feeding behaviour of myceliophagous nematode, *Ditylenchus myceliophagus* Goodey on button mushroom, *Agaricus bisporus* (Lange.) Singer. *Indian J. Nematol.,* 38: 232–235.

Evans, A.A.F. and Fisher, J.M. 1969. Development and structure of population of *Ditylenchus myceliophagus* as affected by temperature. *Nematologica,* 15: 395–402.

Fei, X., Qi Zhi, L., Lin Lin, L. and Dao Wei, Y. 2012. Effect of nematodes in edible mushroom substance on hypha growth of *Auricularia auricula* and its quantity threshold. *Acta- Agriculturae- Zhejiagensis,* 24: 615–619.

Fiori, G., Saglini, V., Bertini, F., Domenighetti, G. and Mombelli, G. 1987. Intoxication severe al'insecticide oraganophosphore thyonazine- deux cas avec development d'un syndrome de detresse respriratoire de l'adulte (SDRA). *Schweizerische-Medizinische- Wochenschrift,* 117: 399–401.

Fisher, J.F. and Evans, A.A.F. 1967. Penetration and feeding by *Aphelenchus avenae. Nematologica,* 13: 425–428.

Grewal, P.S. 1989a. Nematicidal effects of some plant-extracts to *A. composticola* (nematode) infesting mushroom, *Agaricus bisporus. Revue. Nematol.,* 12: 317–322.

Grewal, P.S. 1989b. Effects of leaf matter incorporation on *Aphelenchoides composticola* (Nematoda), mycofloral composition, mushroom compost quality and yield of *Agaricus bisporus. Ann. Appl. Biol.,* 115: 299–312.

Grewal, P.S. 1991. Effect of *Caenorhabditis elegans* (Nematoda: Rhabditidae) on the spread of the bacterium *Pseudomonas tolaasii* in mushrooms (*Agaricus bisporus*). *Ann Appl. Biol.,* 118: 47–55.

Grewal, P.S. and Richardson, P.N. 1991. Effects of *Caenorhabditis elegans* (Nematoda: Rhabditidae) on the yield and quality of cultivated mushroom *Agaricus bisporus. Ann. Appl. Biol.,* 118: 581–594.

Hechler, H.C. 1962. The development of *Aphelenchus avenae* Bastian, 18665 in fungus culture. *Proc Helminth Soc. Wash.,* 29: 162–167.

Hesling, J.J. 1966. Preliminary experiment on the control of mycophagous eelworms in mushroom beds with a note on their swarming. *Plant Pathology,* 15: 163–166.

Hesling, J.J. 1972. Nematode pests of mushroom. *In:* Webster, J.M. (*ed.*), Economic Nematology. Academic press, London.

Hesling, J.J. 1977. *Aphelenchoides composticola*. CIH Description of Plant-parsitic nematoes Set 7, N0. 92. Commonwealth Institute of Helminthology, St. Albans, Herts, England.

Hesling, J.J. 1979. The role of eelworm nematodes in the culture of mushroom in UK. *Part 1. Mushroom Journal*, 83: 423–429.

Hesling, J.J. and Kempton, R.J. 1969. The control of nematodes in mushroom compost. Proc 5th British Insecticide Fungicide Conference, Brighton, pp. 185–188.

Ingratta, F.J. and Olthoff Th, H.A. 1978. The influence of saprophagous nematodes on the production of *Agaricus brunnescens* (*bisporus*). *Mushroom Sci.,* 10: 397–405.

Jairajpuri, M.S. 1968. Some studies on the morphology and biology of *Aphelenchus avenae* Bastian, 1865 (Nematoda: Aphelenchoidea). *Labdev J. Sci. Technology,* 6B: 27–34.

Juhl, H. 1966. Nematoder I champignonkulturer. *Tidsskr PlAvl.,* 69: 511–531.

Kaufman, T.D., Bloom, J.R. and Lukezie, F.L. 1984. The effects of free living nematodes and compost moisture on growth and yield in *Agaricus brunnescens*. *Can. J. Microbiology,* 30: 503–506.

Kermarrec, A. 1973. Investigations on the enemies of cultivated mushrooms. II. Preliminary studies on the utilization of predatory fungi in biological control of nematodes of cultivated mushrooms. *Annales de Zoologigie Ecologie Animals,* 5: 465–480.

Keshari, N. and Kanwar, R.S. 2015. Prey capturing and prey preference of *Fictor composticola,* Khan *et al*. *Paper Presented at National Symposium on Nematode management: A Challenge to Indian Agriculture in the Changing Climate.* Yashwantrao Chavan Academy of Development Administration, Pune.

Khanna, A.S. 1991. Yield losses in white button mushroom due to nematode infestation. *Nematol. Medit.,* 19: 35–36.

Khanna, A.S. 1993. Estimation of yield losses in white button mushrooms due to nemic infestation. *Curr. Nematology,* 4: 173–176.

Khanna, A.S. 1994. Nematode problems in mushroom management. *In*: Nair, M.C. and Luludes Gokulpelen, C. (*eds.*), Advances in Mushroom Biotechnology, Scientific publishers, Jodhpur.

Khanna, A.S. and Sharma, N.K. 1988. Pathogenicity of various *Aphelenchoides* species on *Agaricus bisporus*. *Indian Phytopathology,* 41: 472–473.

Khanna, A.S. and Sharma, N.K. 1989. Host range of *Aphelenchoides agarici*. *Indian J. Nematol.,* 19: 32–34.

Khanna, A.S. and Sharma, N.K. 1990. Efficacy of nematophagous fungi against myceliophagous nematodes *in vitro*. *Proc. National Seminar on Bioagents in Nematode Management, IARI, New Delhi,* pp. 83–84.

Khanna, A.S. and Sharma, N.K. 1994. Efficacy of neem (*Azadirachta indica*) and Ak (*Calotropis procera*) leaf powder as additives in management of mushroom nematodes. *Indian J. Hill Farming,* 7: 114–115.

Khanna, A.S., Chandrana, R. and Kumar, S. 2006. Sources of nematode inoculum in white button and milky mushroom. *Indian J. Nematol.,* 36: 267–269.

Khanna, A.S., Grewal, P.S., Sharma, N.K. and Sohi, H.S. 1988. Screening of some plant extracts for their nematicidal properties against *Aphelenchoides composticola* infesting mushroom. *Indian J. Hill Farming,* 1: 41–44.

Khanna, A.S., Thapa, C.D. and Sharma, V.P. 1996. The effect of plant extract material on the mycelial growth and sporophore production of *Agaricus bisporus* (Lange) Singer and nematode population of *Aphelenchoides composticola*. *Indian J. Hill Farming,* 9: 56–59.

Lin Lin, L., Qi Zhi, L., Fei, X., DaoWei, Y. and Peng Hua, B. 2011. The nematode quantity in edible mushroom substance an dits amplification and impact on the mushroom mycelium growth. *Acta- Agriculturae- Zhejiagensis,* 23: 1157–1161.

Mc Leod, R.W. and Khair, G.T. 1978. Control of *Aphelenchoides composticola* in mushroom compost with nematicides. *Ann. Appl. Biol.,* 88: 81–88.

Mc Leod, R.W. and Nair, G.T. 1981. The effects of saprophagous nematodes on mushroom yields. *Mushroom Science,* 11: 605–608.

Qi, Z.L., ZiHua, L., Yong, B., DaoWei, Y. and Yong, T. 2011. Preliminary study on the harmfulness of nematode (*Trilabiatus* spp.) in edible mushroom mycelium. *Acta- Agriculturae- Zhejiagensis,* 23: 335–339.

Rao, M.S. and Pandey, M. 1991. Comparative efficacy of karanj leaf and carbofuran on the management of *Aphelenchoides composticola. Indian J. Nematol.,* 21: 158–159.

Sanderson, F.R., Marshall, J.W., Ovenden, G.E. and Stengs, H.M. 1981. Problem in New Zealand mushroom production associated with bacterial feeding nematodes (*Mesorhabditis* sp.). *Mushroom Science,* 11: 609–619.

Seth, A. and Sharma, N.K. 1986. Five new species of genus *Aphelenchoides* (Nematoda: Aphelenchida) infesting mushroom in northern India. *Indian J. Nematol.,* 16: 205–215.

Sharma, N.K., Thapa, C.D. and Damanjeet 1984. Estimation of loss in mushroom yield due to myceliophagus nematode *Aphelenchoides sacchari. Indian J. Nematol.,* 14: 121–124.

Sharma, N.K., Thapa, C.D. and Kaur, D. 1985. Pathogenicity and identity of *Ditylenchus myceliophagus* "Indian population" infesting *Agaricus bisporus* (Lange) Sing. in India. *Indian J. Nematol.,* 15: 233–234.

Sharma, N.K., Thapa, C.D. and Nath, A. 1981. Pathogenicity and identity of myceliophagous nematodes infesting *Agaricus bisporus* (Lange) Sing. in H.P. (India). *Indian J. Nematol.,* 11: 230–231.

Sohi, H.S., Grewal, P.S. and Seth, A. 1987. Preliminary studies on the effect of plant extracts on the mycelial growth and yield of *Agaricus bisporus* (Lange) Sing. and fungal flora in compost. *Proc. Int. Sym Scientific and Technical Aspects of Cultivating Edible Fungi. The Pennsylvania State Univ, Univ Park, USA.* Elsevier Science Publishers, BV Amsterdam, pp. 121–128.

Thapa, C.D., Jandaik, C.L. and Sharma, N.K. 1983. Comparative resistance in some cultivated edible fungi against two mushroom nematodes *Aphelenchoides sacchari* and *D. myceliophagus* occurring in India. *Paper Presented at Symposium on Science and Cultivation Technology of Edible Fungi.* Regional research Lab., Jammu and Kashmir, India.

Thapa, C.D., Sharma, N.K. and Nath, A. 1981. Mushroom nematodes and their sources of contamination in mushroom beds in India. *Mush. Sci.,* 11: 443–448.

Vats, R., Bajaj, H.K. and Mann, S.S. 2002a. Assessment of losses in button mushroom due to myceliophagous nematodes. *Paper presented at National Symposium on Biodiversity and Management in Nematodes in Cropping Systems for Sustainable Agriculture,* Agriculture Research Station, Durgapura, Jaipur.

Vats, R., Bajaj, H.K. and Mann, S.S. 2003a. Integrated Management of *Aphelenchoides composticola* in button mushroom with neem seed kernel water extract and achook. *In:* Singh, R.V., Pankaj, Dhawan, S.C., and Gaur, H.S. (*eds.*), *Proceedings of the National Symposium on Biodiversity and Management of Nematodes in Cropping Systems for Sustainable Agriculture, 11-13 Nov., 2002, Agriculture Research Station, Durgapura, Jaipur.*

Vats, R., Bajaj, H.K. and Rana, R.S. 2002b. Pathogenicity of a myceliophagous nematode, *Aphelenchus avenae* on button mushroom. *Paper Presented at III Indian Mushroom Conference-2002, 6-7 March, 2002. T.N.A.U. Coimbatore.*

Vats, R., Bajaj, H.K. and Rana, R.S. 2003b. Management of a myceliophagous nematode, *Aphelenchoides composticola* Franklin in *Agaricus bisporus* (Lange) Singer using non chemical methods. *In:* Current Vistas in Mushroom Biology and Production. Proceedings of the Third Indian Mushroom Conference-2002. Updhyaya, R.C., Singh, S.K. and Rai, R.D. (*eds.*), Mushroom Society of India, pp. 210–213.

Vats, R., Kanwar, R.S. and Bajaj, H.K. 2001. Biology of *Seinura paratenuicaudata* on *Aphelenchus avenae*. *Paper Presented at National Congress on Centenary of Nematology in India:* Appraisal and Future Plans, I.A.R.I., New Delhi.

Vats, R., Kanwar, R.S. and Bajaj, H.K. 2004. Biology of *Seinura paratenuicaudata* Geraert. *Nematol. Medit.,* 32: 117–121.

Vats, R., Surjit, S. and Bajaj, H.K. 2006. Pathogenicity of *Paraphelenchus* spp. on milky mushroom (*Calocybe indica*) and its multiplication on other fungi. *Indian J. Nematol.*, 36: 146–147.

Wyatt, I.J. 1977. Principles of insecticide action on mushroom cropping: Incorporation into compost. *Ann. Appl. Biol.,* 85: 375–388.

Wyatt, I.J. 1978a. Principles of insecticide action on mushroom cropping: Incorporation into casing. *Ann. Appl. Biol.*, 88: 89–103.

Wyatt, I.J. 1978b. Effects of compost insecticides on mushroom yield. *Mushroom J.,* 70: 327–333.

Section-4
Applied Mushroom Biology

4.1

Utilization of Waste Through Mushroom Cultivation

SHWETA KULSHRESHTHA[1*]

ABSTRACT

Mushrooms possess versatile properties. Besides their nutritional and medicinal properties, mushrooms are grown for cleaning up the wastes disposed in the environment by agricultural and agro-industrial activities. Mushrooms have a variety of enzymes which can be used to degrade lignocellulosic wastes and remediate them from the environment. Wastes can be used in various ways to cultivate mushroom. Wastes can be used as substrate for mushroom cultivation. These are used alone or in combination with different other wastes. The wastes can be used as supplement for increasing the nutritional content of mushroom. The wastes can also be used as wetting agent to maintain the moisture content. Here, in this chapter, the role of different wastes as substrate, supplement and wetting agent is highlighted.

Key words: Mushroom cultivation, Supplement, Substrate, Wetting agent, Mushroom enzymes.

1. INTRODUCTION

A mad rat race has been started among the nations over the globe for growth and development which lead to the establishment of industries and urban areas. India has a good industrial infrastructure for producing basic need of the people and for making export quality material. These industries produce the products for satisfying local demand and for export at national

[1] Amity Institute of Biotechnology, Amity University Rajasthan, Jaipur.
**Corresponding author:* E-mail: shweta_kulshreshtha@rediffmail.com

and international level. These industries are also the source of earning foreign currencies. Therefore, it is true that no one can ignore their importance. It is also true that most of the industries are not adopting proper disposal practices. After processing the resources, residues in solid, liquid and gaseous form are being discharged into the environment. This resulted in over exploitation of resources which further causes pollution and disturbances in the ecological balance. Industries generate hazardous wastes comprising organics, inorganics, heavy metals, etc. in very high amount. Many organic contaminants are carcinogenic, toxic, and recalcitrant and have a tendency to accumulate in the environment (Chaudhry and Chapalamadugu, 1991). These pose noteworthy health risks to human and other living organisms and therefore, need to be tackled in a safe manner. Industrial and environmental biotechnology is going to develop "clean technologies" which gives emphasis on the maximum production, reduced waste generation and conversion of waste in some useful form. Physicochemical technologies are difficult to adopt because of their cost; requirement of skilled persons and difficult execution. Bioremediation is considered as a useful technology for the removal of waste from the environment.

2. BIOPERSPECTIVES OF MUSHROOM CULTIVATION

The cultivation of edible mushroom on agricultural and industrial wastes may thus be a value added process capable of converting these discharges, which are otherwise considered to be wastes, into foods and feeds (Bisaria *et al.*, 1997). Besides producing nutritious mushroom, it offers economic incentives for agribusinesses (Bisaria *et al.*, 1987). Mushroom production is the biggest solid-state-fermentation industry in the world (Moore and Chiu, 2001). China and the United States of America (U.S.) are the world's largest producers of mushrooms. During the last few years, research is going on to cultivate new varieties of mushroom like *Morchella, Hericium etc.*

Edible mushrooms are protein rich food and can be used to cure protein deficiency based diseases like Kwashiorkor and marasmus. These have been considered as the source of single cell protein and can be cultivated in fermenter for the production of single cell protein (Kurtzmam, 1981; Alofe *et al.*, 1998). The mushrooms either do not possess or very low amount of fat, lipids, sugars and cholesterol. In the mushroom fruit bodies, high quantities of fibers and amino acids threonine, tyrosine and phenylalanine are present (Khurana *et al.*, 1983).

The nutrient composition of mushroom fruit bodies is influence by a number of factors such as composition of the media or substrate used for its cultivation. Any nutrient supplementation also affects the nutritional composition of mushroom (Okafor and Ejiofor, 1990; Sanni, 1991). The study conducted by Akinyele *et al.* (2011) stated that the crude fibre content of the fermented mushroom was lesser than the unfermented samples. Wheat bran and rice bran contain considerable amount of carbohydrate or easily utilizable sugars which stimulate the cells to express many hydrolytic enzymes and encourage growth initiation and increase protein concentration (Bisaria *et al.*, 1997; Mabrouk and Ahwanyi, 2008; Kadiri, 1999; Akinyele, 2003). Belewu and Belewu (2005) also reported that supplementing the substrate with protein source will lead to the increase content of crude protein in the mushroom fruit body. Despite of little differences in the composition of mushroom, these are good source of nutraceuticals. Hence, these can be cultivated not only for remediating wastes from the environment but also for solving the problem of shortage of food. In the future, mushroom cultivation can solve two major problems simultaneously.

3. BIODEGRADATION BY MUSHROOM AND ROLE OF ENZYMES

Mushrooms, a macrofungi, possess a variety of enzymes due to which it has ability to degrade different pollutants. The mushrooms also have the ability to remove pollutants by biosorption process. Hence, it is considered as a tool of mycoremediation (Kulshreshtha *et al.*, 2014; Kulshreshtha, 2018 a,b). Many types of agricultural and agro-industrial wastes can be used as bioprospective agent for mushroom cultivation. Thusly, they involve in removal of waste which is also known as mycoremediation. According to the requirement, mushroom produces different enzymes for the degradation of pollutants because of the differences in the chemical groups amenable to degradation. The enzymes produced by mushroom are hydrolases (Arboleda *et al.*, 2011), laccases, esterases (Hashimoto *et al.*, 2010) and phenol oxidases (Cohen *et al.*, 2002; Stajia *et al.*, 2009). The mushroom enzymes are capable of degrading a variety of pollutants or substrates in both acidic and alkaline condition. These enzymes carried out specific reaction, unlike chemicals, for the degradation of pollutants. These enzymes, when active under controlled conditions, avoid the formation annoying products.

Mushroom, a ligninolytic fungi, can be used for remediation of pollutants due to having extracellular peroxidases and laccases which is able to oxidize recalcitrant pollutants *in vitro*. Novotný *et al.* (2004) reported the activities of Mn-dependent peroxidase (MnP), lignin peroxidase and laccase (LAC) in various species of ligninolytic fungi grown in liquid media and soil and their effect on degradation of polycyclic aromatic hydrocarbons (anthracene and pyrene), a polychlorinated biphenyl mixture (Delor 106) and a number of synthetic dyes.

Fungi degrade the lignocellulosic substrate by means of hydrolytic extracellular enzymes like cellulases, pectinases and xylanases, along with oxidases and peroxidases. Lignin is also oxidized and degraded by a ligninase system, which is composed of three enzyme *i.e.,* lignin peroxidase (LiP), manganese dependent peroxidase (MnP) and laccase. Cellulose is hydrolyzed by endoglucanases, cellobiohydrolases and β-glucosidases. These enzymes are typically induced by their substrates. Hemicellulases are degraded by xylanases which act synergistically with cellulases to achieve hydrolysis.

Mushrooms (*Phanerochaete chrysosporium*, *Trametes versicolor* etc.) typically secrete one or more of the three principal ligninolytic enzymes (Hatakka, 1994) *i.e.*, lignin peroxidase, Mn-dependent peroxidase and phenol oxidase (laccase) (Thurston, 1994; Orth and Tien, 1995). There are certain other enzymes which are also produced by mushroom such as peroxidases, glyoxal oxidase and superoxide dismutase and other enzymes used in lignocellulose degradation pathways (*e.g.*, glucose oxidase and aryl alcohol oxidase), however, their role in lignin degradation is poorly understood (Pointing, 2001).

4. BIODEGRADATION POTENTIAL OF MUSHROOM

The biodegradation term is used to describe the ultimate degradation and recycling of an organic molecule to its mineral constituents. Alexander (1965) described the principle of microbial infallibility which focuses on the fact that – there is no natural organic compound is totally resistant to biodegradation if environmental conditions are favourable. Biodegradation involves complex mechanisms due to the interference of other biological systems is involved.

For bioremediation purpose, waste is used in mushroom cultivation process. This waste is used in a number of ways which support mushroom growth. The mushroom mycelium runs on the waste and degrades it. Thusly, mushroom cultivation gives mushroom fruit body for consumption purpose and degrades different types of agro-industrial waste and solves the problem of their accumulation.

In mushroom cultivation process, waste can be used as substrate, compost for casing, supplement for increasing the nutritional content and biological efficiency, and wetting agent.

4.1. Waste as Substrate

Wastes, discharge from agricultural field, are collected for mushroom cultivation. These wastes are wetted for overnight in formalin containing solution and then, sterilized after filling in suitable container in the autoclave with free flowing steam for two hours. After sterilization, wastes are inoculated with 2% spawn and incubate at appropriate temperature and humidity conditions. Mushroom fruit body appears in the form of pin heads which acquire their specific shape on maturation. The mature fruit bodies can be harvested. Thusly, mushroom can be cultivated on the agricultural or agro-industrial waste. Wastes used for the cultivation of mushroom are shown in Table 1. The most commonly used agricultural wastes for mushroom cultivation are wheat straw, paddy straw, rice straw, banana leaves, tea leaves, saw dust, coffee straw and cotton straw.

Table 1: Agricultural wastes for cultivation of mushroom.

Sl. no.	*Agricultural waste*	*Strains*	*References*
1.	Rice straw, wheat straw, cotton straw, tea leaves, banana leaves.	*Pleurotus* sp.	Kamthan (2017)
2.	Wheat straw	*Agaricus bisporus*	Kamthan (2017)
3.	Rice bran, coffee pulp	*Lentinula edodes*	Kamthan (2017)
4.	Tea leaves	*Volvariella* sp.	Kamthan (2017) and Yang *et al.* (2016).
5.	Sawdust	*Ganoderma* sp.	Kamthan (2017)
6.	Olive mill wastes	*Pleurotus* sp.	Mansour-Benamar *et al.* (2013).

In certain cases, a combination of agricultural wastes for the cultivation of mushrooms was used. It is also reported that combination can increase the biological efficiency rather than the using waste alone. Mushroom strains vary in their ability of producing lignolytic and cellulolytic enzymes in the substrate which further affect their substrate utilizing ability. This results in low yield and biological efficiency. The combination provides suitable condition for the growth of mushroom and increases their yield and biological efficiency. The combinations of agricultural and agro-industrial wastes are shown in Table 2.

Mushroom and their combinations can be modified by fermentation or supplemented with some supplements in order to increase biological efficiency and yield. Buckweed was fermented by *Rhizopus oligosporus* and then used for the cultivation of *Pleurotus* sp. showed increase in biological efficiency (Lee *et al.*, 2018). Switch grass (*Panicum virgatum*, 99%) with 1% ground limestone and a mixture of pasteurized cottonseed hulls (75% dry wt.), 24% chopped wheat straw, and 1% ground limestone supplementation were used for the cultivation of *Pleurotus cornucopiae*. Results show an increase in biological efficiency due to supplementation (Royse *et al.*, 2004).

Invasive weeds can also be used as supplement in mushroom cultivation. These invasive weeds may or may not be native to a place but have the tendency to spread very rapidly in the environment and cause damage to the crop, human and environment. These invasive weeds are noxious element that needs to be removed. To solve this problem, many researchers have used invasive weeds for mushroom cultivation as shown in Table 3.

Table 2: The combination of the substrates used for mushroom cultivation.

Sl. no.	*Substrate (in combination)*	*Strain*	*Effect*	*References*
1.	Date palm wastes mixed with other agricultural wastes.	*Pleurotus ostreatus*	Ratio of 25 (date palm): 75 (agro -waste) showed the best results.	Alananbeh *et al.* (2014)
2.	Corncob with bran and olive press cake.	*Hypsizygus marmoreus*	Biological efficiency of 85.6% .	Akavia *et al.* (2016).
3.	Vegetable and agro-wastes.	*Pleurotus sapidus*	Vegetable waste alone give low biological efficiency.	Singh and Singh (2012).
4.	Leaf, pseudo-stem and pseudo-stem + leaf) and banana cultivars - *Musa* spp.	*Pleurotus ostreatus*	Used successfully with 70% biological efficiency.	De Carvalho *et al.* (2012).
5.	Maize cob, rice bran, wheat bran and cotton waste.	*Pleurotus* and *Lentinus*	Used successfully with 65.3% biological efficiency.	Oyedele *et al.* (2018).
6.	Date-palm leaves mixed with other agro-wastes.	*Pleurotus ostreatus*	Used successfully with 84.5% biological efficiency	Alananbeh *et al.* (2014).
7.	Barley straw + wheat bran and wood chips + soybean powder + rice bran treatments.	*Pleurotus eryngii*	4.64% protein content.	Kamthan (2017)
8.	Wheat straw + wheat bran + soybean powder treatment.	*Pleurotus eryngii*	13.66% protein content.	Kamthan (2017)
9.	Soybean straw + wheat straw.	*Pleurotus sajorcaju*	87.3% biological efficiency.	Kamthan (2017)
10.	Soybean straw + saw dust.	*Pleurotus sajorcaju*	43.8% biological efficiency.	Kamthan (2017)
11.	Corn cob + sugarcane bagasse.	*Pleurotus ostreatus* *Pleurotus cystidiosus*	High content of protein, ash and mineral (Ca, K, Mg, Mn, and Zn).	Kamthan (2017)

4.2. Mushroom as Supplement

As stated in the aforesaid paragraph, wastes can be used as supplements to increase the biological efficiency and yield of mushroom. Agro-industrial wastes like wheat bran, cereal bran or rice bran can be used as supplement for growing a number of edible varieties of mushroom. These supplements are organic compound and act as nitrogen source. However, they may interfere with the mushroom mycelial growth, their productivity and biological efficiency. The strain of mushroom and its growth stage is the decisive factor for quantity and type of bran supplement (Moda *et al.*, 2005). A detailed description of effect of different supplement on the growth of mushroom is given in Table 4.

Table 3: Effect of using weed as substrate for mushroom cultivation.

Sl. no.	*Substrate*	*Strain*	*Effect*	*References*
1.	Invasive weeds species (*Lantana camara, Prosopis juliflora, Parthenium hysterophorus.*	*Pleurotus* spp.	Crude protein ranged from 40.51- 41.48% for *P. florida* grown on *P. hysterophorus* and *L. camara*. Lowest crude protein content (30.11%) was recorded for *P. ostreatus* grown on wheat straw.	Mintesnot *et al.* (2014).
2.	Dry weed plants, *Leonotis sp, Sida acuta, Parthenium argentatum, Ageratum conyzoides, Cassia sophera, Tephrosia purpurea* and *Lantana camara. Leonotis* sp.	*Pleurotus ostreatus*	*T. purpurea* was the least suited weed for oyster mushroom cultivation. Problem with second flush of mushroom.	Das and Mukherjee (2007).
3	*Chenopodium album* alone or mix with cotton waste.	*Pleurotus* spp.	Cotton waste showed better biological efficiency as this substrate took minimum 20.5 days which was followed by *Chenopodium album* alone and mixture of cotton waste + *Chenopodium album* which relatively taken 24.5 days to reach maximum growth level.	Islam *et al.* (2016).

4.3. Waste as Wetting Agent in Mushroom Cultivation

The maintenance of moisture in the mushroom bed is a great challenge. The mushroom mycelial growth, productivity and biological efficiency are severely affected by the moisture of the substrate bed. Over-moisture and lack of moisture conditions are not favourable for mushroom growth. Therefore, wetting agent is used instead of water. In a study conducted by Kalmis *et al.* (2008) olive mill effluent was used as wetting agent instead of water. It was thought that olive mill effluent contains significant amount of carbon and energy which provides an economic wetting agent for mushroom cultivation. Mushroom cultivated on wheat straw was wetted with water and olive mill effluent and results were compared in terms of biological efficiency and yield. It was suggested that the use of olive mill effluent up to 25% is suitable as moisturizer for mushroom bed (Kalmis *et al.*, 2008).

Table 4: The effect of growth supplement on mushroom growth

Sl. no.	*Wastes as supplement*	*Mushroom*	*Effect*	*References*
1	Vegetable and agro-wastes like pea pod shell, cauliflower leaves, radish leaves, brassica straw in various combinations of paddy straw.	*Pleurotus sapidus*	Failed to grow on these vegetable wastes separately. Protein content in the fruit bodies was found to be higher in the mushroom grown on paddy straw mixed with vegetable wastes than that obtained with paddy straw alone.	Singh and Singh (2012).
2	Vegetable meal, soybean meal.	*Pleurotus* spp.	Increase the protein content	Zied *et al.* (2011).
3	Cereal bran, enriched with minerals or vitamins.	*Agaricus* spp.	Increase protein content and BE	Burton *et al.* (2015).
4	Microbes from SMs were isolated and inoculated.	*Pleurotus* sp.	Increase protein content and BE	Carrasco *et al.* (2018).
5	Rice bran, maize powder, and wheat bran.	*Calocybe indica*	Primordia initiation was observed between 13.5 and 19.3 days. 30% maize powder supplement was effective for producing viable fruiting bodies with highest yield and biological efficiency.	Alam *et al.* (2010).

5. TOXICITY /GENOTOXICITY ASSAYS: AN IMPORTANT CONSIDERATION WHILE USING WASTES

Besides degradation, mushrooms have the ability to adsorb pollutants from the environment (Kulshreshtha *et al.*, 2014; Kulshreshtha, 2018 a,b). If any pollutant is absorbed by the fruiting bodies of mushroom, then these fruiting bodies cannot be consumed. A detailed description of mutagenicity in mushroom is shown in Table 5.

It is necessary to make specific recommendations about health risks associated with mushroom, cultivated on waste. Further information is needed about the nutritional and toxic/ mutagenic compounds of these mushrooms, ignorance of this will cause the associated health related problems.

The use of mushrooms for edible purpose depends on the toxicity in fruiting bodies. Mushroom species, cultivated on handmade paper and cardboard industrial waste, do not possess toxicity or mutagenicity (Kulshreshtha *et al.*, 2010; Kulshreshtha *et al.*, 2011; Kulshreshtha *et al.*, 2013). Aqueous extract of *P. florida* and *P. citrinopileatus* cultivated on handmade paper and cardboard industrial waste, did not show mutagenicity with either *Salmonella typhimurium* strain TA 98 or TA 100. Its mutagenicity was not found to increase with S9 mix (Kulshreshtha *et al.*, 2010). Similarly, aqueous extract of *P. citrinopileatus*, did not show mutagenicity with *Salmonella typhimurium* strain TA 100 in the presence or absence of S9 mix. However, aqueous extracts of this mushroom, collected from all wastes including control, were mutagenic with *Salmonella typhimurium* strain TA 98 even at low dose level. Moreover, mutagenicity was found to increase by the addition of S9 mix. In contrast, this mutagenicity was found to be reduced on boiling the aqueous extract of mushroom (Kulshreshtha *et al.*, 2013).

Table 5: Mutagenicity of naturally occurring and cultivated mushroom species analyzed by Ames test.

Sl. no	*Mushroom Species*	*Mutagenicity*	*Mutagenicity assays*	*References*
1.	*Lactarius* (*L. necator, L. torminosus,L. helvus* and *L. rufus*) and bolete (*Boletus edulis*), (*Agaricus bisporus*) and shiitake (*Lentinus edodes*).	The mushrooms were mutagenic to tester strains sensitive to base-pair substitution mutagens, and *L. necator, L. rufus* and *B. edulis* also had some frameshift activity.	Ames *Salmonella/* microsome test system.	Wright *et al.* (1982).
2.	Isovelleral (5) from *Lactarius* sp., agaritine (3) from *Agaricus bisporus* and related sp. and beta-nitramino-alanine (7) from *Agaricus silvaticus*.	Isovelleral was found to be mutagenic. *A. bisporus* was weakly mutagenic toward all three strains. Agaritine was weakly active only with TA2637 strain.	*Salmonella/* microsome assay, with strains TA98, TA2637 and TA100.	Sterner *et al.* (1982).
3.	*Agaricus bisporus*	Agaritine is not the principal mutagenic component in the mushroom.	Ames test	Papapara-skeva-petrides *et al.* (1993).

6. CONCLUSIONS

In this chapter, methods of using waste for mushroom cultivation are discussed. However, the use of these mushrooms for edible purpose depends on the toxicity in fruiting bodies. Thus, there is a need for further research towards the exploitation of potential of mushroom as bioremediation tool and its safety aspects for consumption. Mycoremediation through mushroom cultivation is an easily accessible, economical, and simple solution that will help alleviate two of the world's major problems *i.e.,* waste accumulation and shortage of proteinaceous food, simultaneously.

7. REFERENCES

Akavia, E., Beharav, A., Wasser, S.P. and Nevo, E. 2009. Disposal of agro-industrial by-products by organic cultivation of the culinary and medicinal mushroom *Hypsizygus marmoreus. Waste Manag.,* 29*(5)*:1622–1627. doi: 10.1016/ j.wasman.2008.10.024.

Akinyele, B.J. 2003. *In vitro* nutritional studies on *Volvariella volvacea* (Bull. Ex Fr.) Sing, An edible mushroom. Ph.D. Thesis, Federal University of Technology, Akure, Nigeria, p. 157.

Akinyele, B.J., Olaniyi, O.O. and Arotupin, D.J. 2011. Bioconversion of selected agricultural wastes and associated enzymes by *Volvariella volvacea*: An edible mushroom. *Res. J. Microbiol.,* 6: 63–70.

Alam, N., Amin, R., Khair, A. and Lee, T.S. 2010. Influence of different supplements on the commercial cultivation of milky white mushroom. *Mycobiology*, 38(3): 184–188. doi:10.4489/MYCO.2010.38.3.184.

Alananbeh, K.M., Bouqellah, N.A. and Al Kaff, N.S. 2014. Cultivation of oyster mushroom *Pleurotus ostreatus* on date-palm leaves mixed with other agro-wastes in Saudi Arabia. *Saudi J. Biol. Sci.,* 21(6): 616–25. doi: 10.1016/j. sjbs. 2014.08.001.

Alexander, M. 1965. Biodegradation: Problems of molecular recalcitrance and microbial infallibility. *Adv. Appl. Microb.,* 7: 35–80.

Alofe, F.V., Odu, E.A. and Illoh, H.C. 1998. Mushrooms, man and nature. Edible wild mushrooms on decaying wood in Obafemi Awolowo University Campus. *The Nig. Field,* 63: 3–18.

Arboleda Valencia, J.W., Valencia Jiménez, A., Gonçalves de Siqueira, F., Dussan Medina, K., Restrepo Franco, G.M., Filho, E.X., Siegfried, B.D. and Grossi-de-Sa, M.F. 2011. Holocellulase activity from *Schizophyllum commune* grown on bamboo: A comparison with different substrates. *Curr. Microbiol.,* 63(6): 581–587.

Belewu, M.A. and Belewu, K.Y. 2005. Cultivation of mushroom (*Volvariella volvacea*) on banana leaves. *Afr. J. Biotechnol.,* 4(12): 1401–1403.

Bisaria, R., Madan, M. and Vasudevan, P. 1997. Utilization of agro- residues as animal feed through bioconversion. *Biores. Technol.*, 59: 5–8.

Burton, K., Noble, R., Rogers, S. and Wilson, J. 2015. Understanding mushroom nutrition: Project aimed at improving yield, substrate efficiency and utilisation and flavor. M056 Final Report. Agriculture and Horticulture Development Board (AHDB), p. 54.

Carrasco, J., Zied, D.C., Pardo, J.E., Preston, G.M. and Pardo-Giménez, A. 2018. Supplementation in mushroom crops and its impact on yield and quality. *AMB Express,* 8: 146. doi: 10.1186/s13568-018-0678-0.

Chaudhry, G.R. and Chapalamadugu, S. 1991. Biodegradation of halogenated organic compounds. *Microbiol. Mol. Biol. Rev.,* 55(1): 59–79.

Das, N. and Mukherjee, M. 2007. Cultivation of *Pleurotus ostreatus* on weed plants. *Bioresour. Technol.,* (14): 2723–2726.

De Carvalho, C.S.M., De Aguiar, L.V.B., Sales-Campos, C., Minhoni, M.T.A and Andrade, M.C.N. 2012. Applicability of the use of waste from different banana cultivars for the cultivation of the oyster mushroom. *Braz. J. Microbiol.,* 43(2): 819–826. doi: 10.1590/S1517-83822012000200048.

Hashimoto, K. Kaneko, S. and Yoshida, M. 2010. Extracellular carbohydrate esterase from the basidiomycete *Coprinopsis cinerea* released ferulic and acetic acids from xylan. *Biosci. Biotechnol. Biochem.,* 74: 1722–1724.

Hatakka, A. 1994. Lignin-modifying enzymes from selected white-rot fungi: Production and role in lignin degradation. *FEMS Microbiology Reviews,* 13: 125–135.

Islam, W., Riaz, A., Zaynab, M. and Zujian, Wu. 2016. Efficacy of different weeds and agricultural substrates for cultivation of *Pleurotus ostreatus* (Jacq. fr.). *Asian J. Agri. Biol.,* 4(4): 108–113.

Kadiri, M. 1999. Physiological studies of some Nigerian mushrooms. Ph.D. Thesis, University of Ibadan, Ibadan, Nigeria.

Kalmis, E., Azbar, N., Yildiz, H. and Kalyoncu, F. 2008. Feasibility of using Olive Mill Effluent (OME) as a wetting agent during the cultivation of oyster mushroom, *Pleurotus ostreatus*, on wheat straw. *Bioresour. Technol.,* 99(1): 164–169. DOI: 10.1016/j.biortech.2006.11.042.

Kamthan, R. and Tiwari, I. 2017. Agricultural wastes- Potential substrates for mushroom cultivation. *Eur. Exp. Biol.,* 7(5): 31. doi: 10.21767/2248-9215.100031.

Khurana, M.L., Kaul, T.N. and Kachroo, J.L. 1987. Nutritive value of wild edible mushrooms for Kashmir. Indian Mushroom Science-II. *Proceedings of the International Conference on Silver Jubilee Symposium on Science and Cultivation Technology of Edible Fungi Held on 9-11th September, 1983 at Srinagar. RRL, Jammu and Kashmir (CSIR),* pp. 184–189.

Kulshreshtha, S. 2018. Mushroom biomass and spent mushroom substrate as adsorbent to remove pollutants. *In*: Crini, G. and Lichtfouse, E. (*eds.*), green adsorbents for pollutant removal. *Environmental Chemistry for a Sustainable World,* 19, Springer, Cham.

Kulshreshtha, S. 2018. Removal of pollutants using spent mushrooms substrates. *Environ. Chem. Lett.* https://doi.org/10.1007/s10311-018-00840-2

Kulshreshtha, S., Mathur, N. and Bhatnagar, P. 2011. Pros and cons of *P. florida* cultivation for managing waste of handmade paper and cardboard industries. *The IIOAB J., Spl. Issue*, 2(1): 45–48.

Kulshreshtha, S., Mathur, N. and Bhatnagar, P. 2014. Mushroom as a product and their role in mycoremediation. *AMB Express,* 1(4): 29. doi: 10.1186/s13568-014-0029-8.

Kulshreshtha, S., Mathur, N., Bhatnagar P. and Jain, B.L. 2010. Bioremediation of industrial wastes through mushroom cultivation, *J. Environ. Biol.,* 31: 441–444.

Kulshreshtha, S., Mathur, N., Bhatnagar, P. and Kulshreshtha, S. 2013. Cultivation of *Pleurotus citrinopileatus* on handmade paper and cardboard industrial wastes. *Ind. Crops and Products,* 41: 340–346.

Kurtzmam, R.H. Jr. 1981. Mushrooms: Single cell protein cellulose. *Ann. Rep. Ferment. Process,* 3: 305–395.

Lee, T.K., Nguyen, T.T.H., Park, N., Kwak, S.H., Kim, J., Jin, S., Son, G.M., Hur, J., Choi, J.I. and Kim, D. 2018. The use of fermented buckwheat to produce L-carnitine enriched oyster mushroom. *AMB Express,* 8(1):138. doi: 10.1186/s13568-018-0664-6.

Mabrouk, E.M. and Ahwany, M.D. 2008. Production of mannanase by *Bacillus amyloliquefaciens* 10A1 cultured on potato peels. *Afr. J. Biotechnol.,* 7: 1123–1128.

Mansour-Benamar, M., Savoie, J.M. and Chavant, L. 2013.Valorization of solid olive mill wastes by cultivation of a local strain of edible mushrooms. *C. R. Biol.,* 336(8): 407–15. doi: 10.1016/j.crvi.2013.07.004.

Mintesnot, B., Ayalew, A. and Kebede, A. 2014. Evaluation of biomass of some invasive weed species as substrate for oyster mushroom (*Pleurotus* spp.) cultivation. *Pak. J. Biol. Sci.,* 17(2): 213–9.

Moda, E.M., Horii, J. and Spoto, M.H. 2005. Edible mushroom *Pleurotus sajor-caju* production on washed and supplemented sugarcane bagasse. *Sci. Agric.,* 62: 127–132.

Moore, D. and Chiu, S.W. 2001. Filamentous fungi as food. *In:* Exploitation of filamentous fungi. Pointing, S.B. and Hyde, D. (*eds.)*, Fungal Diversity Press, Hong Kong.

Novotný, C., Svobodová, K., Erbanováa, P., Cajthamla, T., Kasinatha, A., Lang, E. and Šašeka, V. 2004. Ligninolytic Fungi in bioremediation: Extracellular enzyme production and degradation rate. *Soil Biology and Biochemistry,* 36(10): 1545–1551.

Okafor, N. and Ejiofor, A.O. 1990. Rapid detoxification of cassava mash by a yeast simultaneously producing linamarase and amylase. *Proc. Biochem.,* 25: 82–86.

Orth, A.B. and Tien, M. 1995. Biotechnology of lignin degradation. *In*: Esser, K. and Lemke, P.A. (*eds.), The Mycota. II. Genetics and Biotechnology*, Springer, Berlin, pp. 287–302.

Oyedele, O.A., Adeosun, M.V. and Koyenikan, O.O. 2018. Low cost production of mushroom using agricultural waste in a controlled environment for economic advancement. *Int. J. Waste Resour.*, 8(1): 329. doi: 10.4172/2252-5211.1000329.

Papaparaskeva-Petrides, C., Ioannides, C. and Walker, R. 1993. Contribution of phenolic and quinonoid structures in the mutagenicity of the edible mushroom *Agaricus bisporus*. *Food Chem. Toxicol.*, 31(8): 561–567.

Pointing, S.B. 2001. Feasibility of bioremediation by white-rot fungi. *Applied Microbiology and Biotechnology,* 57: 20–33.

Sanni, M.O. 1991. Gari processing in Ibadan metropolis: Factors controlling quality at the small scale level. *Processing of the 19th Symposium of the International Society of Tropical Roots Crops, (TRC'91), Accra, Ghana*, pp. 256–259.

Singh, M.P. and Singh, V.K. 2012. Biodegradation of vegetable and agro-wastes by *Pleurotus sapidus*: A novel strategy to produce mushroom with enhanced yield and nutrition. *Cell. Mol. Biol. (Noisy-le-grand),* 58(1): 1–7.

Sterner, O., Bergman, R., Kesler, E., Magnusson, G., Nilsson, L., Wickberg, B., Zimerson, E. and Zetterberg, G. 1982. Mutagens in larger fungi. I. Forty-eight species screened for mutagenic activity in the *Salmonella*/microsome assay. *Mutat. Res.,* 101(4): 269–81.

Thurston, C.F. 1994. The structure and function of fungal laccases. *Microbiology,* 140: 19–26.

Von Wright, A., Knuutinen, J., Lindroth, S., Pellinen, M., Widén, K.G. and Seppä, E.L. 1982. The mutagenicity of some edible mushrooms in the Ames test. *Food Chem. Toxicol.,* 20(3): 265–267.

Yang, D., Liang, J., Wang, Y., Sun, F., Tao, H., Xu, Q., Zhang, L., Zhang, Z., Ho, C.T. and Wan, X. 2016. Tea waste: An effective and economic substrate for oyster mushroom cultivation. *J. Sci. Food Agric.,* 96(2): 680–4. doi: 10.1002/jsfa.7140.

Zied, D.C., Savoie, J.M. and Pardo-Giménez, A. 2011. Soybean the main nitrogen source in cultivation substrates of edible and medicinal mushrooms. *In*: *Soybean and Nutrition*. doi: 10.5772/17692.

4.2

Mushroom Cultivation: A Boon for Empowerment of Women Entrepreneurship

MONIKA THAKUR[1]*

ABSTRACT

With the growing population and fast depletion of natural resources, it became necessary to explore the possibilities of using newer untapped resources. In the present scenario people are more aware of functional foods and with this the demand for mushroom as food is rising. It provides ample opportunities for empowerment of women. In the last decade, there is almost stabilization and/or stagnation in agricultural production. It is therefore, high time to divert resources and energy to develop alternative source of high quality food and also create employment. Mushroom cultivation serves both the purposes: it is high quality food and through this enterprise employment can be created. Therefore, it is essential to evolve a development approach by empowerment of women entrepreneurs in mushroom cultivation. Mushroom cultivation is a full-fledged entrepreneurship and is 'women-friendly' operation. Women as mushroom entrepreneurs address diverse problems and promise benefits like healthier lives, nutritious food, enhanced well-being, sustained livelihood, financial independence, employment and mitigation of environment related issues. Mushroom cultivation can be undertaken on the homestead and can provide additional income enhancing family welfare without sacrificing their household responsibilities. Therefore, mushroom cultivation is one such activity that can play a pivotal role in the empowerment of women.

[1] Amity Institute of Food Technology, I-1, Fourth floor, Amity University, Uttar Pradesh, Sector 125, Express highway, Noida, Uttar Pradesh – 201303.
**Corresponding author:* E-mail: monika.harsh05@gmail.com

Key words: Mushroom cultivation, Women empowerment, Sustainable livelihoods, Non-green revolution.

INTRODUCTION

Agriculture continues to remain a major sector in Indian economy; generally it provides 60% of employment and continues to be the primary source of living for the 70% of the population. Entrepreneurship is an innovative and dynamic process and acts as an economic force in rural development. 'Women entrepreneurs' play an important role in economy of the country and in present scenario a large percentage of entrepreneurs in developing countries are undertaken by women. In developing country, like India, where the economic status of women is very pathetic especially in rural areas and opportunities of earning are very less, the economic independence of women is very important. Women entrepreneurs also help in achieving the concept of food security. The women entrepreneurs may be mobilized and may lead to the nation towards the path of progress and prosperity.

MUSHROOM CULTIVATION

Mushrooms are non-photosynthetic group belongs either to ascomycetes or basidiomycetes. They can be saprophytic, parasitic and/or mycorrhizal and secure nutrition by degrading lignocellulosic materials through a variety of enzymes. They can convert the lignocellulosic wastes through solid state fermentation technology into high quality protein in the form of mushroom fruiting bodies which can be consumed directly as vegetable (Chang, 1998). In the recent past, it has been realized that mushroom nutraceuticals which are a source of biologically active compounds of medicinal value, have been used in prevention and treatments of various human diseases (Chang and Buswell, 1999). These bioactive compounds have proven anticancer, antiviral, immunopotentiating, hypocholesterolemic and hepatoprotective properties and therefore regular intake of mushrooms (fruiting bodies or mycelium) can help people maintaining good health. Moreover, mushroom cultivation reduces environmental pollution as it utilizes the agricultural and forest waste converting it into useful matter (Chang, 2007). Therefore, mushrooms are a good source of nutraceuticals, environment remediators and income sources for people especially the women folk.

Being agriculture based country; India offers vast potential for the cultivation of mushrooms due to the presence of abundant raw material and congenial climatic conditions for its growth. With the passage of time mushroom cultivation is becoming more and more popular all over, especially in rural areas for income generation amongst rural, poor and landless women. In addition to the commercial benefits, it is also a perfect health food as it enriches the diet with proteins, vitamins and minerals and has an established nutraceutical potential. Mushroom cultivation is a very efficient system for recycling the waste materials and involves no wastes from production to consumption.

With commercial benefits, it is also a perfect health food by enriching the diet with proteins, vitamins, minerals, fiber and also has immense nutraceutical potential. Mushroom cultivation is compatible with other agricultural and horticultural activities. Many studies have revealed that the cultivation of mushrooms not only improve standard of living but also improve the social status (Kumar and Bindra, 1994; 1997; Prashad *et al.*, 2012; Mishra, 2008). Dutt (2014) studied the strategies for setting and running of small scale enterprises by women entrepreneurs of Kangra district of Himachal Pradesh. Mushroom cultivation was one of the major entrepreneurial activities among other activities selected by women entrepreneurs. Therefore, to build a strong economically forward nation, women folk of that nation are to be profitably exploited and act as a boon for the empowerment of women in the society.

Indian Mushroom Industry

Mushrooms have been collected from the wild and cultivated artificially for human food and for medicinal usage for thousands of years. Gradually the cultivation has assumed the status of an industry. Mushroom industry is now divided into three main interrelated segments: cultivated edible mushrooms; medicinal mushrooms and wild mushrooms (Chang, 2006).

Different types of mushrooms are cultivated in India for commercial purposes. Twelve different species are commonly grown for food and medicinal purposes globally. These include the common mushroom (*Agaricus* spp.), Shiitake (*Lentinus edodes*), Oyster (*Pleurotus* spp.), Paddy straw (*Volvariella volvacea*), Lion's Head (*Hericium* spp.), Black ear (*Auricularia* spp.), Reishi (*Ganoderma*) Maitake (*Grifola frondosa*), Winter mushroom (*Flammulina velutipes*), White jelly fungus (*Tremella* spp.), White milky mushroom (*Calocybe indica*), Nameka (*Pholiota* spp.), and Shaggy mane mushrooms (*Coprinus* spp.). Although cultivation technology is known for number of mushrooms, but the mushroom species cultivated in India on a commercial scale as white button mushroom (*Agaricus bisporus*), Oyster mushroom (*Pleurotus* spp.), Paddy straw mushroom (*Volariella* spp.), *Calocybe indica,* Shiitake (*Lentinus edodes*), Black ear mushroom (*Auricularia* spp.), Winter mushroom (*Flammulina velutipes*) etc. (Dhar, 1997). Besides these cultivated mushrooms various wild edible mushrooms having very good commercial value are species of *Morchella, Helvella, Hericium, Sparassis, Hydnum, Trappeinda, Clavaria, Ramaria, Boletus, Albatrellus, Cordyceps, Lactarius and Rusulla* (Lakhanpal, 2000). All these wild edible mushrooms are consumed as well as sold fresh and collected and dried for sale in local markets for their indigenous use in medicines. Morels are presently the costliest and also most sought after edible mushrooms with great commercial potential (Lakhanpal *et al.*, 2010; Thakur and Lakhanpal, 2014).

Mushroom Market: Production and Marketing

Mushrooms business has a huge export potential and mushroom farming today is being practiced in more than 100 countries. India produced all types of cultivated mushrooms (Wakchaure, 2011). Indian exports mushrooms under three major categories: Fresh mushrooms, dried mushrooms (including wild harvested mushrooms) and processed mushrooms (including preserved – value added products). In view of highly perishable nature, the fresh mushrooms have to be processed to extend their shelf life for off season use. Many value-added products have been produced from freshly harvested mushrooms as ketchup, murabba, candy, chips and pickles or by dehydrating fleshy harvesting mushrooms into dehydrated form and making soup powder, biscuit, nuggets etc. (Mehta *et al.*, 2011).

Mushrooms are highly perishable food and require immediate marketing or processing. Once the body matures, degradation process starts and it becomes inconsumable after some time. Marketing of mushrooms in India is not yet organized and growers face great difficulty in selling the products. Basically, this hurdle deters entrepreneurs from undertaking and then continuing mushroom cultivation. Some women organizations or women entrepreneurs coordinate the marketing of mushrooms. This step itself will be a great entrepreneurship. Women are a serious lot with utmost sincerity, devotion, dedication, selflessness and sacrifice and they can really organize for coordinating and expediting marketing of mushrooms. Therefore, there must be sincere and smart efforts to improve marketing of mushrooms.

Mushrooms Cultivation and Mushrooms Growing Units in India

Previously mushroom cultivation was only confined to hilly and colder regions of Northern and Southern parts of India. Nowadays, it has spread along the whole country. In many of the states like Jammu & Kashmir, Haryana, Uttar Pradesh, Madhya Pradesh, Rajasthan, Gujarat, Maharashtra, Goa, Karnataka, Kerala, Tamil Nadu, Andhra Pradesh, Orissa, West Bengal, there are well established mushroom growing units (Dhar, 1997). North Eastern states and upper North Western Himalayan states, the mushrooms are liked by people but their consumption is confined to mushrooms harvested from the wild. Lakhanpal *et al.* (2010) reported that people not only consume them fresh but collect and dry them for use during winters.

The government agencies, NGO's and even small enterprises have always been on a look out for ventures which can supplement their income with some extra effort and training. Mushroom cultivation offers a good promise in this regard. Even when taking care of household chores and family affairs, women and other people can undertake mushroom cultivation after giving Training and guidance from the Department of Horticulture, Shimla, Directorate of Mushroom Research and Training, Solan, and some private agencies. With this training, the women can undertake mushroom cultivation and supplement their family income, substantially and provide the family a wholesome diet for all around health care. This venture does not require high expenditure or sophisticated and elaborate establishments, but this activity is a good means to diversify the income opportunities for women to raise their productivity. Thus, cultivation of mushrooms has many advantages. Each operation in mushroom cultivation is a full-fledged enterprise and those who can acquire technical skills proficiently can also create facilities to supply compost, spawn, and mycelium to the mushrooms growers.

Mushroom cultivation industry has a very good scope in India with comparatively cheap labour and abundant raw material. Now, this mushroom industry has flourished well, but still the supply-demand gap still exists, and this can only be reduced by encouraging more entrepreneurship and women are good stakeholders. There are a number of Government Department and Agencies supporting this venture *viz.* Agricultural Department, Horticultural Department, District Rural Development Agronomy (DRDA), National Bank of Agricultural and Rural Development (NABARD), Indian Council of Agricultural Research (ICAR), (Food and Agricultural Organization (FAO), Directorate of Mushroom Research (DMR) Solan and various NGO organizations, but there needs to be established a good rapport between these organizations and the users. With so many organizations promising help and guidance, the mushroom industry in India should have been far more ahead than other countries, but the industry is still expanding. The grass root level work for where and what needs to be investigated has to be worked out necessarily, if the entrepreneurship is really to be boosted in mushroom cultivation.

Mushroom Cultivation Role in Sustainable Livelihoods

Entrepreneurship acts as an economic force in the Nation's development. Empowerment is more than ever before linked to entrepreneurship and act as a strategic development intervention that could accelerate the rural development process. According to Khanka (2000), a women entrepreneur is 'a confident, innovative and creative woman capable of achieving economic independence individually or in collaboration generates employment opportunities for others through initiating, establishing and running an enterprise by keeping pace with her family and social life'.

Mushroom cultivation apart from being a source of food, can contribute to sustainable livelihood and economic empowerment for rural people - especially women. It can alleviate poverty and strengthen livelihoods by providing a nutritious source of food and by creating a reliable source of income. Mushroom cultivation is a viable and attractive activity for both rural farmers and poor urban dwellers, since it does not require access to land. Generally, it is a small-scale activity and growing does not require any significant capital investment (Sathiabama, 2010). Mushroom substrate can be prepared from any clean agricultural waste which is rich in lingocellulose. The substrate for cultivation could be easily gathered from the local fruit/vegetable market, which is used as their packaging material. Mushroom cultivation can be a part-time activity, and can well be integrated into their daily household routine. Additionally, mushroom cultivation also helps in improving the sustainability of small farming systems as the recycled organic matter, can be returned to the land as fertilizer (Doraiswamy, 2002). Mushroom cultivation through the provision of income generation, nutrition supplement with great nutraceutical potential, and trade can strengthen livelihood assets, which can not only reduce vulnerability to poverty, but enhance an individual's and a community's capacity to act upon economic growth (Ekpo and Aluko, 2002).

There are many opportunities to, and benefits of, increasing food and income security through incorporating mushroom into livelihood strategies. Many case studies of successful outcomes from growing mushrooms to strengthen the livelihood demonstrate the benefits arising from mushroom production in terms of income, food security and consumption of healthy food (Marshal and Nair, 2009).

Not even cultivated but even wild edible mushrooms have their important contribution in making sustainable livelihoods of rural people. Various wild edible mushrooms constitute a multi-million dollar industry. Picking and selling of these wild fungi provides supplement income to the people as morels (Lakhanpal *et al.*, 2010; Lakhanpal *et al.*, 2016) and *Cordyceps* (Sharma, 2004). However, the subsequent processing, packaging and marketing of mushrooms is similar for both cultivated and wild harvested ones.

Marshall and Nair (2009) reported that there are several reasons why mushrooms are the most effective way of achieving our goal of empowerment of women:

(i) The labour required to grow mushrooms is not overly physical, therefore, there is no need of manpower.

(ii) Growing of mushrooms can be done on a small piece of land. Even landless people can do mushroom cultivation.

(iii) A mushroom growing requires relatively low set up costs.

(iv) Mushrooms take a short period to yield (3 months), and after this period the next lot of mushrooms can be planned straight away.

(v) Mushroom cultivation is not an extremely complex process; therefore the operation will not require a lot of training, which would be an extra cost.

(vi) Mushrooms have numerous health benefits.

Mushroom cultivation and even the wildly harvested mushrooms can directly improve livelihoods through economic, nutritional and medicinal contributions as mentioned below:

(a) Nutritional value

Mushrooms are the fungi that have been used as food since time immemorial. Nutritionally they are a valuable source of healthy food, which is low in calories, rich in carbohydrates, essential amino acids, fiber, important vitamins and minerals. The consumption of mushrooms can make a valuable addition to the often unbalanced diets of people in developing countries. Fresh mushrooms have high water content, so drying them is an effective way to both prolong their shelf-life and preserve their flavour and nutrients. Mushrooms are a good source of vitamin B, C and D, including niacin, riboflavin, thiamine, and folate. The vitamins present are well retained both after cooking and processing. They also provide various minerals including potassium, phosphorus, calcium, magnesium, iron, selenium and copper. They provide carbohydrates, but are low in fat and fiber, and contain no starch. Furthermore, edible mushrooms are an excellent source of high quality proteinaceous food. Modern mushroom culture produces more protein per unit area of land than any other form of agriculture, because they can be grown vertically rather than horizontally.

(b) Medicinal value

Mushrooms have also been used in medicine for centuries in the orient but their potential as 'Health potentiators' and 'Elicitors of immune system' is recent. In the last two decades there has been an upsurge in the use of mushrooms as nutraceuticals and many edible species have been thoroughly investigated and authenticated for medicinal use. The species that have been properly analyzed for medicinal value are: *Ganoderma lucidum* (Reishi), *Lentinus edodes* (Shiitake), *Grifola frondosa* (Maitake), *Agaricus blazei* (Himematsutake), *Cordyceps sinensis* (Caterpillar fungus), *Pleurotus ostreatus* (Oyster mushroom), *Hericium erinaceus* (Lion's mane) and *Morchella* species (morels). There are many more cultivated and wild edible mushroom species that have been analyzed for their nutritional and nutraceutical components but to a lesser extent than those listed above. The active constituents found in mushrooms are polysaccharides, dietary fibers, oligosaccharides, triterpenoids, peptides, proteins, alcohols, phenols, mineral elements (zinc, copper, iodine, selenium and iron), vitamins and amino acids etc. (Lakhanpal and Rana, 2005; Falandysz, 2008; Lakhanpal *et al.*, 2016). These have been found to boost the immune system, have anti-cancerous properties, act as anti-hypercholesterolaemic and hepato-protective agents, show anti-HIV activity and anti-viral activity, ameliorate the toxic effect of chemo- and radiotherapy and many more (Chang and Buswell, 1996; Lakhanpal and Rana, 2005; Thakur and Lakhanpal, 2015). Many commercial products from these mushrooms are available in the market. Prominent among these are: Lentinan from *Lentinus edodes*, Concord Sunchih and Reishi Plus from *Ganoderma lucidum*, Grifon from *Grifola frondosa* and Didanosine from *Cordyceps militaris*. The myconutraceuticals have been found to relieve the stress and pressure of modern society by stimulating both basic and secondary immune responses of the immune system (Chang and Buswell, 1996). There are more than 15000 species of mushrooms recorded from all over the worlds out of which around 5000 are reported to be edible (Lakhanpal and Rana, 2005).

(c) Income benefits

Mushroom cultivation activities can play an important role in supporting the local economy by contributing to subsistence food security, nutrition, and medicine; generating additional employment and income through local, regional and national trade; and offering opportunities for processing enterprises (such as pickling, drying, and various value added dehydrated products) (Marshall and Nair, 2009; Thakur, 2015; Thakur, 2016; Thakur and Lakhanpal, 2015).

Economic Returns from Mushroom Cultivation

Growing any particular type of mushroom in a particular season and stopping the mushroom activity in rest of the season may not lure other villagers; therefore, farmers would like to choose the entrepreneurship that is quite rewarding throughout the year. Hence, the continuous cultivation and production of different mushrooms depending upon the season is certain to increase the economic return of mushroom entrepreneur, by following round the year cultivation of different edible mushrooms. The women entrepreneurs can achieve much more economic returns on a regular basis which will help in their empowerment. Although, it is very difficult to give a generalized cost benefit ratio for all the regions due to variation in the cost of land, labour, electricity, raw materials and market price. However, certain guidelines can be given which should be considered for calculating the cost of cultivation. Major inputs required are substrate and spawn. Cheaper substrates like paddy straw, maize and cotton stalks, waste material from fruits and vegetables from the local markets can be used either alone/ in combination. White button mushrooms (*Agaricus bisporus*) are one of the most widely cultivated mushrooms in the world, contributing 31% of the world's mushroom production. Its cultivation gives high returns to the entrepreneurs whether they are cultivated seasonally / round the year.

Mushroom Cultivation Role in Empowerment of Women in India

For sustainable development, each member of the society should get equal opportunities, so as to unfold one's inner potential. Infact the progress of a nation depends mainly upon its human resources. Women constitute about half the percentage of the country and are indeed an integral part of our society and the growth would not be possible unless the women are brought into the mainstream of national development (Khurana, 2002). Trade of cultivated commercial mushrooms and wild harvested ones can provide a readily available and important source of cash income – for men, women, old, and disabled alike. Emancipation of women in any society is an important indicator of economic growth. The role played by women in rural mushroom production can be very significant. In spite of their poor education and social pressures, the women trainees can learn the process of cultivation and postharvest technique more efficiently. Certain parts of the mushroom cultivation process, such as filling substrates in containers and harvesting, are ideally suited for women's participation. Several programmes by different organizations like – NABARD and NHB have enhanced women's empowerment through mushroom production by giving them the opportunity to gain farming skills, financial independence and self-respect. Mushroom cultivation represents a very suitable and empowering income generating option for women in particular, because it can be combined with traditional domestic duties and can be undertaken at home. There are various institutes and voluntary organizations all over India which offer extension services, spawn and training programme for mushrooms growers. Kaul (1978) reported that women can be actively involved in mushroom cultivation and they can gain financial independence and employment through entrepreneurship. Khare (1996) and Savithiri *et al.* (1996) also reported that mushroom cultivation and empowerment of rural women in India are mutually interdependent. Mushroom cultivation can establish a relationship by removal of poverty and achieving higher quality of life. Various case studies are there which had shown the role of women upliftment in India *viz.* Kotla-Mubarakpur – A slum of Delhi (Kumar and Bindra, 1994, 1997) SC/ST rural women of four districts of Karnataka state (Prashad *et al.*, 2012), Krishi Vigyan Kendra, Bhadrak (2012) and Janki Sree Kumar in Kerala (Harish and George, 2012) as mushroom growers. Mishra (2008) studied the entrepreneurship development of farm through mushroom cultivation in Pipili Block of Khurda district of Orissa. The study revealed that 70% of women mushroom growers had improved their occupation by 'improvement in standard of living' (36%), 'better saving' (30%), 'knowledge and attitude' (24%) and social status (10%). Thus the study reveals that the cultivation of mushrooms not only improves standard of living but also improves the social status of women.

Mushroom cultivation is attractive for the resource-poor for two reasons: firstly, mushroom cultivation can be done on any scale. The initial financial outlay to establish a basic cultivation system need not be very great, and substrate materials are often free. Secondly, compared to many agricultural and horticultural crops, mushroom production systems have a short turn around. A harvestable crop can be produced and sold to meet the increasing demand by private entrepreneurs and in the wake of commercialization. The awareness of this activity and its importance has already been reached the doors of the average housewives and can be seen by large number of them getting trained year after year.

Das (2000) observed that the common for success of women entrepreneurs. Promoting and encouraging women participation in the society showed various benefits for socio-economic development as follows (Thakur and Sharma, 2013). Some of the common reasons have been enlisted as:

(i) Economic independence: Empowerment of women will enable them to improve their incomes, health and food production, benefiting their families and communities overall.
(ii) Fully incorporating the contributions of women will tap underutilized resources when harvested from wild.
(iii) Empowerment of women entrepreneurship doesn't only contribute to women welfare but also immensely towards the National Economic Growth.
(iv) Increasing women empowerment will increase the creativity, innovation and knowledge system in family and society.
(v) Social identity.
(vi) Increasing self – confiden**ce** by improving decision making capabilities.
(vii) Status in family and society.

Health benefits for women

Mushroom is considered to be a complete, health food and suitable for all age groups, from child to aged people (Manikandan, 2011). For thousands of years, mushrooms have been revered for their immense health benefits and extensively used in folk medicines. Specific bioactive components in mushrooms are responsible for improving women's health in many ways. These bioactive components include: polysaccharides, triterpenoids, low molecular weight proteins, immune-modulating compounds, etc. which are very necessary for women (Wasser and Weis, 1999; Chang and Miles, 2004; Thakur, 2012; Thakur, 2015; Lakhanpal *et al*., 2016). Some potential health benefits for women are:

(i) Breast & prostate cancer

Mushrooms are very effective in preventing cancer of breast and prostate due to presence of Beta-glucans and conjugated linoleic acid having anti-carcinogenic effects. Out of these two, linoleic acid is particularly helpful in suppressing effects of estrogen. This estrogen is the prime reason for breast cancer in women after menopause. The Beta-glucans, on the other hand, inhibit growth of cancerous cells in cases of prostate cancer. Selenium in mushrooms is very effective in inhibiting cancerous cells (Isaacs, 2007). The Japan Cancer Association proved that *Agaricus blazei* is effective against Ehrlich's ascites carcinoma, sigmoid colon cancer, ovarian cancer, breast cancer, lung cancer, and liver cancer, as well as against solid cancers (Mizuno, 1996; 1999; 2000 and 2002; Yuen and Gohel, 2005), Shin *et al.* (2010) studied the dietary mushroom intake mushroom intake and the risk of breast cancer based on hormone receptors status.

(ii) Good for cardiovascular health

Mushrooms are very good food for heart. Mushrooms have zero cholesterol, fats and very low carbohydrates. Besides having minimal sodium concentration, mushrooms are rich in potassium. Portobello mushrooms are specifically one of the best sources of potassium. One can also get twenty to forty percent of your daily copper requirement from a single serving of mushrooms. Both potassium and copper are essential for cardiovascular health. In fact, mushrooms have an impressive effect on the cardiovascular system and many varieties (maitake, shiitake and *Cordyceps* etc.) have been reported to reduce total cholesterol levels, reduce the bad cholesterol (low density lipoproteins) and triglycerides, decrease platelet binding and reduce arterial pressure (Kino *et al.*,1986; Zhang and Xia, 1990; Zhu *et al.*, 1998; Nakamura *et al.*, 1999).

(iii) Anti-diabetic active components

Mushrooms have anti-diabetic active compounds can be an ideal low energy diet for diabetic patients. These have low calorific value, no cholesterol, very low carbohydrates, high proteins, vitamins and minerals. Moreover, they contain natural insulin and enzymes which help breaking down of sugar or starch of the food. Basically they are ideal food for diabetic patients (Perera and Li, 2011; Manikandan, 2011).

(iv) Immunomodulatory properties

Scientists have noted that mushrooms have definite primary and secondary physiological effects on the human immune system Ergothioneine, a powerful antioxidant present in *Boletus, Flammulina* and *Pleurotus* mushrooms are very effective in giving protection from free radicals as well as boosting up immune system. This antioxidant is also necessary for healthy eyes, kidneys, bone marrow, liver and skin (Manikandan, 2011). Mushrooms contain natural antibiotics which inhibit microbial and other fungal infections. They also help in healing ulcers, wounds and protect them from infections. A good combination of vitamins especially B and C are found in mushrooms which also strengthens the immune system (Isaacs, 2007). Wasser and Weis (1999) have reported that cellular components and secondary metabolites of many mushrooms affect the immune system of the host and therefore could be used to treat a variety of diseases. Some well-known medicinal mushrooms with benefits for the immune system include Reishi (*Ganoderma lucidum*), Shiitake (*Lentinus edodes*), Maitake (*Grifola frondosa*), and Caterpillar fungus (*Cordyceps sinensis*) can directly stimulate both basic (lymphocytes, neutrophils etc.) and secondary immune responses (immunoglobulins Ig E, Ig A, Ig G) of the immune system. This stimulus can increase production of immune defenders such as cytokines and macrophages, which play vital roles in recognizing and removing foreign antigens, as well as releasing chemical mediators including interleukin-1 (Kino *et al.,* 1989; Zhang and Xia, 1990; Zhu *et al.,* 1998; Nakamura *et al.*, 1999; Chang and Buswell, 1999; Zhuang and Mizuno, 1999).

(v) Anti-ageing properties

Falandysz (2008) reported that mushrooms are a good source of selenium, having very good anti-cancerous properties. Many wild edible mushrooms *e.g., Boletus edulis, Albatrellus, Ganoderma lucidum* have very good concentration of selenium. Mushrooms are also rich in riboflavin along with selenium. These chemicals by strengthening the immune system, prevents the action of free radicals in the body, consequently reducing the ageing process. The polysaccharides from mushrooms are also potent scavengers of super oxides increased free radicals. These antioxidants prevents the action of free radicals in the body, consequently reducing the ageing process (Manikandan, 2011).

(vi) Ovulation in women

Chen *et al*. (2010) reported that the mushroom extract of Maitake (*Grifola frondosa*) is a treatment of the common female disorder polycystic ovary syndrome. Maitake extracts improve ovulation in women.

(vii) Pregnant and Lactating mothers

Mushrooms are rich in vitamin B, iron, other minerals (Se, P and Ca), and antioxidants which are essential nutrients during pregnancy. Mushrooms are a good source of zinc, which is important for normal fetal growth and development during pregnancy. Decoction of morels (*Morchella* spp.) is also given to pregnant and lactating mothers as it provides strength and warmness to the body and considered nutritious (Lakhanpal *et al.*, 2010).

(viii) Other benefits

Mushrooms are the second known source (after cod liver oil) to contain vitamin-D in edible form. They are rich in calcium (good for bones), iron (benefits in anemia), potassium (very good for lowering blood pressure), copper (anti-bacterial) and selenium (very good for health of bones, teeth, nails, hair and as an antioxidant). So, mushrooms can be the best choice for vegetarians to obtain selenium (Lakhanpal *et al.*, 2010; Lakhanpal *et al.*, 2016). Mushrooms are also considered anti-stress food.

FUTURE PROSPECTS

With the constant increase in population, humanity is passing today through a difficult phase as the demands are multiplying. The agricultural production is almost static and therefore the global food and nutritional security of growing population are showing a great challenge. It is imperative therefore to develop alternative sources of employment and quality food production. Undoubtedly, mushroom farming looks not only promising but also enterprising and adaptable. Though it is not a substitute for cereal food, nevertheless it is a very good supplement and health food. Mushroom cultivation has grown up in almost all parts of world and during last three decades, the world mushroom production achieved a growth rate of about 10%. Owing to the increasing demand by private entrepreneurs in the wake of commercialization, the awareness of this activity and its importance has already reached the doors of the average housewives and can be seen by a large number of them getting trained year after year. Women empowerment shall also lead to provide significant benefits including healthier lives enhanced well-being, increased knowledge and capacities, enhanced productivity, sustained livelihood and greater rights.

Mushroom cultivation being an indoor activity, labour intensive and high profit venture provides ample opportunities for gainful employment of small farmers, landless labourers and especially women. It is hoped that the avocation of mushroom farming will become a very important cottage industry activity in the integrated rural development programme, which will lead to the economic betterment of not only small farmers but also of landless labourers and other weaker sections of communities. The advantages of mushroom cultivation can be summarized as:

(i) Wastes (agricultural / organic) are largely burnt by the farmers, which causes air pollution. However, these raw materials can actually be used for the cultivation 'of mushrooms. This kind of bioconversion activity can greatly reduce environmental pollution. Therefore, reduction of environmental pollution by bioconversion of vast quantities of organic wastes into mushrooms is a gainful proposition.

(ii) Mushroom cultivation is a labour intensive activity. Therefore, it will serve as means of generating employment, particularly for rural women and youths in order to raise their economic and social status. It will also provide additional work for the farmers during winter months when the farming schedule is light.

(iii) It will provide the people with an additional vegetable of high quality, and enrich the diet with high quality proteins, minerals and vitamins which can be of direct benefit to the human health and fitness. The extractable bioactive compounds from medicinal mushrooms would elicit the human immune systems and improve the quality of life.

(iv) Mushroom cultivation is a cash crop. The harvested fruiting bodies can be sold in local markets for additional family income or exported for an important source of foreign exchange that will definitely improve the economic standards of the people.

(v) Some high temperature loving mushrooms *e.g., Volvariella volvacea* (Straw mushrooms) and *Pleurotus sajor-caju* (Oyster mushrooms) are relatively fast growing organisms and can be harvested in 3 to 4 weeks after spawning. It is a short return agricultural business and can be of immediate benefit to the community.

(vi) Cost of production is low. Majority of the mushrooms can be sun-dried and even pickled.

Therefore, the significant impact of applied mushroom biology on human welfare has been named as a "Non -green revolution" (Chang, 1999). There is significance of mushrooms in our lives towards alleviating poverty, enhancing human health, and arresting environmental degradation. Mushrooms can serve as agents for promoting equitable economic growth in society. They are a unique group of fungi through which we can pilot a non-green revolution in less developed countries, and in the world at large. They demonstrate great potential for generating a great socio-economic impact in human welfare, at local, national and international levels.

CONCLUSIONS

Mushroom cultivation is an ideal activity for women, which can make a valuable contribution to sustainable livelihood of both rural and urban women, because it is highly compatible with their other livelihood activities, requiring minimal physical and financial inputs and resources. Mushroom cultivation aims to provide women empowerment with self-sufficiency, entrepreneurship skills, and overall economic independence. Participating women can derive benefit by improving their health, wealth and the quality of life of their families. Empowerment of women entrepreneurship not only contributes to the welfare of women but also towards the Government's objectives of fighting poverty. Mushroom cultivation is a low input activity and the productivity is both ways: locally and in the international market. Mushrooms can be cultivated on both small and large scale with a potential for commercial enterprises. Indeed, the basic requirements are good spawn source, suitable substrate, the means to sterilize it, a few bags for inoculation and a clean, dark room for cultivation. Thus, mushroom cultivation with empowerment of women in India improves the quality of life of rural women with aims at self-sufficiency rather than dependency. In a developing country like India, the two basic issues are providing quality food and solving unemployment besides the environmental issues and these can be resolved by involving rural women in mushroom cultivation as entrepreneurship programme. Besides, this venture assures a clean environment and can prove a boon for the empowerment of rural women entrepreneurship.

REFERENCES

Chang, S.T. 1998. Development of novel agroscience industries based on bioconversion technology. *In*: Chou, C.H. and Shao, K.T. (*eds.*), *Frontiers in Biology: The Challenges of Biodiversity*. Academia Sinica, Taipei, pp. 217–222.

Chang, S.T. 1999. Global impact of edible medicinal mushrooms on human welfare in the 21st century: Non green evolution. *International Journal of Medicinal Mushrooms,* pp. 1–7.

Chang, S.T. 2006. The world mushroom industry: Trends and technological development. *International Journal of Medicinal Mushrooms,* pp. 297–314.

Chang, S.T. 2007. Mushroom cultivation using the "ZERI" principle: Potential for application. *International Journal of Brazil Applied Microbiology*, pp. 33–34.

Chang, S.T. and Buswell, J.A. 1996. Mushroom nutriceuticals. *World Journal of Microbiology and Biotechnology*, pp. 473–476.

Chang, S.T. and Buswell, J.A. 1999. *Ganoderma lucidum* (Curt.: Fr.) P. Karst. (Aphyllophoromycetideae) – A mushrooming medicinal mushroom. *International Journal of Medicinal Mushrooms*, pp. 139–148.

Chang, S.T. and Miles, P.G. 2004. *Mushroom Cultivation, Nutritional Value, Medicinal Effects and Environmental Impact.* CRC Press.

Chen, J.T., Tominaga, K., Sato, Y., Anzai, H. and Matsuoka, R. (2010). Maitake mushroom improves ovulation in women's health study. *The Journal of Alternative and Complementary Medicine*, pp. 1295–1299.

Das, M. 2000. Women entrepreneurs from India: Problems, motivations and success factors. *Journal of Small Business and Entrepreneurship*, pp. 67–84.

Dhar, B.L. 1997. Mushroom industry in India. *In:* Rai, R.D., Dhar, B.L. and Verma, R.N. (*eds.*), *Advances in Mushroom Biology and Production*. National Research Center for Mushrooms, Solan, HP. Nirmal Vijay Printers, pp. 369–378.

Doraiswamy, S. 2002. Career option for woman in science engineering. *Conference of Women Scientist and Technologists Role in National Development*. Vigyan Bhawan, Department of Biotechnology, New Delhi, pp. 49–51.

Dutt, B. 2014. Strategies for setting and running of small scale enterprises by women entrepreneurs of Kangra District of Himachal Pradesh. *Journal of Agricultural Extension and Rural Development,* 6(1): 35–43.

Ekpo, E.N. and Aluko, A.P. 2000. Cultivation strategies and nutritive values of edible mushrooms (*Lentinus tuber-regium*) as a component of sustainable livelihood. *Proceeding* 28th*Annual Conference Forestry Association of Nigeria*.

Falandysz, J. 2008. Selenium in edible mushrooms. *Journal of Environmental Health*, pp. 256–299.

Harish, S. and George, A. 2012. Young entrepreneur: Janaki Sreekumar: Pioneering mushroom farming in Kerala. Retrieved from: *www.yentha.com/.../young-entrepreneur-janaki-sreekumar-pioneering-mushroom-far*. Accessed on 13th December, 2012.

Isaacs, T. 2007. Mushrooms for health immune system and cancer. Retrieved from: *www.disabled-world.com/artman/publish/mushrooms.shtml* (Accessed 05 December 2012).

Kaul, T.N. 1978. Mushroom cultivation and rural development. *Indian Journal of Mushroom Science*, pp. 1–12.

Khanka, S.S. 2000. *Entrepreneurship Development*. S. Chand & Company Ltd. India.

Khare, N. 1996. Mushroom growing in Raipur for empowerment of rural women. *In*: Kumar, S. and Bindra Ban, V. (*eds.*), *Mushroom Growing in India for Empowerment of Rural Women*. Swastik Printer, New Delhi, pp. 71–74.

Khurana, A. 2002. Science and engineering education and woman empowerment.*Conference of Women Scientist and Technologists Role in National Development, Vigyan Bhawan, Department of Biotechnology, New Delhi*, pp. 34–36.

Kino, K.Y., Yamaoka, K., Watanabe, J., Shimizu, K. and Tsunoo, H. 1989. Isolation and characterization of a new immune-modulatory protein Ling Zhi-8 (LZ-8) from *Ganoderma lucidum*. *Journal of Biological Chemistry*, pp. 472–478.

Kino, T., Tabata, H., Ukai, S. and Hara, C. 1986. A minor protein- containing galactomannan from a sodium carbonate extracts of *Cordyceps sinensis*. *Carbohydrate Research,* pp. 189–197.

Krishi Vigyan Kendra Bhadrak ZPD – VII. 2012. Retrieved from: *http//kvkbhadrakzpdvii.org/case-study-01.html.* (Accessed 05 December 2012).

Kumar, S. and Bindra Ban, V. 1994. Mushroom cultivation for small scale farmers – A farming approach. *Mushroom Research*, 103.

Kumar, S. and Bindra Ban, V. 1997. Mushroom cultivation and upliftment of women in India. *In*: Rai, R.D., Dhar, B.L. and Verma, R.N. (*eds.*), *Advances in Mushroom Biology and Production* . National Research Center for Mushrooms, Solan, HP. Nirmal Vijay Printers, pp. 399–406.

Lakhanpal, T.N. 2000. Wild and cultivated mushrooms. *In:* Verma, L.R. (*ed.*), *Natural Resources and Development in Himalaya*. Malhotra Publishing House, New Delhi, pp. 402–417.

Lakhanpal, T.N. and Rana, M. 2005. Medicinal and nutraceutical genetic resources of mushrooms. *Plant Genetic Resources – Characterization and Utilization,* pp. 1–17.

Lakhanpal, T.N., Sai, I. and Thakur, M. 2016. Biodiversity of wild edible, mycorrhizal and nutraceutically important mushrooms. *In:* Sharma, S., Singh, M., Prasad and Rathore, H. (*eds.*) Mushroom of India, III Delhi, and Institute of Rural Development. IIT Delhi.

Lakhanpal, T.N., Shad, O. and Rana, M. 2010. *Biology of Indian Morels.* I.K. International Publishing House, p. 266.

Manikandan, K. 2011. Nutritional and medicinal values of mushrooms. *In*: Singh, M., Vijay, B., Kamal, S. and Wakchaure, G.C. (*eds.*), *Mushroom: Cultivation, Marketing and Consumption*. Directorate of Mushroom Research, Solan, H.P., pp 11–14).

Marshal, E. and Nair, N.G. 2009. Make money by growing mushrooms. *FAO,* pp. 1–51.

Mehta, B.K., Jain, S.K., Sharma, G.P., Doshi, A. and Jain, H.K. 2011. Cultivation of button mushroom and its processing: An techno-economic feasibility. *International Journal of Advanced Biotechnology and Research*, pp. 201–207.

Mishra, S. 2008. Entrepreneurship development for farm women through mushroom cultivation. *International Journal on Rice*, pp. 68–71.

Mizuno, T. 1996. A development of antitumor polysaccharides from mushroom fungi. *Journal of Food and Food Ingredients,* pp. 69–85.

Mizuno, T. 1999. The extraction and development of antitumor active polysaccharides from medicinal mushrooms in Japan. *International Journal of Medicinal Mushrooms*, pp. 19–30.

Mizuno, T. 2000. The development of an antitumor BRM (Biological Response Modifier) from *Phellinus linteus* (Berk. et Curt.) Teng Mushroom. *International Journal of Medicinal Mushrooms*, pp. 21–34.

Mizuno, T. 2002. Medicinal properties and clinical effects of *Agaricus blazei* Murill. *International Journal of Medicinal Mushrooms*, 4.

Nakamura, K., Yamaguchi, Y., Kagota, S., Shinozuka, K. and Kunitomo, M. 1999. Activation of Kupffer cell function by oral administration of *Cordyceps sinensis* in rats. *Japan Journal of Pharmacy,* pp. 505–508.

Perera, P.K. and Li, Y. 2011. Mushrooms as a functional food mediator in preventing and ameliorating diabetes. *Functional Foods in Health and Disease,* pp. 161–171.

Prashad Raghu, K.P., Gangadhrappa, N.R. and Prasad Madhu, V.L. 2012. Mushroom cultivation using agricultural wastes as a sustainable livelihood alternative for resource poor SC/ST rural women. *Journal of Rural Development*, pp. 173–180.

Sathiabama, K. 2010. Rural women empowerment and entrepreneurship development. Retrieved from: *https//ideas.repec.org/p/ess/wpaper/id2475.html.* (Accessed 01 December, 2013).

Savithiri, A, Solundaradjane, S. and Shanthakumar, H. 1996. *In*: Kumar, S. and Bindra Ban, V. (*eds.*). Mushroom growing in India for empowerment of rural women. Swastik Printer, New Delhi, pp. 64–66.

Sharma, S. 2004. Trade of *Cordyceps sinensis* from high altitudes of the Indian Himalaya: Conservation and biotechnological priorities. *Current Science,* pp. 1614–1619.

Shin, A., Kim, J., Lim, S.Y., Kim, G., Sung, M.K., Lee, E.S. and Ro, J. 2010. Dietary mushroom intake and the risk of breast cancer based on hormone receptor status. *Nutrition Cancer*, 62(4): 476–83.

Thakur, M. 2012. Mushrooms: A prominent source of nutraceuticals. *J. Food Process Technol.,* 3(10): 177–178

Thakur, M. 2015. Wild mushrooms as natural untapped treasures. *In: Natural Products: Recent Advances* (Chauhan Ashok K., Pushpangadan Palpu and Varughese George) (*eds.*), Write and Print Publications.

Thakur, M. 2016. Significance of prebiotics, probiotics and synbiotics as health potentiators. *Food Sci. Res. J.,* 7(2): 327–334

Thakur, M. and Lakhanpal, T.N. 2014. Qualitative phytochemical screening, total phenolic content and *in-vitro* antioxidant activity in methanolic extracts of *Morchella esculenta* Fr. *In: Proceedings of the 8th International Conference on Mushroom Biology and Mushroom Products (ICMBMP8),* pp. 215 –220.

Thakur, M. and Lakhanpal, T.N. 2015. Mushroom cultivation and women entrepreneurship: Prospects and Challengers. *In:* Vasudevan Padma, Sharma Satyawati, Sharma, V.P. and Verma Monica. Women, Technology and Development. III Delhi, Narosa Publishers, pp 220–240.

Thakur, M. and Sharma, N. 2013. Women empowerment as agricultural entrepreneurs - An Economic Force in Rural Development. (Aggarwal, M. (*ed.*)). *Changing Status of Women in India: Issues and Challenges*. Chapter 7. Kanishka Publishers, pp. 79–99.

Wakchaure, G.C. 2011. Production and marketing of mushrooms: Global and national scenario. *In:* Singh, M., Vijay, B., Kamal, S. and Wakchaure, G.C. (*eds.*). *Mushroom: Cultivation, Marketing and Consumption*. Directorate of Mushroom Research, Solan, H.P., pp 15–22.

Wasser, S.P. and Weis, A.L. 1999. Medicinal properties of substances occurring in higher basidiomycete mushrooms: Current perspectives. *International Journal of Medicinal Mushrooms*, pp. 31–62.

Yuen, J.W. and Gohel, M.D. 2005. Anticancer effects of *Ganoderma lucidum*: A review of scientific evidence. *Journal of Nutrition Cancer*, pp. 11–7.

Zhang, Z. and Xia, S. S. 1990. *Cordyceps sinensis* - 1 as an immunosuppressant in hetero suppressant in heterotrophic heart allograft in rats. *Journal of Tougjichi Medicine*, 100–103.

Zhu, H., Jia-Shi, Halpern, G. and Jones, K. 1998. The scientific rediscovery of an ancient Chinese herbal medicine: *Cordyceps sinensis*. *Journal of Alternative Complementary Medicine,* pp. 289–303.

Zhuang, C. and Mizuno, T. 1999. Biological responses from *Grifola frondosa* (Dick.: Fr.) S. F. Gray - Maitake (Aphyllophoromycetideae). *International Journal of Medicinal Mushrooms*, pp. 317–324.

4.3

Reoccurrence of Genus *Morchella* from District Faizabad, Uttar Pradesh (India)

SIDDHANT[1*], C.S. SINGH[2] AND R.S. KANAUJIA[3]

ABSTRACT

Genus Morchella is commonly known as morel, dryland fish, pine cone, guchchi and sponge mushroom. It is easy to spot due to its distinctive honey-combed upper portion which is composed of a network of ridges and pits. The present paper reports its reoccurrence in district Faizabad, Uttar Pradesh (India).

Key words: Morels, *Morchella* sp., Ascomycetes, Faizabad, Uttar Pradesh.

INTRODUCTION

Morchella, the true morels belonging to class Ascomycetes, are amongst the most highly prized fungi in the world (Pegler, 2003). The average morels are sold at a price of Rs. 10000-15000 per Kg (Prasad *et al.*, 2002) that makes a good income generating source for livelihood (Singh and Rawat, 2000; Semwal *et al.*, 2014). All the true morels are known to edible (Arora, 1986) with excellent flavor and taste. They contain 42% protein on dry weight basis. They are also rich in minerals (Gençcelep *et al.*, 2009) and vitamins *viz.*, thiamine, riboflavin, niacin, pantothenic acid, pyridoxine and cyanocobalamin (Samajpati, 1978). Not only morels but even their mycelia have same nutritive value (Hayes and Hadded, 1976). Apart from this, morel mycelium has been reported to contain compounds with anti Mau *et al.*

[1] Department of Botany, Durgesh Nandini Degree College, Faizabad (U.P.), India.
[2] Department of Botany, K.S. Saket P.G. College, Ayodhya, Faizabad (U.P.), India.
[3] Indo-Canadian Farms, Pune (Maharashtra), India.
**Corresponding author:* E-mail: siddhant.ani@gmail.com

(2004), like antioxidant (Mau *et al.*, 2004), immune enhancer (Duncan *et al.*, 2002), anti-inflammatory tumor activity (Nitha *et al.*, 2007) and free radical scavenging activity (Shagug and Masaphy, 2010). Occurrence of 18 species of *Morchella* are reported from 28 countries, wherein altogether 14 species are reported to be edible and used as food, and 5 are used medicinally (FAO, 2004). In India, majority of *Morchella* species are reported from Northern Western Himalayan regions especially Jammu and Kashmir (Ghosh and Pathak, 1962; Waraitch, 1976; Kaul, 1981; Khan *et al.*, 2009; Kumar and Sharma, 2010), Himachal Pradesh (Sohi *et al.*, 1965; Lakhanpal and Shad, 1986; Manikandan *et al.*, 2011) and hilly tract of Uttar Pradesh (Hennings, 1901; Theissen, 1911; Negi, 2006; Vishwakarma *et al.*, 2011) which are known to be main *Morchella* producing sites in India (Shad, 1989; Jandaik and Sharma, 1995). However, occasional reports on collection of *Morchella* have also been documented. Wakode (1983) collected some specimen of morels from Amravati, Maharashtra during rainy season while Kaviyasaran *et al.* (2006) reported *Morchella esculenta* from Vattakanal falls of Kodaikanal hills, Tamil Nadu. Likewise, occurrence of *Morchella vulgaris* from Assam forest (Bhattacharya and Baruah, 1953), *M. deliciosa* from Amritsar, Punjab (Purkayastha and Chandra, 1985), *M. conica* from Maharashtra (Ghurde and Wakode, 1981) and Madhya Pradesh (Rai *et al.*, 1999), *M. esculenta* from Darjeeling, West Bengal (Acharya *et al.*, 2005) and Mount Abu, Rajasthan (Paliwal *et al.*, 2013) have been given by different workers.

Recently, *Morchella* species is reported from District Faizabad (Siddhant *et al.*, 2014) which is one of the 75 districts of Uttar Pradesh state in Northern India. Earlier, there is no report in respect of incidence of *Morchella* from this region. It may be first report from plains of Uttar Pradesh state too. The present communication reveals reoccurrence of Genus *Morchella* from same region.

MATERIALS AND METHODS

The specimens of *Morchella* were collected during December 04 to 27, 2014 from district Faizabad which lies between the parallels of 26°47'N to 26°78'N latitude and 82.08'E to 82.13'E longitude having an average elevation of 97 metres above the sea level. The macroscopic features of the fruiting bodies (*e.g.,* shape, size and colour of sporophore) were noted in the field while the microscopic features (*e.g.,* spore shape) were noted in laboratory. The specimens were photographed in natural habitat. It was identified following Arora (1986). Later these were sun-dried to use as a reference.

RESULTS AND DISCUSSION

Eight *Morchella* specimens were collected during December 04 to 27, 2014 near railway station area, Ayodhya, Faizabad.

Apothecia

The specimens were stipitate and characterized by honeycombed fertile cap (Fig. 1). The fruiting body was 11-12.0 cm high. It was hollow from the base of the stipe to the top of the pileus with cottony fibers (Fig. 2). The pileus rises continuously from stipe. Pileus subglobose, ovoid, or elongated, occasionally attenuated upwards but obtuse at the apex, 6-7 cm high, 4.5-5 cm wide. Pileus was longer than stipe. Pits rounded, irregular, occasionally longitudinally elongated. Cream to yellowish when younger, becoming brownish or blackish when dried (Fig. 3) stipe stout, relatively short and narrower than cap. Up to 4.8-5.0 cm in high and 2.0 cm broad, delicately pubescent, the base was wrinkled, cream coloured. Asci 200-250 µm long and 12-20 µm broad, with 8 spores. Spores 12-14 µm × 20-24 µm, elliptical and smooth, creamy to yellowish in mass. Odor pleasant.

Habit and Habitat

All fruiting bodies were solitary on humus rich soil near fern, cactus and other ornamental plants (Fig. 4). Earlier in 2013, a *Morchella* specimen was collected from the same place which showed the possibility of mycorrhizal association (Siddhant *et al.*, 2014). There is also evidence that suggests some morels may form mycorrhizal associations (Buscotand and Kotte, 1990).

Comments

The above collections are typical of *Morchella* as described by Arora (1986). Apothecia terrestrial and differentiated into pileus and stipe. Interior hollow. The pileus is uniformly shaped, honeycombed with ridges and pits. It appears to be pitted inwards. Stipe is well developed and wrinkled at base.

CONCLUSIONS

Although, the appearance of *Morchella* sp. is uncommon in plains of Uttar Pradesh, its repeated occurrence in same locality of Ayodhya during two consecutive years indicates its new probable permanent habitat that is to be confirmed on the basis of results found in the coming years.

ACKNOWLEDGEMENTS

This finding is dedicated to Late Dr. Harish Chandra Sachan, Sr. Consultant, Noida whose honesty, dedication and hard work will be an inspiration for us.

REFERENCES

Acharya, K., Bhattacharya, M., Pradhan, R. and Rai, M. 2005. Wild edible Halvellaceae from Darjeeling Himalaya. *J. Mycopathological Research,* 43(1): 135–136.

Arora, D. 1986. Mushroom demystified: A comprehensive guide to the fleshy fungi. Berkeley, CA: Ten Speed Press, pp. 785–793.

Bhattacharya, B. and Baruah, H.K. 1953. Fungi of Assam. *J. Univ. Gauhati,* 4: 287–312.

Buscot, F. and Kotte, I. 1990. The association of *Morchella rotunda* (Pers.) Boudier with roots of *Picea abies* (L.) Karst. *New Phytologist,* 116: 425–430.

Duncan, C.J.G., Pugh, N., Pasco, D.S. and Ross, S.A. 2002. Isolation of a galactomannan that enhances macrophage activation from the edible fungus *Morchella esculenta. J. Agric. Food Chem.,* 50(20): 5683–5685.

FAO 2004. Wild edible fungi: A global overview of their use and importance. *In:* Boa, E. *(eds.), Non Wood Forest Products*, FAO Publication, Rome, 17: 157.

Gençcelep, H., Uzunm, Y., Tunçtürk, Y. and Demirel, K. 2009. Deter-mination of mineral contents of wild-grown edible mush-rooms. *Food Chem.,* 113(4): 1033–1036.

Ghosh, R.N. and Pathak, N.C. 1962. Fungi of India-1 *Morchella*, *Verpa* and *Helvella*. *Bull. Nat. Bot. Gardens (Lucknow),* 71: 1–19.

Ghurde, V.R. and Wakode, D.D. 1981. A new report of *Morchella* from Central India. *Indian J. Mycol. Pl. Pathol.,* 11: 314–315.

Hayes, W.A. and Hadded, N. 1976. The food value of cultivated mushrooms and its importance to the mushroom industry. *Mush. J.,* 40: 104–106.

Hennings, P. 1901. *Fungi Indiae Orientalis-II. Hedw.,* 40: 323–342.

Jandaik, C.L. and Sharma, S.R. 1995. Present status of *Morchella* in India. *In:* Chadda and Sharma *(eds.)*, *Advances in Horticulture*. Malhotra Publishing House, New Delhi, pp. 171–194.

Kaul, T.N. 1981. Common edible mushrooms of Jammu and Kashmir. *Mush. Sci.,* 11(1): 79–82.

Kaviyasaran, V., Kumar, M., Shiva, R. and Natarajan, K. 2006. *Morchella esculenta*-A new record from South India. *Mush. Res.,* 15(1): 87–88.

Khan, M., Kumar, S., Hamal, I.A. and Koul, S. 2009. Wild edible plants of Sewa catchment area in Northwest Himalaya. *J. Plant Devel. Sci.,* 1(1 and 2): 1–7.

Kumar, S. and Sharma, YP. 2010. Pezizales from Jammu Province of Jammu and Kashmir-1, India. *Indian J. Mush.,* 28(1 and 2): 23–32.

Lakhanpal, T.N. and Shad, O.S. 1986. Studies on wild edible mushrooms of Himachal Pradesh (N.W. Himalayas)-II. Ecological relationships of *Morchella* species. *Indian J. Mush.,* 12–13: 15–20.

Manikandan, K., Sharma, V.P., Kumar, S., Kamal, S. and Shirur, M. 2011. Edaphic conditions of natural sites of *Morchella* and *Phellorinia. Mush. Res.,* 20(2): 117–120.

Mau, J.L., Chang, C.N., Huang, S.L. and Chen, C.C. 2004. Antioxidant properties of methanolic extracts from *Grifola frondosa*, *Morchella esculenta* and *Termitomyces albuminosus* mycelia. *Food Chem.,* 87(1): 111–118.

Negi, C.S. 2006. Morels (*Morchella* spp.) in Kumaun Himalaya. *Natural Product Radiance,* 5(4): 306–310.

Nitha, B., Meera, C.R. and Janardhanan, K.K. 2007. Anti-inflammatory and antitumor activities of cultured mycelium of morel mushroom *Morchella esculenta. Curr. Sci.,* 92: 235–239.

Paliwal, A., Boha, A., Pillai, U. and Purohit, D.K. 2013. First Report of *Morchella*- An edible morel from Mount Abu, Rajasthan. *Middle-East J. Sci. Res.,* 18(3): 327–329.

Pegler, D.N. 2003. Useful fungi of the world: Morels and truffles. *Mycologist,* 17(4): 174–175.

Prasad, P., Chauhan, K., Kandari, L.S., Maikhuri, R.K., Purohit, A., Bhatt, R.P. and Rao, K.S. 2002. *Morchella esculenta* (Guchhi): Need for scientific intervention for its cultivation in Central Himalaya. *Curr. Sci.,* 82(9): 1098–1100.

Purkayastha, R.P. and Chandra, A. 1985. *Manual of Indian Edible Mushrooms.* Today and Tomorrow's Printers and Publishers, New Delhi, p. 226.

Rai, M.K., Singh, J. and Acharya, D. 1999. *Morchella conica*: A new report from Madhya Pradesh. *Mush. Res.,* 8: 61–62.

Samajpati, N. 1978. Nutritional value of some Indian edible mushrooms. *Mush. Sci.,* 10(2): 695–703.

Semwal, K.C., Stephenson, S.L., Bhatt, V.K. and Bhatt, R.P. 2014. Edible mushrooms of the Northwestern Himalaya, India: A study of indigenous knowledge, distribution and diversity. *Mycosphere,* 5(3): 440–461.

Shad, O.S. 1989. Biological studies on *Morchella* species (Morels) in Himachal Pradesh. Ph.D. Thesis, Himachal Pradesh University, Shimla- 171005.

Shagug, G.J. and Masaphy, S. 2010. Free radical scavenging activity of culinary medicinal morel mushroom, *Morchella* Dill. ex Pers. (Ascomycetes): Relation to color and phenol contents. *Int. J. Med. Mush.,* 12(3): 299–307.

Siddhant, Singh, R., Singh, C.S. and Kanaujia, R.S. 2014. A report on occurrence of *Morchella* sp. from District Faizabad, Uttar Pradesh. *Trends in Biosciences,* 17(7): 2473–2474.

Singh, S.K. and Rawat, G.S. 2000. Morel mushroom industry in India. *Plant Talk,* 21: 36–37.

Sohi, H.S., Kumar, S. and Seth, P.K. 1965. Some interesting fleshy fungi from Himachal Pradesh-I. *J. Indian Bot. Soc.,* 54: 69–74.

Theissen, F. 1911. Fungi aliquot *Bombayensis,* a revised edition, *Blatter* collections. *Ann. Mycol.,* 9: 153–159.

Vishwakarma, M.P., Bhatt, R.P. and Gairola, S. 2011. Some medicinal mushrooms of Garhwal Himalaya, Uttarakhand. India. *Int. J. Med. Arom. Plants,* 1(1): 33–40.

Wakode, D.D. 1983. Distribution of *Morchella* in time and space with special reference to Melghat forest. *In: Silver Jubilee Symposium on 'Science and Cultivation of Edible Fungi' Regional Research Laboratory, Sanat Nagar, Srinagar,* p. 87 (Abst.).

Waraitch, K.S. 1976. The genus *Morchella* in India. *Kavaka,* 4: 69–76.

Subject Index

R

S

T

V

W

X

Z